ATTP 3-18.11 (FM 3-05.211)
AFMAN 11-411(I)
NTTP 3-05.26M

Special Forces Military Free-Fall Operations

October 2011

DISTRIBUTION RESTRICTION: Distribution authorized to U.S. Government agencies and their contractors only to protect technical or operational information from automatic dissemination under the International Exchange Program or by other means. This determination was made on 1 September 2011. Other requests for this document must be referred to Commander, United States Army John F. Kennedy Special Warfare Center and School, ATTN: AOJK-CDI-SF, 3004 Ardennes Street, Stop A, Fort Bragg, NC 28310-9610, or by e-mail to AOJK-DT-SF@soc.mil.

DESTRUCTION NOTICE: Destroy by any method that will prevent disclosure of contents or reconstruction of the document.

FOREIGN DISCLOSURE RESTRICTION (FD 6): This publication has been reviewed by the product developers in coordination with the United States Army John F. Kennedy Special Warfare Center and School foreign disclosure authority. This product is releasable to students from foreign countries on a case-by-case basis only.

Headquarters, Department of the Army

Foreword

This publication has been prepared under our direction for use by our respective commands and other commands as appropriate.

BENNET S. SACOLICK
Major General, USA
Commanding
U.S. Army John F. Kennedy Special
 Warfare Center and School

T. C. HANIFEN
Major General, USMC
Director
Naval Expeditionary Warfare Division (N85)

ATTP 3-18.11, C1

Change 1
Army Tactics, Techniques, and Procedures
No. 3-18.11

Headquarters
Department of the Army
Washington, DC, 1 February 2013

Special Forces Military Free-Fall Operations

1. Change Army Tactics, Techniques, and Procedures (ATTP) 3-18.11/Air Force Manual (AFMAN) 11-411(I)/Navy Tactics, Techniques, and Procedures (NTTP) 3-05.26M, dated 14 October 2011, as follows:

Remove old pages:	Insert new pages:
iii and iv	iii and iv
ix through xii	ix through xii
14-13 through 14-16	14-13 through 14-16

2. A bar (|) marks changed material.

3. File this transmittal sheet in front of the publication.

DISTRIBUTION RESTRICTION: Distribution authorized to U.S. Government agencies and their contractors only to protect technical or operational information from automatic dissemination under the International Exchange Program or by other means. This determination was made on 1 September 2011. Other requests for this document must be referred to Commander, United States Army John F. Kennedy Special Warfare Center and School, ATTN: AOJK-CDI-SF, 3004 Ardennes Street, Stop A, Fort Bragg, NC 28310-9610, or by e-mail to AOJK-DT-SF@soc.mil.

DESTRUCTION NOTICE: Destroy by any method that will prevent disclosure of contents or reconstruction of the document.

FOREIGN DISCLOSURE RESTRICTION (FD 6): This publication has been reviewed by the product developers in coordination with the United States Army John F. Kennedy Special Warfare Center and School foreign disclosure authority. This product is releasable to students from foreign countries on a case-by-case basis only.

ATTP 3-18.11, C1
1 February 2013

By Order of the Secretary of the Army:

RAYMOND T. ODIERNO
General, United States Army
Chief of Staff

Official:

JOYCE E. MORROW
Administrative Assistant to the
Secretary of the Army
1301403

DISTRIBUTION:
Active Army, Army National Guard, and U.S. Army Reserve: Not to be distributed; electronic media only.

PIN: 102456-001

*ATTP 3-18.11 (FM 3-05.211)
AFMAN 11-411(I)
NTTP 3-05.26M

Army Tactics, Techniques, and Procedures
No. 3-18.11 (FM 3-05.211)

Headquarters
Department of the Army
Washington, DC, 14 October 2011

Special Forces Military Free-Fall Operations

Contents

		Page
	PREFACE	xiv
Chapter 1	MILITARY FREE-FALL PARACHUTE OPERATIONS	1-1
	Characteristics	1-1
	Planning Considerations	1-2
	Phases of Military Free-Fall Operations	1-5
Chapter 2	MC-4 RAM-AIR PERSONNEL PARACHUTE SYSTEM	2-1
	MC-4 Ram-Air Personnel Parachute System Components	2-1
	Donning and Recovering the MC-4 Ram-Air Personnel Parachute System	2-22
Chapter 3	CYBERNETIC PARACHUTE RELEASE SYSTEM	3-1
	General Information on Military CYPRES 2 Models	3-1
	Military CYPRES 2 Principles of Operation	3-2
	Military CYPRES 2 Models	3-3
	Expert CYPRES 2 Model	3-4
	CYPRES Model Identification	3-4
	Components	3-6

DISTRIBUTION RESTRICTION: Distribution authorized to U.S. Government agencies and their contractors only to protect technical or operational information from automatic dissemination under the International Exchange Program or by other means. This determination was made on 1 September 2011. Other requests for this document must be referred to Commander, United States Army John F. Kennedy Special Warfare Center and School, ATTN: AOJK-CDI-SF, 3004 Ardennes Street, Stop A, Fort Bragg, NC 28310-9610, or by e-mail to AOJK-DT-SF@soc.mil.

DESTRUCTION NOTICE: Destroy by any method that will prevent disclosure of contents or reconstruction of the document.

FOREIGN DISCLOSURE RESTRICTION (FD 6): This publication has been reviewed by the product developers in coordination with the United States Army John F. Kennedy Special Warfare Center and School foreign disclosure authority. This product is releasable to students from foreign countries on a case-by-case basis only.

*This publication supersedes FM 3-05.211, 6 April 2005.

Contents

Maintenance	3-8
General Terms	3-9
Modes of Operation	3-10
Mode Determination	3-11
Operating Procedures	3-16
Using Military CYPRES 2 Calculators	3-25

Chapter 4 USE OF OXYGEN IN SUPPORT OF MILITARY FREE-FALL OPERATIONS ... 4-1

Oxygen Handling and Safety	4-1
Physiological Effects of High-Altitude Military Free-Fall Operations	4-2
Oxygen Forms	4-4
Oxygen Requirements	4-5
Oxygen Life-Support Equipment	4-6
The "PRICE" Check	4-27
Oxygen Safety Personnel and Preflight Checks	4-28
Oxygen Handling and Safety	4-32

Chapter 5 EQUIPMENT AND WEAPON RIGGING PROCEDURES ... 5-1

Equipment and Weapon Packing Considerations	5-1
Parachutist and Parachute Load Limitations	5-1
Hook-Pile Tape (Velcro) Lowering Line Assembly	5-3
Combat Packs and Other Equipment Containers	5-5
Parachutist Drop Bag	5-29
Weapon-Rigging Procedures	5-33
Flotation Devices/Life Preservers	5-54

Chapter 6 AIRCRAFT PROCEDURE SIGNALS AND JUMP COMMANDS ... 6-1

Aircraft Procedure Signals	6-1
Jump Commands	6-9

Chapter 7 BODY STABILIZATION ... 7-1

Tabletop Body Stabilization Training	7-1
Main Rip Cord Pull	7-5
Tracking	7-7
Recovery from Instability	7-8

Chapter 8 RAM-AIR PARACHUTE FLIGHT CHARACTERISTICS AND CANOPY CONTROL ... 8-1

Ram-Air Parachute Characteristics	8-1
Ram-Air Parachute Deployment Sequence	8-3
Ram-Air Parachute Theory of Flight	8-6
Canopy Performance Factors	8-7
Parachute Flight Characteristics	8-8
Canopy Control	8-9
Canopy Maneuvers	8-13
Landing Maneuvers	8-19
Landing Approaches	8-20
Turbulence	8-24

Chapter 9 EMERGENCY PROCEDURES FOR MILITARY FREE-FALL OPERATIONS ... 9-1

Chapter 10	**HIGH-ALTITUDE HIGH-OPENING AND LIMITED-VISIBILITY OPERATIONS**	
	Refresher Training	9-1
	Emergency Measures	9-1
	Actions for Dust Devils and Turbulent Air	9-10
		10-1
	Techniques and Requirements	10-1
	Special Equipment	10-2
	Free-Fall Delays	10-8
	Parachute Jump Phases	10-8
	Limited-Visibility Operations	10-11
	Military Free Fall With Night Vision Goggles	10-12
Chapter 11	**MILITARY FREE-FALL DROP ZONE OPERATIONS**	11-1
	Responsibilities	11-1
	Drop Zone Selection Criteria	11-2
	Drop Zone Surveys	11-3
	Drop Zone Personnel Qualifications and Responsibilities	11-4
	Military Free-Fall Drop Zone Markings	11-6
	High-Altitude Release Point and Military Free-Fall Drop Zone Detection	11-8
	Aircraft or High-Altitude High-Opening Team Identification	11-9
	Authentication System	11-9
Chapter 12	**DELIBERATE WATER MILITARY FREE-FALL OPERATIONS**	12-1
	Additional Support Requirements	12-1
	Parachutist Requirements	12-2
	Equipment Requirements	12-2
	Drop Zone Requirements and Markings	12-4
	Parachutist Procedures for Water Jumps	12-5
	Drop Zone Procedures for Pickup of Parachutists and Equipment	12-5
	Night Water Parachute Operations	12-6
	Water Jumps With Combat Equipment	12-6
Chapter 13	**JUMPMASTER RESPONSIBILITIES AND CURRENCY QUALIFICATIONS**	13-1
	Responsibilities	13-1
	Qualifications	13-2
	Cardinal Rules for the Jumpmaster	13-2
	Currency and Requalification Requirements	13-3
Chapter 14	**WEATHER FACTORS FOR THE MILITARY FREE-FALL JUMPMASTER**	14-1
	Criticality of Weather Knowledge	14-1
	Mission Planning Tools	14-2
	Atmosphere	14-5
	Weather	14-7
	Temperature	14-8
	Environmental Effects on Altimeters	14-16
	Air Density Altitude	14-17
	Mapping of Pressure Systems	14-18
	Atmospheric Circulation	14-19

Contents

	Wind Flow Mechanics 14-20
	Land and Sea Breezes 14-23
	Eddy Winds 14-24
	Clouds 14-28
	Thunderstorms 14-29
	Application of General Weather Principles 14-32
	Moon Phases 14-36
	Reference Tables 14-39
Appendix A	MILITARY FREE-FALL CRITICAL TASK LISTS A-1
Appendix B	MILITARY FREE-FALL PARACHUTIST QUALIFICATION AND REFRESHER TRAINING REQUIREMENTS B-1
Appendix C	RECOMMENDED MILITARY FREE-FALL TRAINING PROGRAMS C-1
Appendix D	SUGGESTED MILITARY FREE-FALL SUSTAINED AIRBORNE TRAINING D-1
Appendix E	SAMPLE ACCIDENT REPORT E-1
Appendix F	HIGH-ALTITUDE RELEASE POINT CALCULATION F-1
Appendix G	JUMPMASTER PERSONNEL INSPECTION G-1
Appendix H	SAMPLE AIRCRAFT INSPECTION CHECKLIST H-1
Appendix I	JUMPMASTER AIRCREW BRIEFING CHECKLIST I-1
Appendix J	JOINT PRECISION AIRDROP SYSTEM J-1
	GLOSSARY Glossary-1
	REFERENCES References-1
	INDEX Index-1

Figures

Figure 1-1. Military free-fall operations planning phases 1-6
Figure 2-1. MC-4 RAPPS components 2-1
Figure 2-2. MC-4 RAPPS harness and container assembly components 2-4
Figure 2-3. MC-4 RAPPS assembly components 2-5
Figure 2-4. Location of the three-ring canopy release assembly 2-6
Figure 2-5. Location of the main rip cord handle and cutaway handle 2-7
Figure 2-6. Location of the chest strap, reserve rip cord handle, large equipment attachment ring, and reserve rip cord cable housing 2-8
Figure 2-7. Location of the oxygen fitting block and equipment lowering line attachment V-ring 2-9
Figure 2-8. Location of the main and reserve parachutes in the container 2-9
Figure 2-9. Location of straps 2-10
Figure 2-10. Location of the equipment tie-down loop and main risers 2-10
Figure 2-11. Location of reserve components 2-11
Figure 2-12. MA2-30/A and PA-200 free-fall altimeters 2-12

Figure 2-13. Altimeter setting ... 2-13
Figure 2-14. MA-10 altimeter buttons ... 2-14
Figure 2-15. On/Off buttons ... 2-14
Figure 2-16. Power-saving mode ... 2-15
Figure 2-17. Zeroing the MA-10 altimeter ... 2-15
Figure 2-18. Manual offset ... 2-16
Figure 2-19. Setting the drop zone ... 2-17
Figure 2-20. Replacing batteries in the MA-10 altimeter ... 2-19
Figure 2-21. Jumpsuits ... 2-21
Figure 2-22. Parachutist individual equipment kit ... 2-21
Figure 2-23. Aviator's and MC-4 kit bags ... 2-22
Figure 2-24. Donning the MC-4 Ram-Air Personnel Parachute System ... 2-24
Figure 3-1. Military CYPRES 2 1500 35 A ... 3-3
Figure 3-2. Expert CYPRES 2 ... 3-4
Figure 3-3. Military CYPRES 2 Model 1000 35 A control unit ... 3-5
Figure 3-4. Military CYPRES 2 Model 1500 35 A control unit ... 3-5
Figure 3-5. Military CYPRES 2 Model 2500 29 A control unit ... 3-6
Figure 3-6. Expert CYPRES 2 control unit ... 3-6
Figure 3-7. Military CYPRES 2 control unit ... 3-7
Figure 3-8. Back of Military CYPRES 2 control unit ... 3-7
Figure 3-9. Military CYPRES 2 processing unit ... 3-7
Figure 3-10. Military CYPRES 2 release unit ... 3-8
Figure 3-11. Example of Military CYPRES 2 serial number ... 3-9
Figure 3-12. Example of next required maintenance date for Military CYPRES 2 ... 3-9
Figure 3-13. Power ON sequence for Military CYPRES 2 in default (training) mode ... 3-18
Figure 3-14. Beginning of Military CYPRES 2 self-test countdown in default (training) mode ... 3-18
Figure 3-15. Military CYPRES 2 displaying current barometric pressure in millibars ... 3-18
Figure 3-16. Military CYPRES 2 set in default (training) mode ... 3-18
Figure 3-17. Example of Military CYPRES 2 error code ... 3-19
Figure 3-18. Power ON sequence for Military CYPRES 2 in absolute (operational) mode ... 3-20
Figure 3-19. Beginning of Military CYPRES 2 self-test countdown in absolute (operational) mode ... 3-20
Figure 3-20. Military CYPRES 2 displaying current barometric pressure in millibars ... 3-20
Figure 3-21. Military CYPRES 2 set in absolute (operational) mode ... 3-20
Figure 3-22. Example of Military CYPRES 2 error code ... 3-21
Figure 3-23. First value of 1 chosen for millibar setting ... 3-21
Figure 3-24. Second value of 0 chosen for millibar setting ... 3-21
Figure 3-25. Third value of 1 chosen for millibar setting ... 3-21
Figure 3-26. Final value chosen and Military CYPRES 2 set ... 3-22
Figure 3-27. Power ON sequence for Expert CYPRES 2 in offset mode ... 3-23

Figure 3-28. Expert CYPRES 2 displaying countdown ... 3-23
Figure 3-29. Example of Expert CYPRES 2 error code ... 3-24
Figure 3-30. Expert CYPRES 2 control unit displaying countdown at zero down in offset mode ... 3-24
Figure 3-31. Expert CYPRES 2 set at 120-foot offset ... 3-24
Figure 3-32. Power OFF sequence for CYPRES 2 ... 3-25
Figure 3-33. Military CYPRES Absolute Adjust Circular Calculator (Whiz Wheel) ... 3-26
Figure 3-34. PDA computer with Military CYPRES Absolute Model Calculator software download ... 3-27
Figure 3-35. Military CYPRES Calculator ... 3-27
Figure 3-36. Military CYPRES Absolute Adjust Circular Calculator ... 3-28
Figure 3-37. Value for the Military CYPRES setting is displayed in the CYPRES setting box ... 3-29
Figure 3-38. Usage instructions for the Military CYPRES calculator ... 3-30
Figure 3-39. Instructions and first page of online "Military CYPRES Absolute Adjust Model Calculator" ... 3-30
Figure 3-40. Step 1: Military CYPRES Absolute Adjust Model Calculator ... 3-31
Figure 3-41. Step 2: Military CYPRES Absolute Adjust Model Calculator ... 3-31
Figure 3-42. Step 3: Military CYPRES Absolute Adjust Model Calculator ... 3-31
Figure 3-43. Step 4: Military CYPRES Absolute Adjust Model Calculator ... 3-32
Figure 3-44. Military CYPRES 2 air travel card ... 3-33
Figure 4-1. MBU-12/P pressure-demand oxygen mask components ... 4-8
Figure 4-2. Parachutist Oxygen Mask ... 4-9
Figure 4-3. The improved oxygen harness ... 4-9
Figure 4-4. Complete Parachutist Oxygen Mask with MICH ... 4-10
Figure 4-5. Advanced combat helmet accessory rail connector with oxygen single-strap and double-strap kits ... 4-11
Figure 4-6. Parachutist Oxygen Mask and HS-57 quick disconnect ... 4-12
Figure 4-7. Fitting the MBU-12/P oxygen mask ... 4-13
Figure 4-8. Properly fitted mask ... 4-14
Figure 4-9. Parachutist Oxygen Mask with bayonet connectors and taped straps ... 4-15
Figure 4-10. The 106-cubic-inch PBOS with the quick-disconnect oxygen hose ... 4-17
Figure 4-11. AIROX VIII assembly ... 4-17
Figure 4-12. Rigging the portable bailout oxygen system with the AIROX VIII assembly to the RAPPS ... 4-19
Figure 4-13. Completed rigging of the portable bailout oxygen system with the AIROX VIII assembly to the RAPPS ... 4-20
Figure 4-14. Completed rigging of the POM/ASFS with the Parachutist Oxygen System assembly to the RAPPS ... 4-21
Figure 4-15. Six-Man Prebreather Portable Oxygen System ... 4-22
Figure 4-16. OXCON rigged in C-130 aircraft ... 4-24
Figure 4-17. Charging assembly looped and taped out of the way of parachutists ... 4-25
Figure 4-18. Side and top view of strap on K-bottle ... 4-25

Figure 4-19. MA-1 Portable Oxygen Assembly 4-26
Figure 4-20. Tie-down assembly and installation 4-27
Figure 4-21. Portable Bailout Oxygen System preflight inspection and operational checklist 4-29
Figure 4-22. Sample prebreather preflight inspection and operational function checklist 4-30
Figure 4-23. Pressure gauge and manual shutoff valve 4-31
Figure 4-24. Removing end plugs and depressing poppets 4-32
Figure 5-1. Stowing the HPT lowering line assembly 5-4
Figure 5-2. H-harness with attaching straps 5-5
Figure 5-3. H-harness attached to the kit bag 5-6
Figure 5-4. Combat pack and frame rigged with the modified H-harness 5-7
Figure 5-5. Improved equipment attachment sling and lowering line (spider harness) 5-8
Figure 5-6. Combat pack and frame rigged with the improved equipment attachment sling 5-9
Figure 5-7. Attaching the lowering line to the combat pack 5-10
Figure 5-8. Attaching the rear-mounted combat pack 5-11
Figure 5-9. Lowering line attached to the lowering line attachment V-ring 5-12
Figure 5-10. Attaching the front-mounted combat pack 5-13
Figure 5-11. Opened SARPELS cargo carrier 5-15
Figure 5-12. SARPELS with folded side flaps 5-15
Figure 5-13. Stowage pockets 5-16
Figure 5-14. Side flap 5-16
Figure 5-15. Top flaps 5-17
Figure 5-16. SARPELS with secured horizontal straps 5-17
Figure 5-17. SARPELS with secured vertical straps 5-18
Figure 5-18. Inserting the white webbing through the parachute harness link ... 5-19
Figure 5-19. Inserting the green 550 cord through the white webbing 5-19
Figure 5-20. Inserting the red 550 cord through the green 550 cord 5-20
Figure 5-21. Inserting the red 550 cord through the grommet 5-20
Figure 5-22. Leg strap cable retainer with buckle and grommet 5-20
Figure 5-23. SARPELS release assembly 5-21
Figure 5-24. Stowage pocket with 8-foot lowering line 5-21
Figure 5-25. Securing the 8-foot lowering line to the cargo carrier 5-22
Figure 5-26. Mounted SARPELS 5-23
Figure 5-27. Single-point release handle 5-23
Figure 5-28. Harness, single-point release (NSN 1670-01-227-7992) 5-24
Figure 5-29. Release handle and D-ring attaching straps 5-25
Figure 5-30. Attaching snap hooks and leg strap release assembly 5-26
Figure 5-31. Rigging the HSPR 5-27
Figure 5-32. Completing rigging the HSPR 5-27
Figure 5-33. Attaching the hook-pile tape lowering line assembly 5-28

Contents

Figure 5-34. Attaching the HSPR-rigged combat equipment 5-29
Figure 5-35. Parachutist with HSPR-rigged combat pack 5-29
Figure 5-36. Compression straps connected and tightened 5-30
Figure 5-37. Loading the drop bag 5-31
Figure 5-38. Drop bag zipped shut with compression straps connected and tightened 5-31
Figure 5-39. Drop bag attaches to the parachutist by standard quick-release
connectors 5-32
Figure 5-40. PDB rigged for rear-mounted jump 5-32
Figure 5-41. PDB rigged for front-mounted jump 5-33
Figure 5-42. Center-mounted weapons harness 5-34
Figure 5-43. Center-mounted weapons harness components 5-35
Figure 5-44. Main lift web attaching points 5-35
Figure 5-45. Attaching horizontal straps to the pile portion weapons harness 5-36
Figure 5-46. Securing triple-fold hook and pile 5-36
Figure 5-47. Securing weapon harness to weapon 5-37
Figure 5-48. Attaching weapon to main lift web attaching points 5-37
Figure 5-49. Sling over chest strap 5-38
Figure 5-50. Attaching straps routed over weapon 5-38
Figure 5-51. M4 carbine-series rifles rigged for jumping 5-40
Figure 5-52. Positioning the weapon on the parachutist 5-41
Figure 5-53. SCAR in folded and open positions 5-42
Figure 5-54. Right-side weapon rigging 5-43
Figure 5-55. M203 rigged for jumping 5-44
Figure 5-56. M14 rigged for jumping 5-45
Figure 5-57. M110 Semi-Automatic Sniper System 5-45
Figure 5-58. MP5 rigged for jumping 5-47
Figure 5-59. M249 and Para M249 squad automatic weapons rigged for jumping 5-48
Figure 5-60. M240G disassembled and packed for jumping 5-49
Figure 5-61. AT-4 and 84-mm Carl Gustaf rigged for jumping 5-50
Figure 5-62. Routing of vertical compression straps 5-50
Figure 5-63. Antiarmor weapon tie-down locations 5-51
Figure 5-64. Parachutist rigged for jumping with an AT weapon mounted
on top of combat pack 5-51
Figure 5-65. Front-mounted weapon with rear-mounted rucksack 5-52
Figure 5-66. Front-mounted weapon with front-mounted rucksack 5-52
Figure 5-67. M224 60-mm mortar rigged for front mount 5-53
Figure 5-68. Left-side mount for M224 60-mm mortar 5-54
Figure 5-69. Underwater demolition team life preservers 5-55
Figure 5-70. Oralock valve 5-57
Figure 5-71. Parachutist with UDT life vest and MC-4 parachute harness 5-59
Figure 6-1. DON HELMETS signal 6-3
Figure 6-2. UNFASTEN SEAT BELTS signal 6-4

Contents

Figure 6-3. EMERGENCY BAILOUT signal ... 6-5
Figure 6-4. MASK signal ... 6-5
Figure 6-5. CHECK OXYGEN signal ... 6-6
Figure 6-6. OXYGEN PROBLEM signal ... 6-6
Figure 6-7. TIME WARNINGS signal ... 6-7
Figure 6-8. WIND SPEED signal ... 6-8
Figure 6-9. GUSTING WINDS signal ... 6-9
Figure 6-10. STAND UP command ... 6-10
Figure 6-11. MOVE TO THE REAR command ... 6-10
Figure 6-12. STAND BY command ... 6-11
Figure 6-13. GO command ... 6-11
Figure 6-14. ABORT command ... 6-12
Figure 7-1. Poised exit position ... 7-1
Figure 7-2. Box man method ... 7-2
Figure 7-3. Diving exit position ... 7-3
Figure 7-4. Stable free-fall position ... 7-4
Figure 7-5. Body turn ... 7-4
Figure 7-6. Gliding ... 7-4
Figure 7-7. Altimeter check ... 7-5
Figure 7-8. Main rip cord pull ... 7-6
Figure 7-9. Tracking position ... 7-7
Figure 7-10. Example of tracking away for separation ... 7-8
Figure 8-1. Shape of the ram-air parachute canopy ... 8-1
Figure 8-2. Structure of the ram-air parachute canopy ... 8-1
Figure 8-3. Components and nomenclature of the ram-air parachute ... 8-2
Figure 8-4. Location of components of the ram-air parachute ... 8-3
Figure 8-5. Detailed lower portion of the ram-air parachute ... 8-4
Figure 8-6. Deployment sequence ... 8-5
Figure 8-7. Cutaway sequence and deployment of the reserve parachute ... 8-6
Figure 8-8. Ram-air parachute theory of flight ... 8-7
Figure 8-9. Applying brakes on the ram-air parachute ... 8-8
Figure 8-10. Controlling ground speed ... 8-10
Figure 8-11. Parachutist guide to good canopy control ... 8-10
Figure 8-12. Holding maneuver ... 8-11
Figure 8-13. Running maneuver ... 8-11
Figure 8-14. Crabbing maneuver ... 8-12
Figure 8-15. Effective canopy range ... 8-12
Figure 8-16. Brake-setting glide angles ... 8-13
Figure 8-17. Full flight ... 8-14
Figure 8-18. Half brakes ... 8-15
Figure 8-19. Full brakes ... 8-16

Contents

Figure 8-20. Stall 8-16
Figure 8-21. Spiral turn 8-17
Figure 8-22. Flat turn 8-18
Figure 8-23. Glide angles for a final approach 8-22
Figure 8-24. Landing approaches 8-22
Figure 8-25. High and low wind patterns 8-23
Figure 8-26. Significant change in wind direction 8-24
Figure 8-27. Adjusting for increase in winds on downwind leg 8-25
Figure 8-28. Adjusting for increase in winds on base leg 8-25
Figure 8-29. Adjusting for decrease in winds on base leg 8-26
Figure 9-1. Emergency preparations before takeoff 9-1
Figure 9-2. Parachutist postopening procedures 9-5
Figure 9-3. Controllability check 9-5
Figure 9-4. Parachutist emergency landing procedures 9-10
Figure 9-5. High-wind landing procedures 9-10
Figure 10-1. Jumper with individual body armor 10-2
Figure 10-2. Compass mounted to high-altitude high-opening navigation board 10-3
Figure 10-3. Navigation aid attaching point 10-4
Figure 10-4. Wilcox Parachutist Navigation Board 10-6
Figure 10-5. Parachutist navigation board closed and open position 10-7
Figure 10-6. Navigation aid attached to parachutist 10-7
Figure 10-7. Wedge formation 10-9
Figure 10-8. Trail formation 10-10
Figure 10-9. Trim tab locations 10-10
Figure 10-10. AN/AVS-6(V)3 10-13
Figure 10-11. AN/PVS-14 10-13
Figure 10-12. AN/PVS-15 10-13
Figure 10-13. Night vision goggle mounts 10-14
Figure 10-14. Bungee position on night vision goggle mount 10-15
Figure 11-1. Military free-fall drop zone markings 11-7
Figure 11-2. Examples of wind socks 11-7
Figure 14-1. Atmosphere 14-6
Figure 14-2. Mercury barometer 14-7
Figure 14-3. Temperature scales 14-10
Figure 14-4. Effect of temperature change on altimeter's indicated altitude (AGL) 14-11
Figure 14-5. Atmospheric pressure change over large distance resulting in false altitude (AGL) readings 14-12
Figure 14-6. **Deleted by Change 1** ....
Figure 14-7. **Deleted by Change 1** ....
Figure 14-8. **Deleted by Change 1** ....
Figure 14-9. Air density variation with temperature change 14-17

Figure 14-10. Pressure systems 14-19
Figure 14-11. General pattern of atmospheric circulation... 14-20
Figure 14-12. Wind shift with altitude increase 14-21
Figure 14-13. Pressure gradient principles 14-22
Figure 14-14. Change in velocity with altitude 14-22
Figure 14-15. Land and sea breezes 14-23
Figure 14-16. Single-obstacle eddy current 14-26
Figure 14-17. Terrain-induced eddy currents... 14-26
Figure 14-18. Tree-line-induced eddy currents 14-27
Figure 14-19. Cloud classification 14-29
Figure 14-20. Air movement beneath a thunderstorm cell... 14-30
Figure 14-21. First gust wind flow 14-31
Figure 14-22. Wind shift as a front passes.. 14-33
Figure D-1. Mock aircraft rehearsal.. D-1
Figure D-2. Actions in free fall and canopy flight.. D-1
Figure D-3. Sample jumpmaster troop briefing D-2
Figure D-4. Emergency procedures D-4
Figure E-1. Sample accident report.. E-2
Figure F-1. Plotting the HARP, free-fall, and canopy drift for a 20,000-foot HALO mission profile F-2
Figure G-1. JMPI without oxygen, weapon, or rucksack.. G-2
Figure G-2. JMPI with oxygen and life preserver G-7
Figure G-3. JMPI for weapon, front-mounted rucksack G-10
Figure G-4. JMPI with the rear-mounted rucksack/parachutist drop bag G-12
Figure H-1. Sample aircraft inspection checklist.. H-1
Figure I-1. Sample jumpmaster aircrew briefing checklist I-1
Figure J-1. Personnel and JPADS combination airdrop operations J-3

Tables

Table 1-1. Minimum and maximum exit and opening altitudes ... 1-3
Table 1-2. Surface interval chart for conducting military free-fall operations after diving ... 1-3
Table 1-3. METT-TC analysis ... 1-4
Table 3-1. CYPRES 2 model identification ... 3-5
Table 3-2. CYPRES 2 power ON self-test error codes in default (training) mode ... 3-17
Table 3-3. CYPRES 2 power ON self-test error codes in absolute (operational) mode ... 3-19
Table 3-4. Expert CYPRES 2 power ON self-test error codes in offset mode ... 3-22
Table 4-1. Supplemental oxygen requirements for MFF parachutists ... 4-6
Table 5-1. Container weight limits ... 5-1
Table 5-2. Parachute load limits ... 5-2
Table 5-3. Weight of parachutist with two equipment loads ... 5-2
Table 5-4. Weight of parachutist with two equipment loads and basic load ... 5-2
Table 5-5. Lift capabilities ... 5-56
Table 6-1. Aircraft procedure signals (oxygen and nonoxygen jumps) ... 6-1
Table 6-2. Aircraft jump commands (oxygen and nonoxygen jumps) ... 6-2
Table 9-1. In-flight emergency procedures and signals ... 9-2
Table 9-2. In-flight emergency procedures ... 9-3
Table 9-3. Emergencies in free fall ... 9-4
Table 9-4. Cutaway procedures ... 9-5
Table 9-5. Malfunction procedures ... 9-6
Table 9-6. Canopy entanglement procedures ... 9-9
Table 10-1. Required free-fall delays ... 10-8
Table 12-1. Wind/sea state observation chart ... 12-4
Table 13-1. Jumpmaster responsibilities ... 13-1
Table 14-1. Military free-fall operations windchill determination ... 14-9
Table 14-2. **Deleted by Change 1**
Table 14-3. Military free-fall lunar data example ... 14-38
Table 14-4. Approximate wind velocity by natural indicators ... 14-39
Table 14-5. Handkerchief angle wind velocity ... 14-39
Table 14-6. Linear measure ... 14-40
Table 14-7. Liquid measure ... 14-40
Table 14-8. Weight ... 14-40
Table 14-9. Square measure ... 14-40
Table 14-10. Cubic measure ... 14-40
Table 14-11. Temperature ... 14-41
Table 14-12. Approximate conversion factors ... 14-41
Table 14-13. Area ... 14-41

Table 14-14. Volume 14-42
Table 14-15. Capacity 14-42
Table 14-16. Statute miles to kilometers and nautical miles 14-42
Table 14-17. Nautical miles to kilometers and statute miles 14-43
Table 14-18. Kilometers to statute and nautical miles 14-43
Table 14-19. Yards to meters 14-44
Table 14-20. Meters to yards 14-44
Table 14-21. Determination of altitude by barometric pressure (in inches of mercury) 14-44
Table C-1. Minimum quarterly training guide C-1
Table C-2. Suggested 10-day combat-ready training program C-3
Table F-1. HAHO K factors for Department of Defense Ram-Air Personnel Parachute Systems F-3

Preface

Department of Defense Directive (DODD) 5100.1, *Functions of the Department of Defense and its Major Components*, tasks the Army to "train and equip, as required, forces for airborne operations, in coordination with the other Military Services, and in accordance with (IAW) joint doctrine" and directs the Army, which has primary responsibility for the development of airborne doctrine, procedures, and techniques, to develop, in coordination with the other Military Services, doctrine, procedures, and equipment that are of common interest.

Department of the Army (DA) memorandum, Subject: Army Military Free-Fall Proponency, dated 23 September 1998, establishes the United States Army Special Operations Command (USASOC) as the proponent for military free-fall (MFF) training, operations, equipment, and doctrine.

United States Special Operations Command (USSOCOM) Directive 10-1, *Organization and Functions, Terms of Reference Roles, Missions, and Functions of Component Commands*, Appendix A, page A-3, paragraph g, establishes the Commander, USASOC, as the lead component for MFF training, doctrine, safety, equipment, and interoperability for USSOCOM Active Army and Reserve forces.

The Commander, United States Army John F. Kennedy Special Warfare Center and School (USAJFKSWCS), Fort Bragg, North Carolina, serves as the USASOC-specified proponent for MFF parachuting training and doctrine.

PURPOSE

Army Tactics, Techniques, and Procedures (ATTP) 3-18.11 presents a series of concise, proven techniques and guidelines that are essential to safe, successful MFF operations. The techniques and guidelines prescribed herein are generic in nature and represent the safest and most effective methodologies available for executing MFF operations.

SCOPE

This ATTP provides a consolidated reference for MFF airborne operations and training and will assist commanders at all levels in preparing special operations forces (SOF) in the execution of MFF airborne operations. These operations may involve the employment of forces from air platforms to meet objectives aground. MFF operations may be in support of or independent from other air or ground operations.

APPLICABILITY

This ATTP applies to Army and USSOCOM MFF-capable units. USSOCOM components are authorized to produce publications to supplement this manual to clarify and amplify the procedures and equipment being utilized to meet the different varieties of equipment being used by SOF. Commanders can request waivers from their Service or component commanders to meet specific operational requirements when methodologies contained in this manual impede mission accomplishment.

When Service publications and USSOCOM publications conflict, USSOCOM publications will take precedence during operations in which USSOCOM units are the supported unit. When conducting Service-pure MFF operations, Services will use their applicable regulations and standing operating procedures (SOPs).

ADMINISTRATIVE INFORMATION

The proponent and preparing agency of this publication is the USAJFKSWCS. Submit comments and recommended changes on DA Form 2028 (Recommended Changes to Publications and Blank Forms) directly to Commander, USAJFKSWCS, ATTN: AOJK-CDI-SF, 3004 Ardennes Street, Stop A, Fort Bragg, NC 28310-9610; by e-mail to AOJK-DT-SF@soc.mil; or by electronic DA Form 2028. This ATTP implements Standardization Agreement (STANAG) 3570, *Drop Zones and Extraction Zones—Criteria and Markings*, dated 26 March 1986. Unless this publication states otherwise, masculine nouns and pronouns do not refer exclusively to men.

Chapter 1

Military Free-Fall Parachute Operations

SOF must conduct a detailed mission analysis to determine an appropriate method of infiltration. MFF operations are one of the many options available to a commander to infiltrate personnel into a designated area of operations (AO). MFF operations are ideally suited for, but not limited to, the infiltration of operational elements, pilot teams, pathfinder elements, special tactics team (STT) assets, and personnel replacements conducting various missions across the operational continuum. A thorough understanding of all the factors impacting MFF operations is essential due to the inherently high levels of risk associated with MFF operations. The objective of this chapter is to familiarize the reader with MFF operations and to outline the planning considerations needed to successfully execute MFF operations.

CHARACTERISTICS

1-1. MFF parachute operations are used when enemy air defense systems, terrain restrictions, or politically sensitive environments prevent low-altitude penetration or when mission needs require a clandestine insertion. MFF parachute infiltrations are conducted using the ram-air personnel parachute system (RAPPS), which is a high-performance gliding system. The RAPPS is a highly maneuverable parachute that has forward air speeds of 20 to 30 miles per hour (mph). The RAPPS can be manually deployed during free fall or with the assistance of a static line, depending on mission and jumper capabilities. The glide capability of the RAPPS provides commanders the means to conduct standoff infiltrations of designated areas without having to physically fly over the target area. This process allows commanders to keep high-value air assets outside the detection and threat ranges of enemy air defense systems or politically sensitive areas.

1-2. MFF parachuting allows SOF personnel to deploy their parachutes at a predetermined altitude, assemble in the air, navigate under canopy, and land safely together as a tactical unit ready to execute their mission. Although free-fall parachuting can produce highly accurate landings, it is primarily a means of entering a designated impact area within the objective area. The following are two basic types of MFF operations:

- High-altitude low-opening (HALO) operations are jumps made with an exit altitude of up to 35,000 feet mean sea level (MSL) and a parachute deployment altitude at or below 6,000 feet above ground level (AGL). HALO infiltrations are the preferred MFF method of infiltration when the enemy air defense posture is not a viable threat to the infiltration platform or when a low opening will not compromise the team's position on infiltration. HALO infiltrations require the infiltration platform to fly within several kilometers of the drop zone (DZ).

- High-altitude high-opening (HAHO) operations are standoff infiltration jumps made with an exit altitude of up to 35,000 feet MSL and a parachute deployment altitude at or above 6,000 feet AGL to 25,000 feet AGL. HAHO infiltrations are the preferred method of infiltration when the enemy air defense threat is viable or when a low-signature infiltration is required. Standoff HAHO infiltrations provide commanders a means to drop MFF parachutists outside the air defense umbrella, where they can navigate undetected under canopy to the DZ or objective area. The most important objective of a HAHO is for team members to land together, even if circumstances force the team to land in an area that might not have been the original landing zone. Sometimes it is necessary to choose an alternate suitable area close to the objective area that provides the advantages of a clandestine insertion.

Chapter 1

1-3. Personnel involved in MFF operations require extensive knowledge of meteorology and navigation. They must be able to conduct realistic premission training, gather information, plan, rehearse, and use the appropriate MFF infiltration technique to accomplish their assigned mission. (Appendix A includes the critical task lists for the MFF basic, advanced, and jumpmaster courses.) Selected units within SOF have the capability to conduct tandem infiltration with large-capacity bundles, personnel, and multipurpose canines, or a combination of two; for example, a jumper, multipurpose handler, and multipurpose canine under one canopy during tandem infiltration.

1-4. When used correctly, MFF infiltrations give commanders another means to move SOF and influence the battlefield. The skills and techniques used in MFF operations are equally applicable to all Special Forces core tasks, especially direct action, special reconnaissance, unconventional warfare, and foreign internal defense.

PLANNING CONSIDERATIONS

1-5. Successful MFF operations depend on thorough mission planning, preparation, coordination, and rehearsals. MFF operations are almost always joint operations that require coordination with an aircrew. Premission planning must include joint briefings and rehearsals between the infiltrating element and the supporting aircrew. Both elements must have a thorough understanding of primary, alternate, and emergency plans. When planning MFF operations, mission planners must consider—

- Mission, enemy, terrain and weather, troops and support available, time available, and civil considerations (METT-TC).
- Ingress and egress routes.
- Suppression of enemy air defenses (SEAD) support.
- Availability of deception air operations in support of actual infiltration.
- Use of commercial airline routes if clandestine infiltration is required in politically sensitive areas.
- In-flight abort criteria.
- En route evasion plan of action for the infiltrating element and the aircrew.
- Availability of aircrews working under arduous conditions in depressurized aircraft at high altitudes.
- Specialized training of personnel and special equipment requirements.
- Currency and proficiency level of the parachutist.
- Drop altitudes requiring the use of oxygen and special environmental protective clothing.
- Limitations on jumping with extremely bulky or heavy equipment. The total combined weight of the parachutist, parachute, and equipment cannot exceed the maximum suspended weight of the canopy (Chapter 5 has more information).
- For joint operations, considerations of different types of parachutes being used.
- Accurate weather data. This information is essential. The lack of accurate meteorological data, such as winds aloft, jet stream direction and velocity, seasonal variances, or topographical effects on turbulence, can severely affect the infiltration's success or the mission's combat effectiveness.
- HAHO standoff operations. Wind, cold, and high-altitude openings increase the probability of physiological stress and injury, parachute damage, and opening shock injuries.
- Minimum and maximum exit and opening altitudes for training (Table 1-1, page 1-3).
- Surface interval after diving operations (Table 1-2, pages 1-3 and 1-4).

Military Free-Fall Parachute Operations

Table 1-1. Minimum and maximum exit and opening altitudes

	Exit Altitude (in Feet)	Opening Altitude (in Feet)
Minimum	5,000 AGL	3,500 AGL
Maximum	35,000 MSL*	25,000 MSL

*NOTE: Openings above 25,000 feet MSL exceed the MC-4, MC-5, and MT-2XX/SL parachute design specifications. The United States Navy (USN) MT-1SS maximum deployment altitude is 12,500 feet MSL.

WARNING

Ascent to altitude after diving increases the risk of decompression sickness because of the additional reduction in atmospheric pressure. The higher the altitude, the greater the risk. (1)

Personnel shall not fly for 72 hours after saturation diving. (2)

Personnel shall wait 48 hours before flying after exceptional exposure. (3)

A diver may need to wait up to 29 hours after a no-compression dive or repetitive dive. (3)

Flying is permitted immediately after oxygen diving unless part of a multiple underwater breathing apparatus dive profile. (4)

References:

(1) USN Diving Manual, Revision 6, paragraph 9-14
(2) USN Diving Manual, Revision 6, paragraph 15-24
(3) USN Diving Manual, Revision 6, Table 9-6
(4) USN Diving Manual, Revision 6, paragraph 18-9

Table 1-2. Surface interval chart for conducting military free-fall operations after diving

Exit Altitude (Maximum)	Oxygen Dive	No-Decompression Dive	Decompression or Repetitive Dive	Exceptional Exposure Dive	Saturation Dive
<13,000 feet MSL	No wait	24 hours	24 hours	72 hours	96 hours
<18,000 feet MSL	No wait	24 hours	36 hours	96 hours	96 hours
<25,000 feet MSL	No wait	24 hours	48 hours	96 hours	120 hours
<35,000 feet MSL	No wait	36 hours	48 hours	120 hours	120 hours

Chapter 1

Table 1-2. Surface interval chart for conducting military free-fall operations after diving (continued)

NOTES:
1. Diving definitions in the table are based on the USN Diving Manual. Listed times include all breathing mixtures.
2. For MFF HAHO operations with opening altitudes above 13,000 feet MSL, 12 hours must be added to the listed times. For MFF HAHO operations with opening altitudes above 18,000 feet MSL, 24 hours must be added to the listed times.
3. When conducting an operation that combines MFF and military scuba diving, the most recent published edition of the USN Diving Manual must first be consulted.

1-6. The successful execution of any operation is directly related to thorough and detailed planning. Mission planning begins with a detailed analysis of METT-TC questions (Table 1-3, pages 1-4 and 1-5), with qualifiers pertaining to MFF operations that the executing element must consider when selecting a method on infiltration.

Table 1-3. METT-TC analysis

Factors	Questions
Mission	• Is the objective located in an area that is conducive to MFF operations? • Is the mission time-critical? • Given the complexity of MFF operations, is time available for the executing element to plan, rehearse, and execute an MFF infiltration? • Is the mission flexible enough to allow for an MFF infiltration window that is dependent on favorable meteorological conditions?
Enemy	• How do the enemy threat, capabilities, disposition, security measures, and air detection or air defense systems affect the method of infiltration? • Does the enemy have the ability to detect or interdict conventional infiltration methods; for example, static-line, waterborne, or air-mobile insertion? • Does the enemy have an air defense system that can be exploited either through gaps in coverage or by SEAD support?
Terrain and Weather	• Is the terrain conducive to an MFF infiltration? • Does the terrain hinder ingress and egress routes? • How does the terrain affect the weather and winds at altitude? • Are there suitable primary, alternate, and contingency DZs available within the objective area (located along the ingress route and in close proximity to one another)? • Are there any storm systems in the AO that might cause unacceptable wind and cloud conditions? • What is the percent of illumination? • Does the executing element have experience navigating under canopy in limited-visibility conditions?

Military Free-Fall Parachute Operations

Table 1-3. METT-TC analysis (continued)

Troops and Support Available	• Does the detachment have the training and experience to successfully execute the selected infiltration method? Is additional training required? • What equipment is required to execute the primary mission? • Does the detachment have the means to infiltrate the required equipment into the AO? • Does the equipment require special rigging? Does it have special handling and storage requirements? • Do overall equipment requirements exceed the suspended weight limitations of the parachute? • Are MFF-capable infiltration aircraft available? • Are SEAD assets available for support if there is a viable air defense threat? • Does the MFF infiltration require additional aircraft to support a deception plan? • Will a reception committee be used on the DZ?
Time Available	• Does the detachment have time to conduct the required training and rehearsals? • How far is it from the high-altitude release point (HARP) to the primary and alternate DZs? • How will unexpected wind conditions at altitude or a low jumper affect the estimated glide distance of the parachute? • Can the detachment make it from the HARP to the primary DZ and complete actions at the assembly area during the hours of darkness? • Will the detachment have the altitude and time to move to an alternate DZ in case the primary DZ is unsuitable or compromised?
Civil Considerations	• Can the operation be executed clandestinely so that the civilian populace is unaware of it? • If the operation is compromised, what will be the repercussions to the local populace? • If the detachment is receiving support from the locals, is there a risk of reprisals against them?

1-7. A thorough METT-TC analysis concentrating on those questions pertaining to MFF operations will determine if MFF infiltration is appropriate. The detachment must then complete the remainder of the mission planning process.

PHASES OF MILITARY FREE-FALL OPERATIONS

1-8. To aid the Special Forces operational detachment in planning and executing, MFF operations are divided into seven phases. Figure 1-1, page 1-6, shows each phase, and the following paragraphs provide the details for each. Graphic Training Aid (GTA) 31-01-003, *Detachment Mission Planning Guide*, has additional information.

Chapter 1

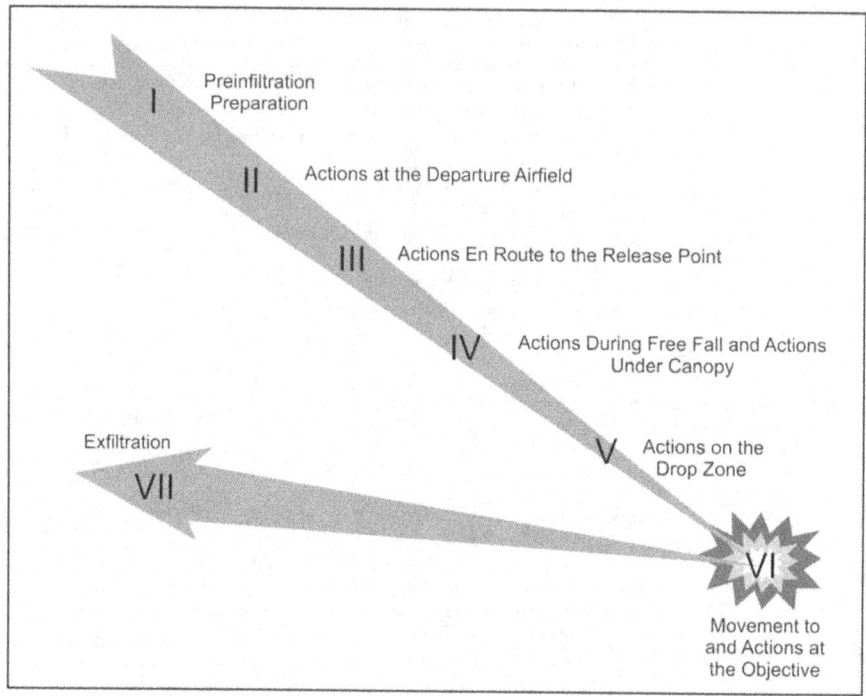

Figure 1-1. Military free-fall operations planning phases

PHASE I: PREINFILTRATION PREPARATION

1-9. Preinfiltration preparation starts with preparing an estimate of the situation. The detachment uses the military decisionmaking process to identify critical nodes in the mission and develop courses of action to address them. During this phase, the detachment will plan the mission, prepare plans and orders, conduct briefbacks, conduct training, prepare equipment, and conduct inspections and rehearsals.

1-10. The air mission brief is one of the key briefings conducted during premission planning. The air mission brief takes place during isolation, before the briefback. The ground commander, primary jumpmaster, and aircrew conduct face-to-face coordination to discuss the following items:
- Flight routes and in-flight checkpoints, to include the point of no return.
- En route mission-abort criteria.
- En route evasion plan of action procedures.
- Emergency landing procedures.
- Actions in the aircraft, including bundles:
  - The earliest possible time the aircraft can be rigged, especially when 6-man oxygen consoles, bundles, and/or rollers are to be used in the mission.
  - The use of the Military Cybernetic Parachute Release System (CYPRES) 2 and restrictions, once armed.
- Call signs and frequencies or visual recognition signals, if used.

1-11. Aircrew attendance at the briefing is mandatory and should include, at a minimum, the aircraft commander, navigator, and the primary loadmaster.

PHASE II: ACTIONS AT THE DEPARTURE AIRFIELD

1-12. In Phase II of infiltration, the detachment moves from the isolation facility to the departure airfield. A pilot-jumpmaster and United States Air Force (USAF) physiological technician briefing normally takes place planeside before loading the aircraft; any changes or updates to the plan are made at this time to discuss any last minute updates to the jump plan. Aircraft is rigged at this point with any additional equipment, such as oxygen consoles and rollers for heavy bundle movement. If in-flight rigging will occur, the MFF rigs, combat packs, weapons, and additional equipment will need to be secured inside the aircraft for departure. If prebreathing is required, Table 4-1, page 4-6, provides start times.

PHASE III: ACTIONS EN ROUTE TO THE RELEASE POINT

1-13. Under tactical conditions, the operational element completely rigs itself, and the jumpmasters make jumpmaster personnel inspections (JMPIs) before the point of no return. This procedure ensures the personnel will exit the aircraft with all their equipment in case of a bailout over enemy territory. All detachment members calibrate their altimeters so that the instruments read distance above the ground at the DZ and Military CYPRES 2 units are set to indicate the barometric pressure at the intended DZ.

1-14. During flight to the HARP, the aircraft commander keeps the jumpmaster informed of the aircraft's position. In turn, the jumpmaster keeps the parachutist updated about the aircraft's location and mission progress. This information is essential. The parachutist must know his relative position along the route so that he can apply the required actions in case of an abort or enemy action. A small dry erase board is located on the aircraft so that all jump team members can read important information, especially when oxygen is being used. Communication is limited to written notes on the dry erase board or predetermined signals from the jumpmaster.

1-15. While in flight, the aircraft commander keeps the MFF jumpmaster informed of changes to the altimeter reading should it be necessary to abort and make an emergency exit. All actions and time warnings issued will be IAW premission briefings and this manual. The pilot will signal the jumpmaster upon arriving at the HARP. The parachutists exit the aircraft on the jumpmaster's command.

PHASE IV: ACTIONS DURING FREE FALL AND ACTIONS UNDER CANOPY

1-16. MFF parachute jumps consist of four phases. They are—
- Exit, delay, and deployment.
- Assembly under canopy.
- Flight in formation.
- Final approach and landing.

PHASE V: ACTIONS ON THE DROP ZONE

1-17. At the DZ, the team leader and team sergeant immediately account for their personnel and all equipment. Infiltrating detachments are especially vulnerable to enemy action during this phase. To minimize the chances of detection, the detachment must clear the DZ as rapidly as possible and move to the preselected assembly area. This area must provide cover and concealment and facilitate subsequent movement to the objective area. Parachutes and air items should be buried or cached. If a reception committee is present, its leader coordinates personnel movement and provides current intelligence on the enemy and battle situation. Finally, the detachment sterilizes the assembly area and begins moving to the objective area.

PHASE VI: MOVEMENT TO AND ACTIONS AT THE OBJECTIVE

1-18. Movement from the DZ to the objective area may require guides. If a reception committee is present, it provides guides to the area or mission support sites where additional equipment brought may be cached. If guides are not available, the detachment follows the preselected route based on detailed intelligence and the patrolling plan developed during isolation. A well-planned route to the objective area must take maximum advantage of cover and concealment and avoid enemy outposts, patrols, civilians, and installations. The detachment carries only mission-essential equipment and supplies (individual equipment, weapons, communications, and ammunition).

PHASE VII: EXFILTRATION

1-19. Exfiltration planning considerations require the same planning, preparations, tactics, and techniques as infiltrations. However, in exfiltration the planners are primarily concerned with recovery methods. Distances involved in exfiltration usually require additional means of transport. Fixed- or rotary-wing aircraft, vehicles, surface craft, submarines, or various combinations of these methods can be used to recover operational elements.

Chapter 2

MC-4 Ram-Air Personnel Parachute System

The evolution of the parachute used in MFF operations has been considerable over the years. This chapter identifies the MC-4 RAPPS components and donning and recovery procedures. There are several RAPPSs used in the Department of Defense that have similar employment and flight characteristics to the MC-4—the USN MT-1SS and MT2-XX/SL, and the United States Marine Corps (USMC) Multimission Parachute System (MMPS), Tandem Offset Resupply Delivery System (TORDS), and MC-5. The MT-2XX/SL and MC-5 RAPPSs have a static-line capability. The MT-1SS has smaller 5-cell main and reserve canopies.

> *Note*: Questions regarding employment of the RAPPS in the static-line configuration should be addressed to USASOC, G-37, Special Skills, Fort Bragg, North Carolina. Technical Manual (TM) 70244A-OI, *Tactics, Techniques, and Procedures Manual for U. S. Marine Corps Military Free-Fall Operations*; and TM 10-1670-287-23&P, *Unit and Direct Support Maintenance Manual Including Repair Parts and Special Tools List for MC-4 Ram-Air Free-Fall Personnel Parachute System*, contain information on repairing and maintaining the MC-4.

MC-4 RAM-AIR PERSONNEL PARACHUTE SYSTEM COMPONENTS

2-1. Figures 2-1 through 2-11, pages 2-1 through 2-11, depict the various components associated with the MC-4 RAPPS.

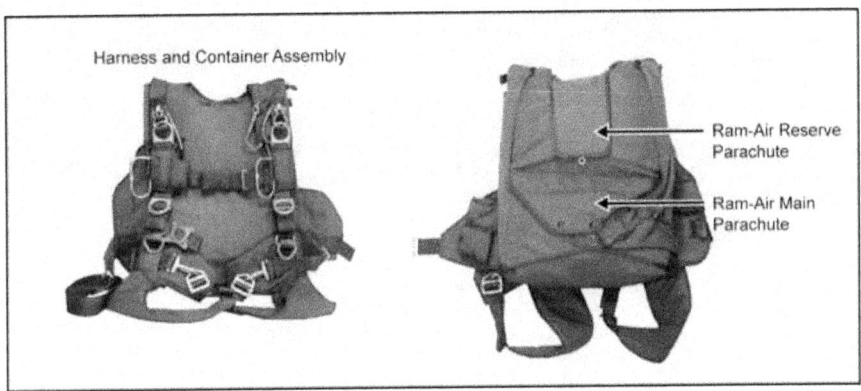

Figure 2-1. MC-4 RAPPS components

Chapter 2

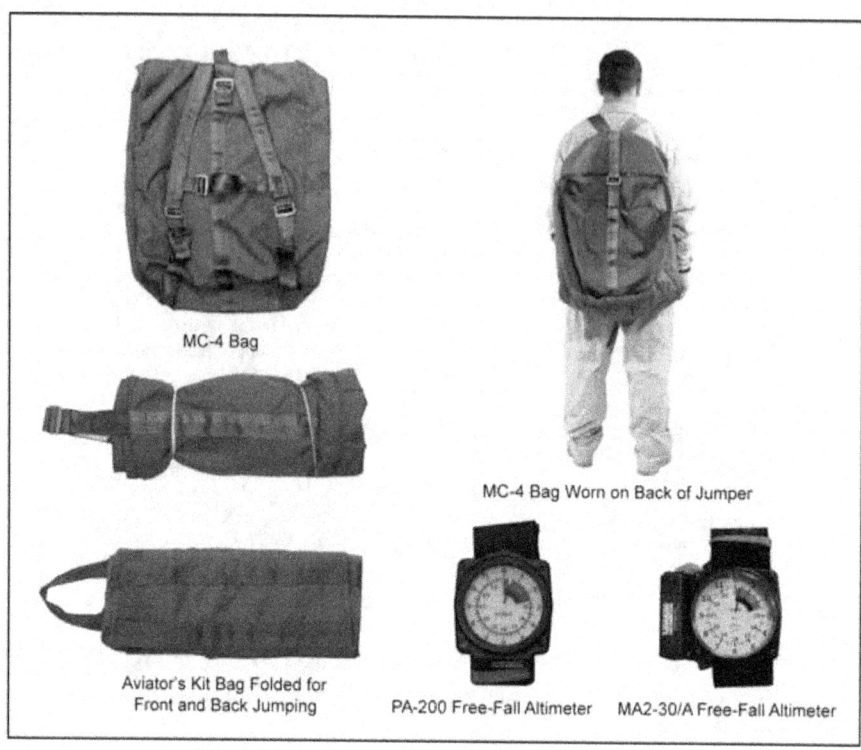

Figure 2-1. MC-4 RAPPS components (continued)

MC-4 Ram-Air Personnel Parachute System

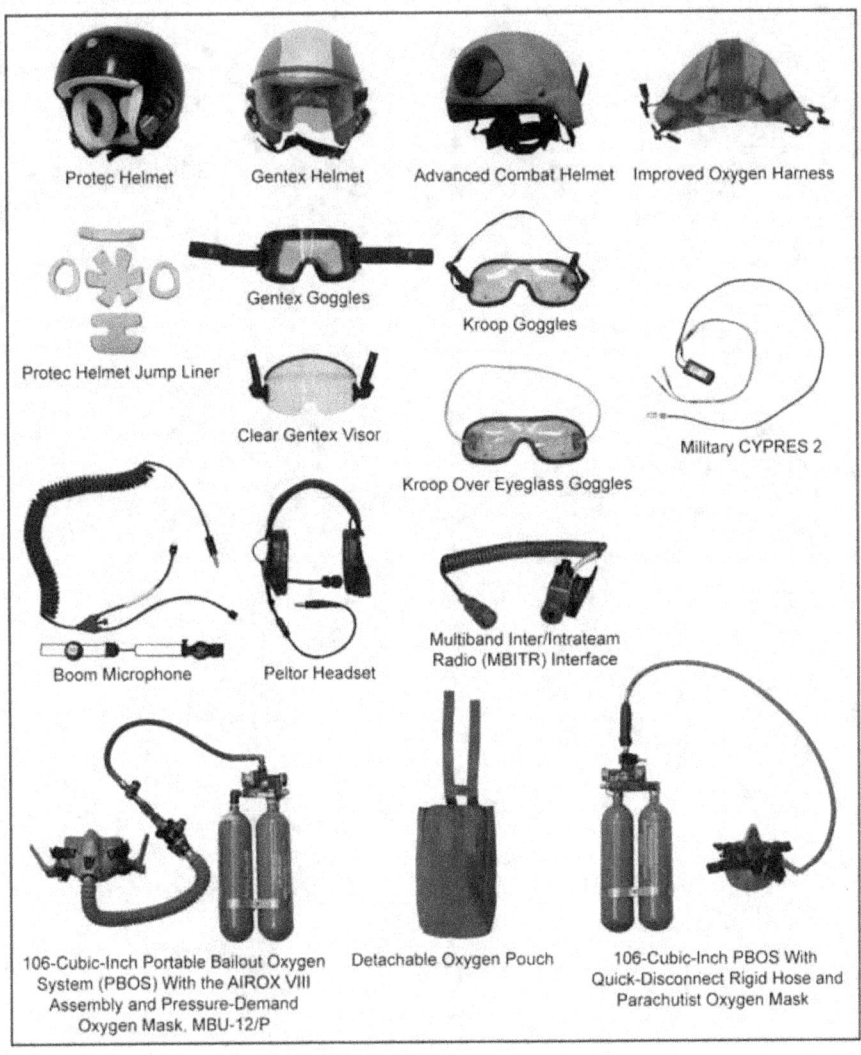

Figure 2-1. MC-4 RAPPS components (continued)

Chapter 2

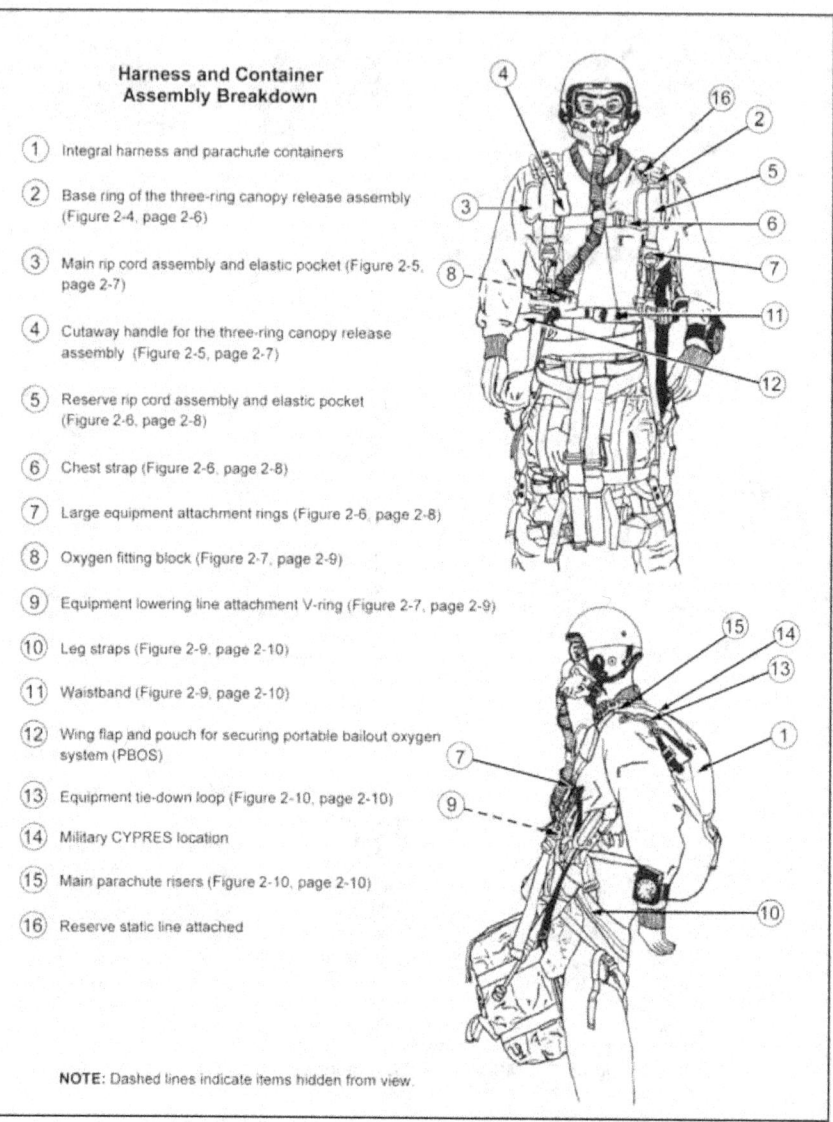

Harness and Container Assembly Breakdown

1. Integral harness and parachute containers
2. Base ring of the three-ring canopy release assembly (Figure 2-4, page 2-6)
3. Main rip cord assembly and elastic pocket (Figure 2-5, page 2-7)
4. Cutaway handle for the three-ring canopy release assembly (Figure 2-5, page 2-7)
5. Reserve rip cord assembly and elastic pocket (Figure 2-6, page 2-8)
6. Chest strap (Figure 2-6, page 2-8)
7. Large equipment attachment rings (Figure 2-6, page 2-8)
8. Oxygen fitting block (Figure 2-7, page 2-9)
9. Equipment lowering line attachment V-ring (Figure 2-7, page 2-9)
10. Leg straps (Figure 2-9, page 2-10)
11. Waistband (Figure 2-9, page 2-10)
12. Wing flap and pouch for securing portable bailout oxygen system (PBOS)
13. Equipment tie-down loop (Figure 2-10, page 2-10)
14. Military CYPRES location
15. Main parachute risers (Figure 2-10, page 2-10)
16. Reserve static line attached

NOTE: Dashed lines indicate items hidden from view.

Figure 2-2. MC-4 RAPPS harness and container assembly components

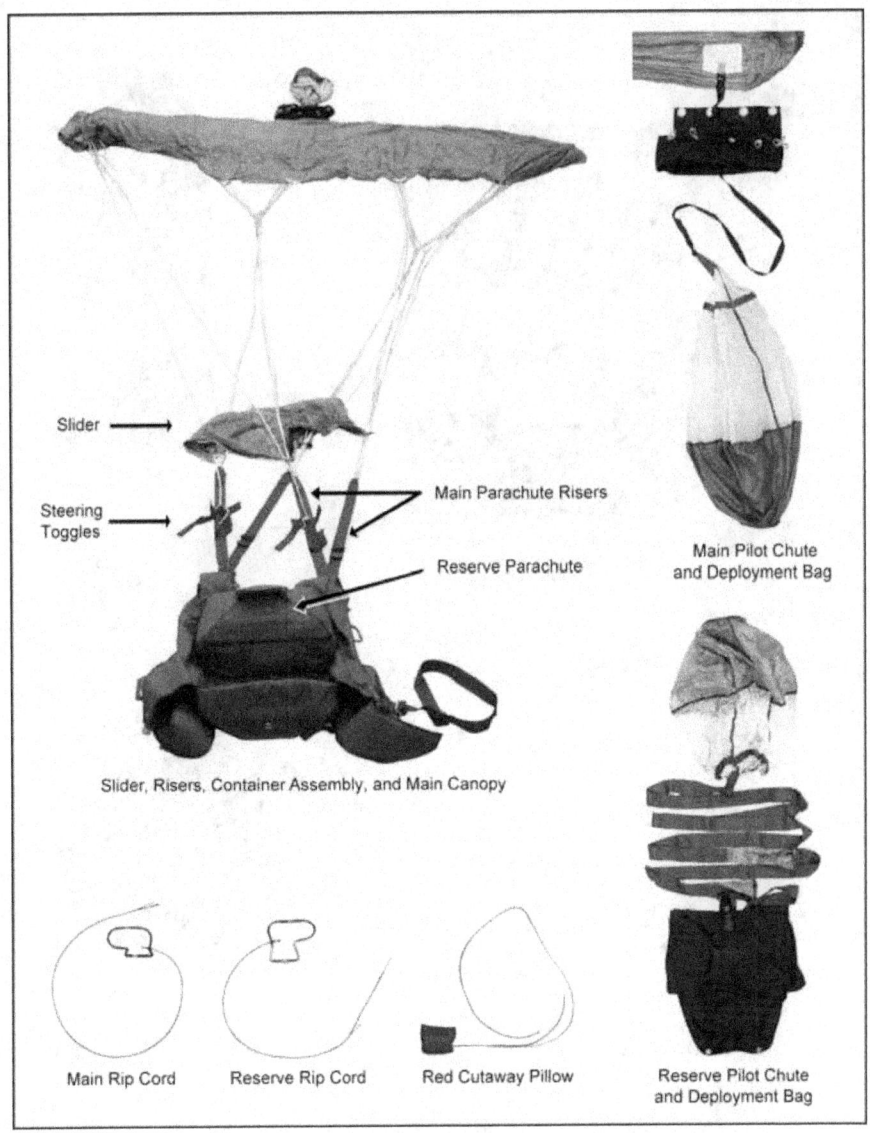

Figure 2-3. MC-4 RAPPS assembly components

Chapter 2

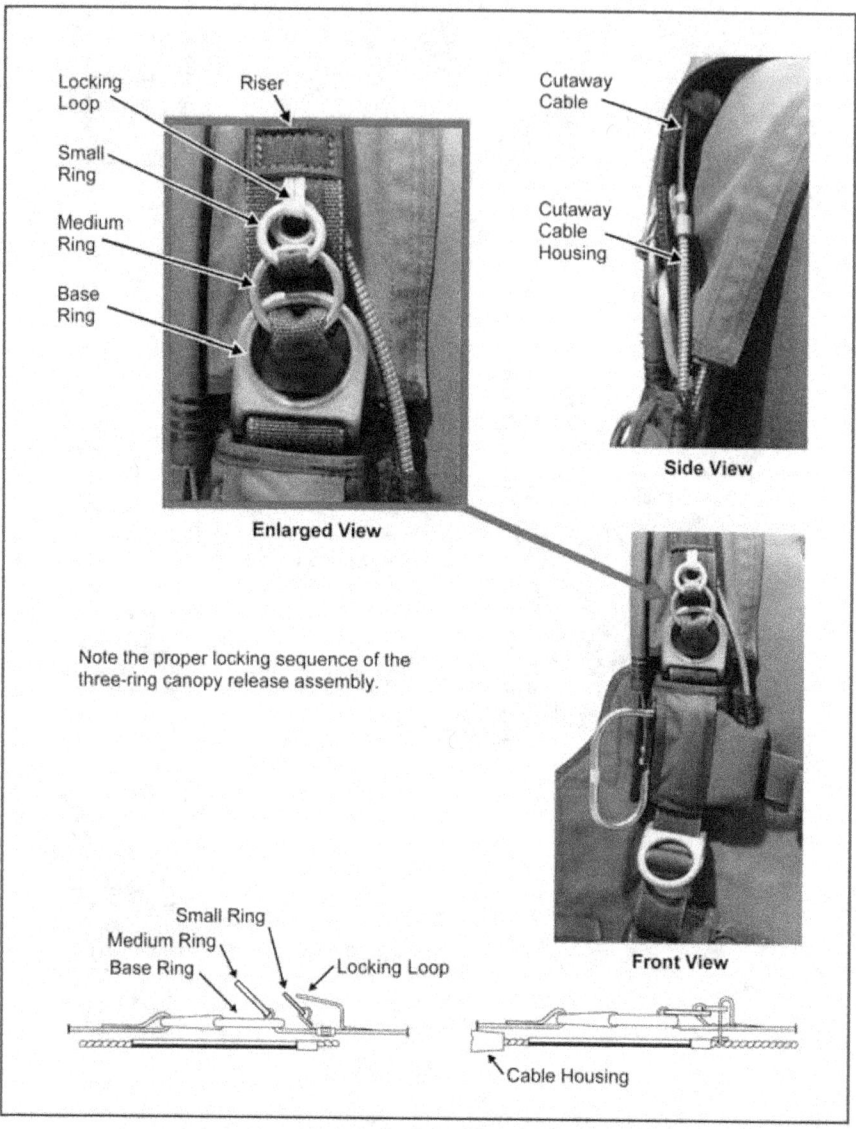

Figure 2-4. Location of the three-ring canopy release assembly

MC-4 Ram-Air Personnel Parachute System

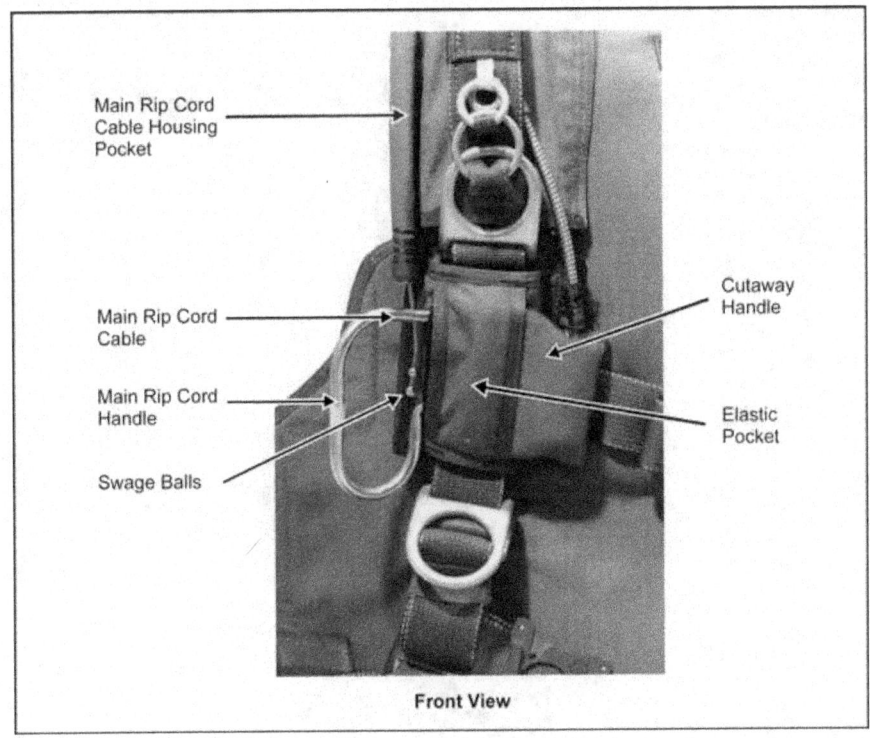

Figure 2-5. Location of the main rip cord handle and cutaway handle

Chapter 2

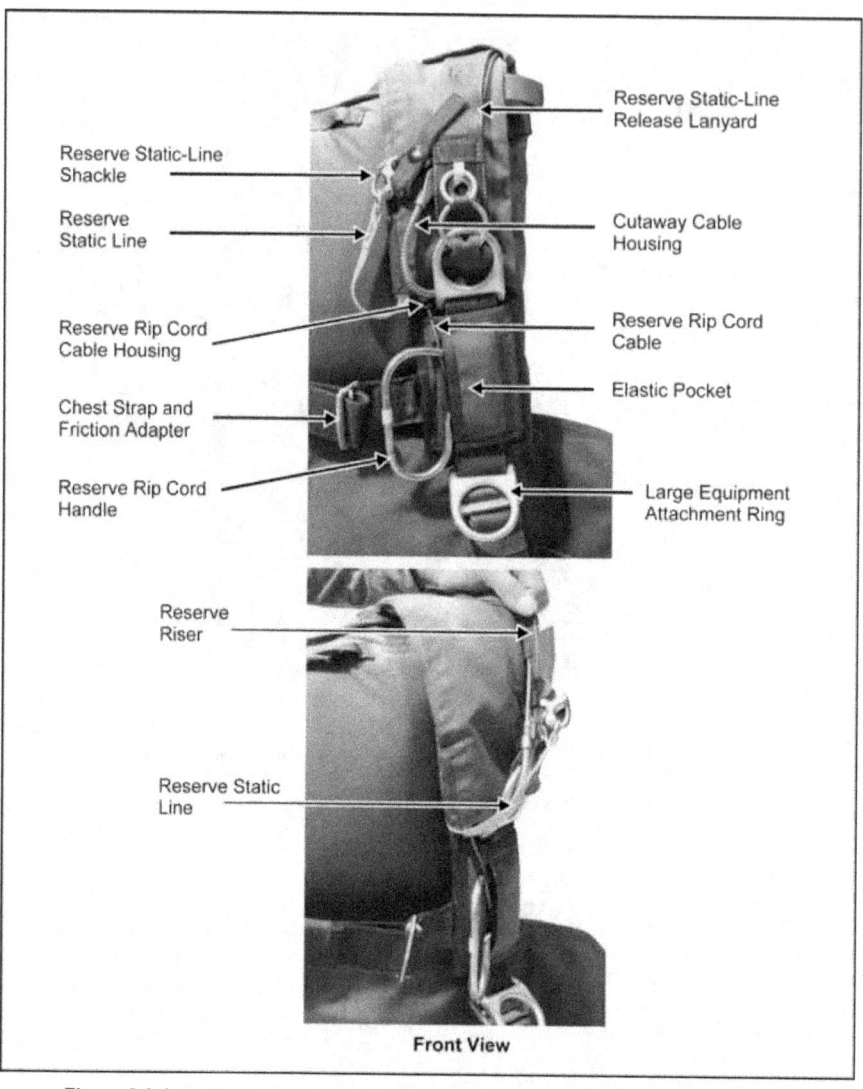

Figure 2-6. Location of the chest strap, reserve rip cord handle, large equipment attachment ring, and reserve rip cord cable housing

MC-4 Ram-Air Personnel Parachute System

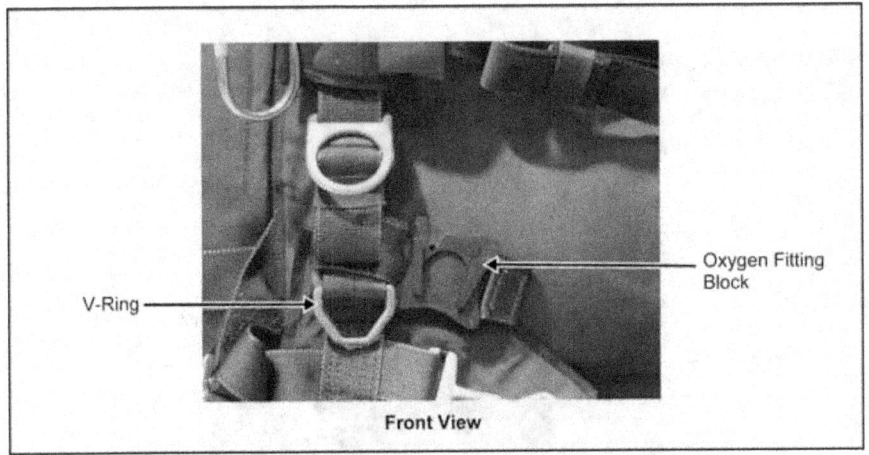

Figure 2-7. Location of the oxygen fitting block and equipment lowering line attachment V-ring

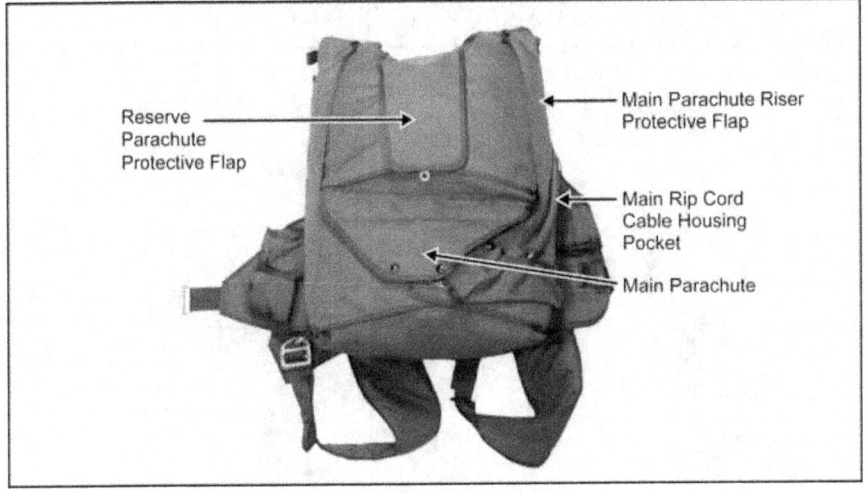

Figure 2-8. Location of the main and reserve parachutes in the container

Chapter 2

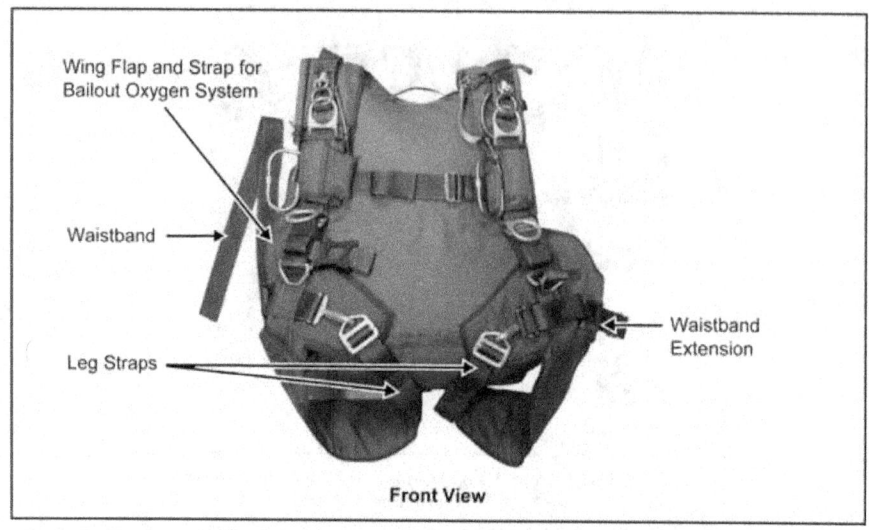

Figure 2-9. Location of straps

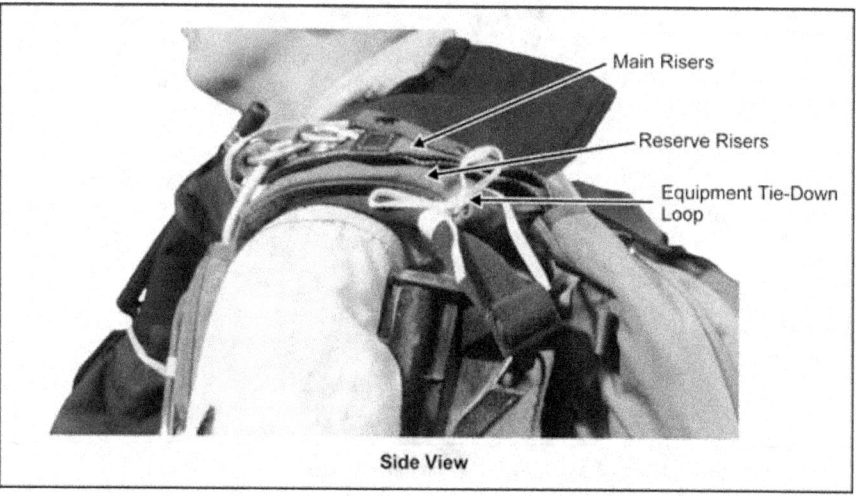

Figure 2-10. Location of the equipment tie-down loop and main risers

MC-4 Ram-Air Personnel Parachute System

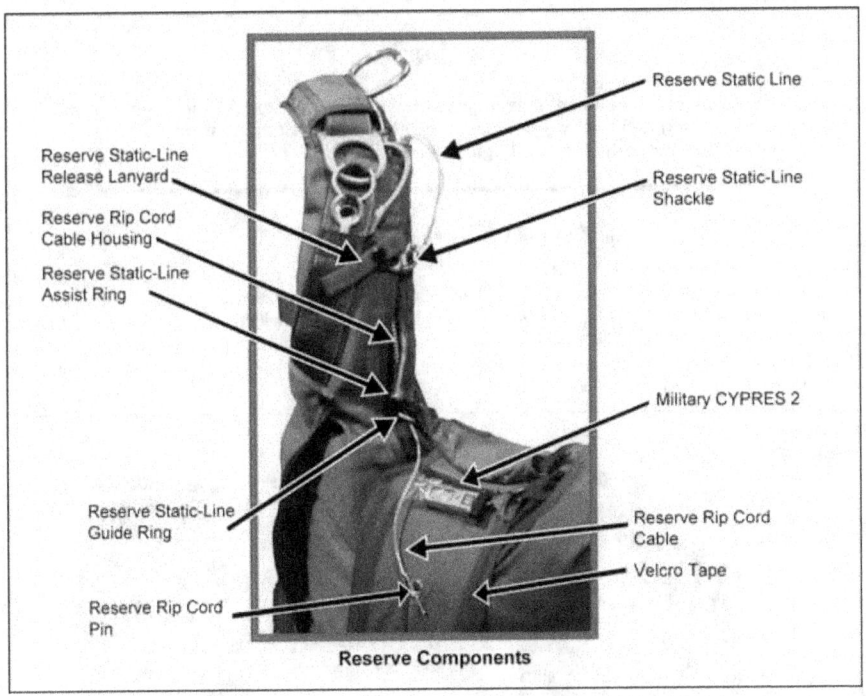

Figure 2-11. Location of reserve components

MILITARY FREE-FALL PARACHUTIST HELMET ASSEMBLY

2-2. MFF parachutists use the following helmets: advanced combat helmet (ACH), the Gentex HGU-55/P helmet, the Gentex lightweight parachutist helmet, the MC-3 helmet (a semirigid, padded leather helmet) used for passengers on tandem jumps, the Protec helmet with free-fall liner, and the Bell motorcyclist helmet (full-face helmet not authorized for MFF). To conduct MFF with oxygen, personnel must wear helmets with bayonet receptacles attached or use the improved oxygen harness (IOH) (skull cap) to attach the oxygen mask. The jumpmaster should have internal earphones and a microphone for communication within the aircraft and while under canopy.

> **WARNING**
>
> The parachutist makes sure that bayonet receivers on his helmet are compatible with the oxygen mask and that the mask fits properly.

Chapter 2

> **WARNING**
>
> The clear full-face shield (issued with the Gentex helmet and jumped with the oxygen mask) may become dislodged in free fall if not properly fitted and tightened.

MA2-30/A AND THE PA-200 FREE-FALL ALTIMETERS

2-3. The parachutist wears the MA2-30/A or the PA-200 altimeter on his left wrist (Figure 2-12). The altimeter shows his altitude above the ground during free fall. The altimeter permits him to determine when he has reached the proper altitude for deploying the main parachute. The altimeter must be transported and stored with care. It must be chamber-tested for accuracy. The altimeter must be rechecked after an unusually hard landing and after accidentally dropping it. If the altimeter is not waterproof, it should be replaced if it has been submerged in water.

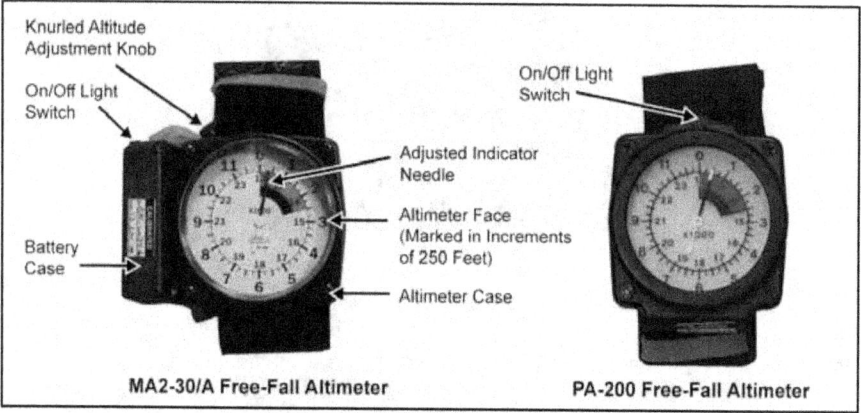

Figure 2-12. MA2-30/A and PA-200 free-fall altimeters

2-4. The computation for the altimeter setting follows. The jumpmaster—
- Applying the formula below, inputs data (Figure 2-13, page 2-13).
- Converts meters to feet if map data is in meters.
- Places departure airfield (DAF) elevation in first block.
- If DZ elevation is lower than DAF, places it in lower block.
- If DZ elevation is higher than DAF, places it in upper block.
- If numbers are the same (+/+ or -/-), then subtracts.
- If numbers are different (+/-), then adds.
- Places total in appropriate block.

The positive or negative sign next to the block identifies the altimeter setting.

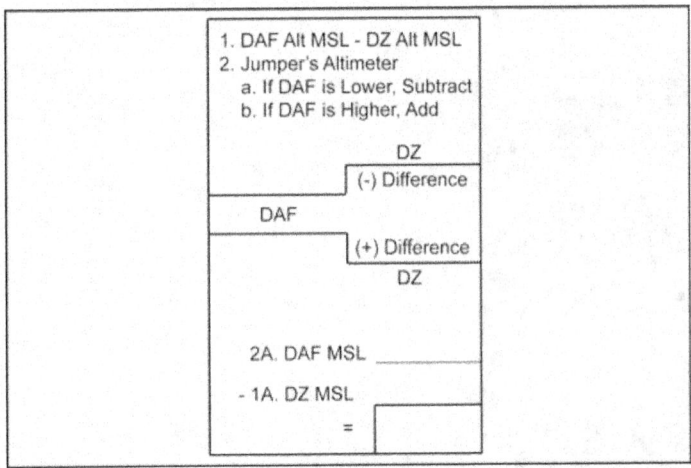

Figure 2-13. Altimeter setting

> **CAUTION**
>
> Special consideration will be given to any obstacles (for example, ridgelines, mountains, towers, and other such items and their elevations) that may be located within 3 nautical miles or 5.5 kilometers of the parachutist's release point or desired impact point.

MA-10 MILITARY ALTIMETER

2-5. The MA-10 altimeter is a solid state electronic device with manufacturer-updatable embedded software and a stepper motor that moves the pointer on an analog display. The 12,000-foot linear scale is capable of reading up to 40,000 feet MSL with the pointer rotating 12,000 feet per revolution (3 1/3 revolutions to 40,000 feet). The face is highlighted with a red warning arc that begins at 2,500 feet. The MA-10 is powered by two 1.5-volt AA lithium batteries; standard 1.5-volt AA batteries may be used with reduced battery life. The MA-10 can be comfortably worn with a Velcro wrist-mount band and is waterproof to a depth of 6 feet for 1 hour. The battery compartment is not waterproof. The aluminum housing measures 3.27 x 3.20 x 1.37 inches and the face has a 2.50-inch dial. The electroluminescence face automatically turns on and provides backlighting during low light conditions. The manufacturer-replaceable lens is protected by a self-adhesive lens protector that can be replaced by a designated parachute rigger. The MA-10 conducts a power-on self-test, checking the pressure sensor, blockage of the filter, stepper motor, battery voltage, and other critical functions. Using the external buttons, the MA-10 can be set in three ways: zeroed to the current location, manually entering the DZ offset, or by calculating the DZ offset entering the DZ altitude and a form of barometric pressure called "altimeter setting" for the DZ. Figures 2-14 through 2-20, pages 2-14 through 2-19, depict the MA-10 altimeter.

Chapter 2

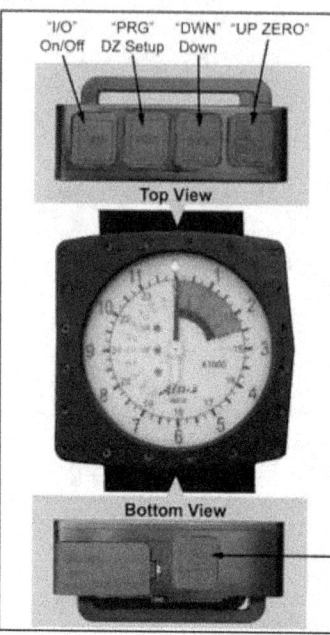

The MA-10 altimeter has a total of 5 buttons.

The single button marked "SEL" near the 6 o'clock position simply activates the other buttons. This eliminates the possibility of accidental button pushes.

Figure 2-14. MA-10 altimeter buttons

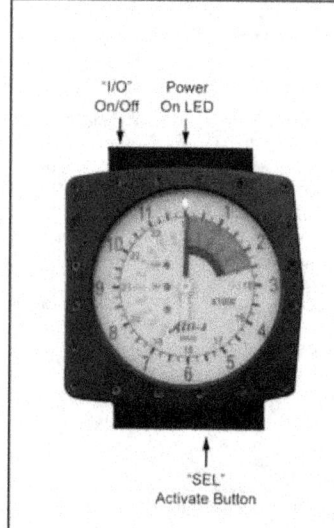

The needle parks at approximately 10,500 feet or -1,500 feet when the unit is off.

To turn the unit on, press and hold the Activate Button marked "SEL" and then press On/Off Button marked "I/O."

The needle will first show battery status:

7–3 (White) = Okay.
2.5–3 (Yellow) = Be ready to change battery.
< 2.5 (Red) = Change battery before next jump.

The needle will move to the current altitude/pressure.

The Power On light-emitting diode (LED) at the 12 o'clock position will be illuminated.

The Power On LED intensity adjusts automatically based on ambient light levels. In very low light conditions, the LED is turned off to prevent loss of night vision and the electroluminescent backlight indicates that the unit is active.

NOTE:
If the needle pauses at the 6 o'clock position, this indicates that the DZ altitude and pressure/elevation have been set. The needle will also pause at the 6 o'clock position if the manual offset has been programmed.

Figure 2-15. On/Off buttons

MC-4 Ram-Air Personnel Parachute System

Power On LED will flash when unit is in power-saving mode.

The system will turn off the motor and backlight if the altitude is below 7,000 feet MSL and there is no significant change in altitude for a period of 30 minutes.

The Power On LED at the 12 o'clock position will flash to indicate power-saving mode.

If altitude activity is sensed, the unit will automatically sweep the pointer one revolution and revert to full function.

If MA-10 has gone to sleep, wake altimeter by pressing one of the top buttons. Do NOT press the "SEL" Activate Button.

Figure 2-16. Power-saving mode

"UP ZERO"

"SEL" Activate Button

For training jumps when the departure airfield and the target DZ are the same location, zero the altimeter when standing on the DZ.

To zero the altimeter, momentarily press the two buttons shown: "UP ZERO" and the "SEL" Activate Button.

Do NOT hold the "UP ZERO" button; this will cause the set altitude to increase.

NOTE: This action will clear any preset DZ altitude and pressure settings.

NOTE: The manual zero altitude will be retained when the unit is turned off. The MA-10 acts like a mechanical altimeter; it will react to barometric changes and will need to be rezeroed when powered back on.

Figure 2-17. Zeroing the MA-10 altimeter

Chapter 2

When the departure airfield and the target DZ are at different altitudes, the DZ Offset may be set manually.

To manually offset the altitude reading, the jumper presses and holds the "SEL" Activate Button and then uses the "UP ZERO" or "DWN" buttons to set the desired altitude.

The rate of pointer movement will speed up (this helps with larger offsets). If the jumper releases the "UP ZERO" or "DWN" button and continues to hold the bottom button, the rate will start slowly again when he presses up or down.

NOTE: This action will clear any preset DZ altitude and pressure settings.

Figure 2-18. Manual offset

MC-4 Ram-Air Personnel Parachute System

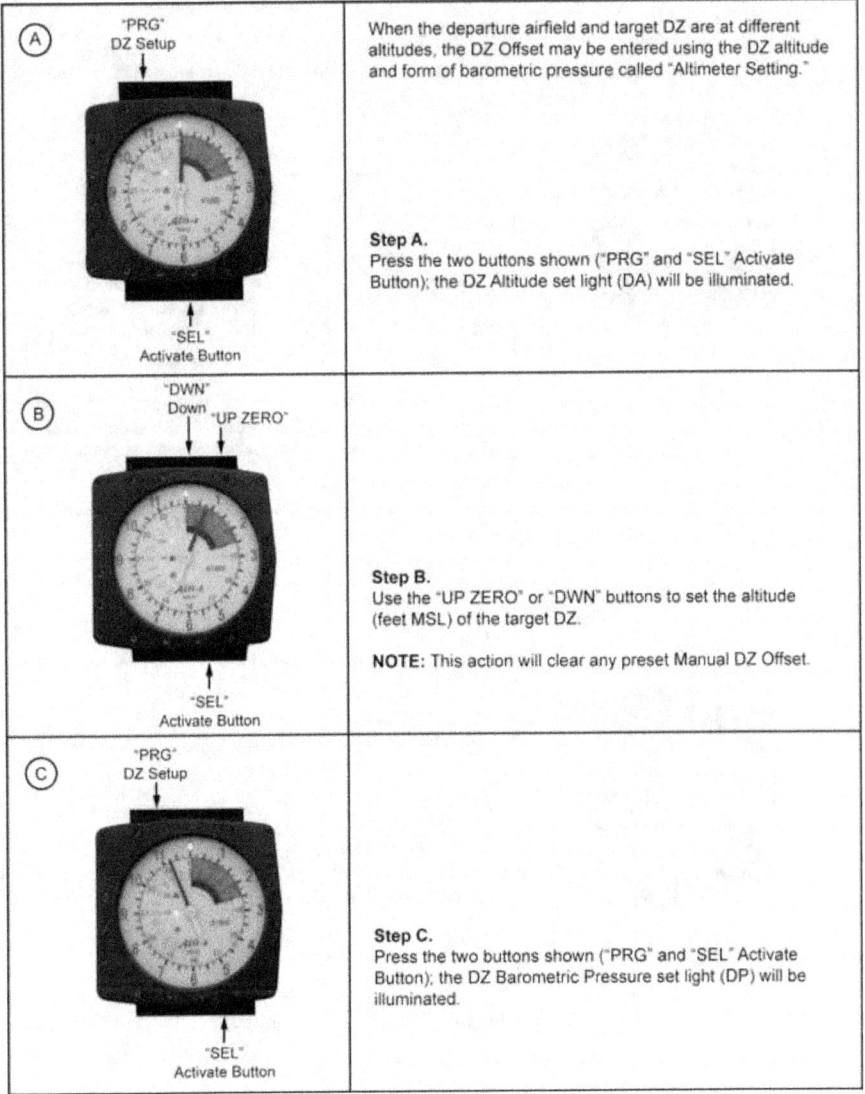

Figure 2-19. Setting the drop zone

Chapter 2

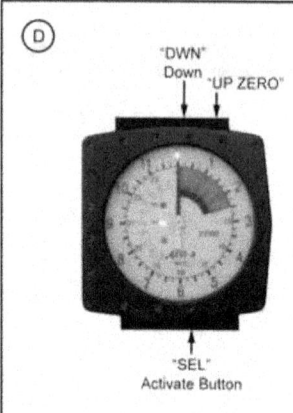

"DWN"
Down "UP ZERO"

"SEL"
Activate Button

Step D.
Use the "UP ZERO" or "DWN" buttons to set the barometric pressure (inches of mercury [in/Hg]) of the target DZ.

Barometric pressure is marked in gray numerals inside the scale.

> **WARNING**
>
> When obtaining the barometric pressure, always request the "altimeter setting" for the DZ. Do not use the actual barometric pressure (station pressure) or sea-level corrected pressure from the DZ.

The current "altimeter setting" for the DZ in inches of mercury (in/Hg) within 100 miles of the intended DZ must be determined by using the most accurate methods available.

If there are no available means to calculate the current "altimeter setting," the combat setting of 29.92 in/Hg will be used.

"PRG"
DZ Setup

"SEL"
Activate Button

Step E.
Press the two buttons shown ("PRG" and "SEL" Activate Buttons).

This completes the DZ setup and the altimeter is in RUN mode. The altimeter displays the DZ Offset between the current altitude and the target DZ.

NOTE: Prior to programming after initial start-up, zero the MA-10 by pressing the "SEL" and "UP ZERO" buttons simultaneously. This step checks the absolute pressure sensor (if the altimeter does not zero, the absolute pressure sensors are faulty).

NOTE: The DZ altitude and pressure settings will be retained when the unit is turned off.

Figure 2-19. Setting the drop zone (continued)

MC-4 Ram-Air Personnel Parachute System

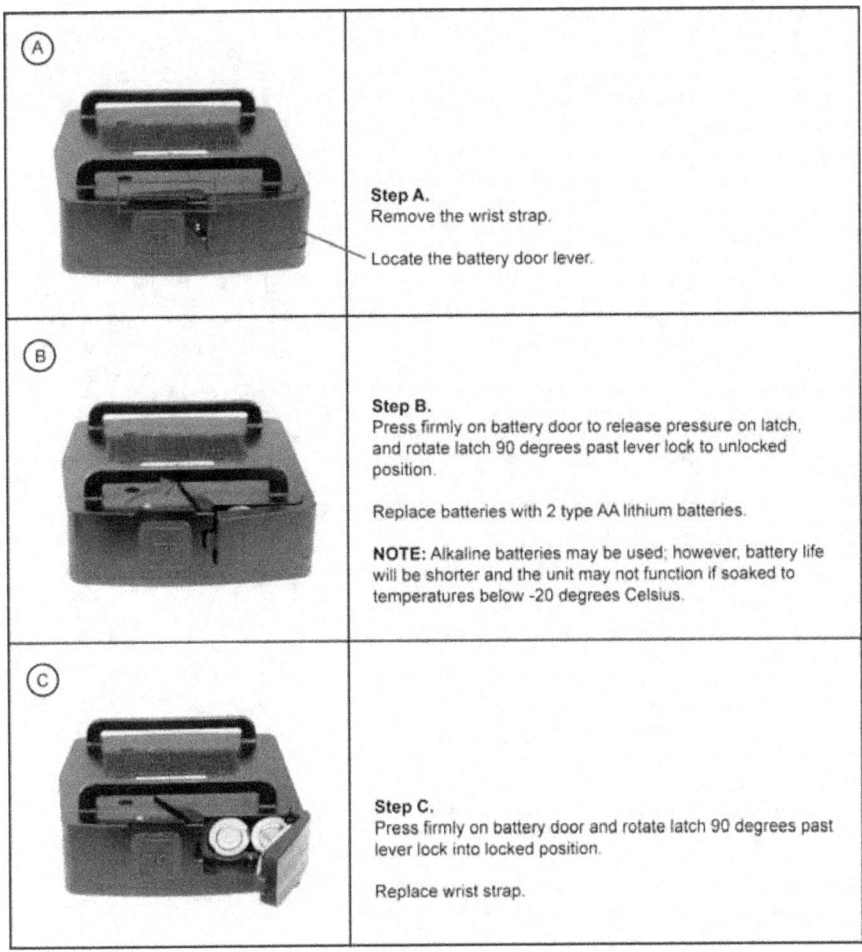

Figure 2-20. Replacing batteries in the MA-10 altimeter

OTHER COMPONENTS AND PROTECTIVE CLOTHING

2-6. The following paragraphs provide basic information about other items required by the MFF parachutist.

Gloves

2-7. Gloves are worn to protect the parachutist's hands from the elements and to prevent injury from the action of the handles, lines, and webbing during canopy deployment and steering. The gloves must be chosen to provide good protection and must allow the parachutist to retain dexterity. Gloves should not interfere with the operation of the handles on the parachute.

Chapter 2

Note: Leather-palmed gloves are required when the temperature at exit altitude is 40 degrees Fahrenheit and below.

Boots

2-8. While boots without speed-lacing hooks are not RAPPS components, they are considered to be mandatory safety equipment.

Eye Protection

2-9. MFF parachutists must use eye protection. Commercial (Kroop) goggles provide a wide field of vision and come in two sizes: regular and a larger box design that fits over standard military eyeglasses. Military-issue sun, wind, and dust goggles are authorized, but not recommended, as they restrict the parachutist's field of vision. Commercial (Kroop) and military-issue goggles are authorized for parachuting with or without an oxygen mask. All lenses used should be clear and relatively free of scratches that might obstruct vision. When oxygen is used, all goggles, no matter what type, will be clear. All goggles shall be made of shatterproof materials.

Note: The clear full-face shield issued with the Gentex helmet is authorized for use only with an oxygen mask.

Military Free-Fall Parachutist Helmet

2-10. The following helmets are authorized for MFF:
- Gentex HGU-55/P.
- Gentex lightweight parachutist helmet.
- MC-3 helmet.
- Protec helmet (only with free-fall liner).
- Bell open-face motorcycle helmet.
- ACH/modular integrated communication headset (MICH).
- Soft-shell helmet (authorized for tandem passenger use only).

Note: The Bell full-face motorcycle helmet is not authorized for MFF operations.

Note: Bayonet receiver assemblies may be installed on most helmets for use with an oxygen mask requiring the use of bayonet lugs. Helmets that cannot be fitted with bayonet receiver assemblies will be fitted with the rail connector and oxygen strap kit or jumper will use the IOH.

Communications Capabilities

2-11. The jumpmaster should have internal earphones and an external microphone or internal oxygen mask microphone for communication. The Gentex lightweight parachutist helmet has bayonet receivers for an oxygen mask and provides each parachutist with communications capability with or without an oxygen mask. The MICH can be used with the ACH. The Peltor headset will sit between the ACH and IOH, if used.

Other Protective Clothing

2-12. Appropriate garments are chosen per individual MFF mission requirements. Jumpsuits (Figure 2-21, page 2-21) with lightweight polypropylene undergarments, insulated undergarments, or insulated overlayers may be necessary depending upon the degree of environmental protection required. The parachutist individual equipment kit (Figure 2-22, page 2-21) worn during USMC MFF operations, is another example of outstanding environmental protection clothing for MFF operations.

MC-4 Ram-Air Personnel Parachute System

Figure 2-21. Jumpsuits

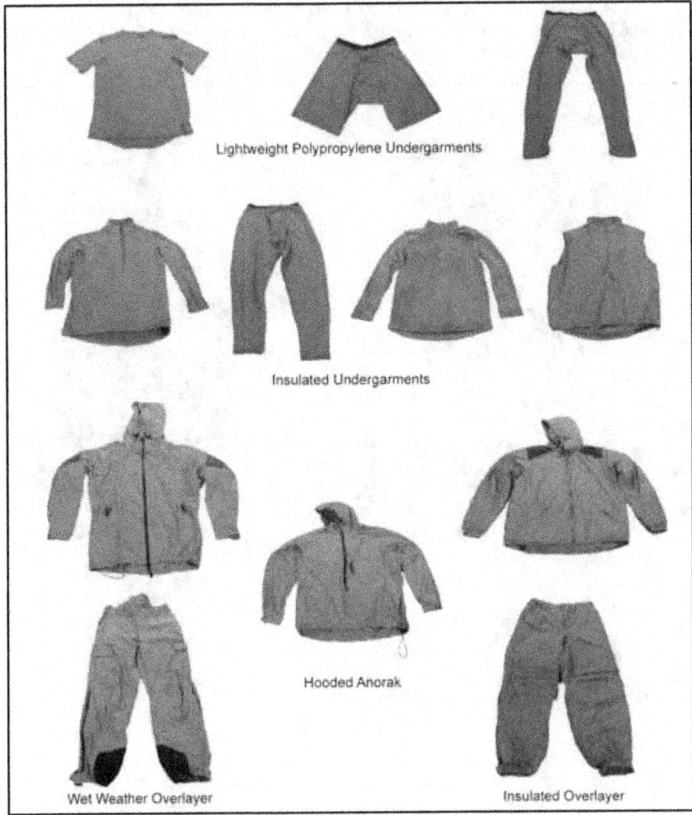

Figure 2-22. Parachutist individual equipment kit

Chapter 2

DONNING AND RECOVERING THE MC-4 RAM-AIR PERSONNEL PARACHUTE SYSTEM

2-13. The buddy system, or the pairing of parachutists, within each operational element provides the most efficient and accurate way for parachutists to don, adjust, and check each other's parachutes. Using the buddy system to properly don and adjust the MC-4 RAPPS provides an additional safety check and prevents unnecessary delays during the JMPI.

PREPARING THE KIT BAGS

2-14. The parachutist will determine if the kit bag will be worn on the front across the groin area, on the rear between the jumper's back and container, or put on like a rucksack on the back. He selects the method and prepares the kit bag as follows:

- *Aviator's kit bag.* The parachutist closes the slide fastener and secures all snap fasteners. If the kit bag is worn rear- or front-mounted, he folds each end of it with one fold toward the center leaving the handles exposed at one end (Figure 2-23).
- *MC-4 kit bag.* The parachutist closes the slide fastener. The kit bag can be worn like a rucksack on the jumper's back (Figure 2-23) under the MC-4 RAPPS. If it is front-mounted, the parachutist rolls it from bottom to top with shoulder straps exposed and places retainer bands on each end (Figure 2-23).

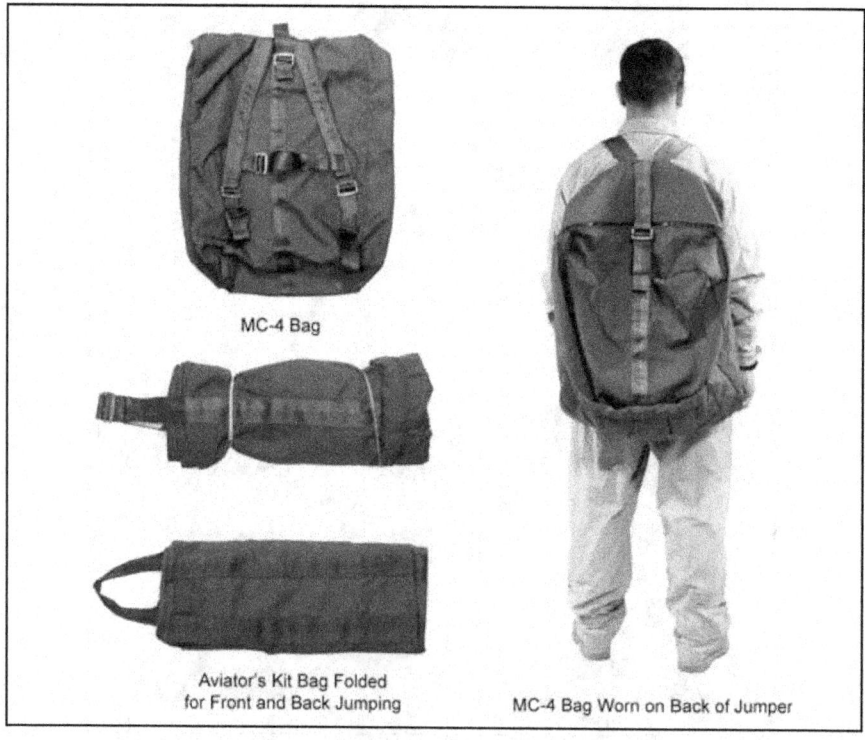

Figure 2-23. Aviator's and MC-4 kit bags

2-22   ATTP 3-18.11/AFMAN 11-411(I)/NTTP 3-05.26M   14 October 2011

DONNING THE MC-4 RAM-AIR PERSONNEL PARACHUTE SYSTEM

2-15. The following are procedures for donning the MC-4 RAPPS (Figure 2-24, page 2-24):

- The parachutist checks the parachute assembly for visible defects, lets out all harness adjustments for ease of donning (Figure 2-24A), and lays the assembly out with the pack tray face down.
- To don the parachute, the parachutist (No. 1) assumes a modified high jumper position. The second parachutist (No. 2) holds the harness container by the main lift webs at the canopy release assemblies and places it on No. 1's back (Figure 2-24B).
- No. 1 remains bent forward at the waist and No. 2 pushes the container high on No. 1's back as No. 1 threads and fastens the chest strap (Figure 2-24C).
- No. 2 prepares the leg straps. No. 2 calls out, LEFT LEG STRAP, and passes it to No. 1. No. 1 repeats, LEFT LEG STRAP, and grasps the left leg strap with one hand. With his other hand, he starts from the saddle and feels the length of the leg strap, removing any twists and turns. He inserts the leg strap through one aviator's kit bag handle (if the kit bag is front-mounted) and fastens the leg strap (Figure 2-24D). He repeats the procedure for the remaining leg strap. He performs the same steps for the MC-4 kit bag by inserting the leg straps through the shoulder straps.
- No. 1 stands erect and checks to make sure the canopy release assemblies are in the hollows of his shoulders by adjusting the main lift webs (Figure 2-24E).
- No. 1 locates the free-running ends of the horizontal adjustment straps and tightens the harness so it fits snugly and comfortably (Figure 2-24F).
- No. 2 then threads the long-running end of the waistband through both kit bag handles (if the kit bag is rear-mounted), and No. 1 fastens the waistband to the waistband extension (Figures 2-24G and H).
- After final adjustment, No. 1 folds all excess straps inward, except for the main lift webs that are folded outward, and secures them using the elastic keepers (Figure 2-24I). No. 1 should be able to stand erect without straining.
- When properly donned, the system should feel snug but not so tight as to restrict movement. The jumper should be able to properly arch, look, reach, and pull the rip cord on the ground before the actual jump.
- No. 1 and No. 2 then change positions and repeat the procedure.
- When both parachutists have donned their parachute assemblies and adjusted their harnesses, they face each other, make a visual inspection of each other, and correct any deficiencies before the JMPI.

RECOVERING THE RAM-AIR PERSONNEL PARACHUTE SYSTEM

2-16. The following procedures are used to recover a RAPPS; the parachutist—

- If jumping oxygen, locks the ON/OFF switch in the OFF position and removes the bailout bottles and pouch from the waistband.

Note: Parachutists do not place the oxygen mask on the ground unprotected during parachute recovery. Moisture from breathing and condensation due to temperature changes will cause dirt and debris to adhere to the mask, interfering with sealing and increasing risk of injury.

- Removes the harness and container and daisy-chains the suspension lines.
- Removes and opens the aviator's kit bag.
- If using the Military CYPRES 2, turns it off.
- Replaces the rip cord in the rip cord cable housing and the rip cord handle in the stow pocket.
- Places the pilot chute next to the kit bag.
- Places the canopy, deployment bag, suspension lines, and risers in the kit bag.

Chapter 2

- Removes the quick-release snap hooks and lowering line quick-ejector snap from the equipment rings on the parachute harness.
- Places the harness and container in the kit bag with the back pad facing up to protect the Military CYPRES 2.
- Finally, places the pilot chute in the kit bag or on top of the MC-4 kit bag and snaps or zips the fasteners.

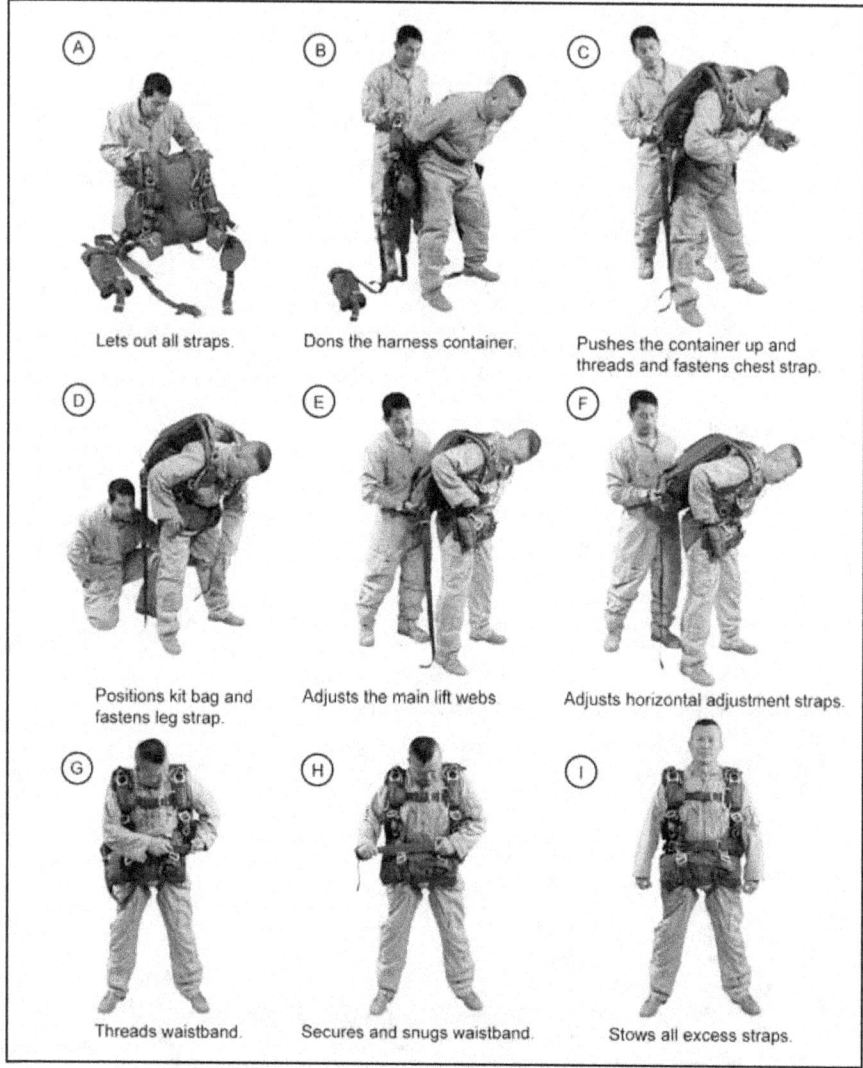

Figure 2-24. Donning the MC-4 Ram-Air Personnel Parachute System

Chapter 3
Cybernetic Parachute Release System

The CYPRES is an electronic automatic activation device (EAAD) designed to cut the loop material that is holding the reserve pilot chute in place, which deploys the reserve in the event that the MFF parachutist meets the criteria that the CYPRES uses to make the decision to fire the release unit.

GENERAL INFORMATION ON MILITARY CYPRES 2 MODELS

3-1. Three Military CYPRES 2 models and one Expert CYPRES 2 model are being used by the U.S. Army as safety devices designed to activate and enable the reserve parachute to deploy in the absence of the parachutist failing to deploy his main parachute or having a malfunction of his main parachute:

- The Military CYPRES 2 is designed specifically for tactical application use. There are three Military CYPRES 2 models: 1000 35 A, 1500 35 A, and the 2500 29 A. All models have two modes of operation—training mode and operational mode:
  - Default (training) mode can be used for nontactical jumps that meet specific parameters; inclusively, the DAF and DZ must be the same location.
  - Absolute (operational) mode can be used for both tactical and nontactical jumps in any scenario.
- The Expert CYPRES 2 is designed for use in authorized nonstandard parachutes and has two modes of operation—training mode and offset mode:
  - Default (training) mode can be used for nontactical jumps that meet specific parameters; inclusively, the DAF and DZ must be the same location.
  - Offset mode can be used when the DAF and DZ are at different altitudes or locations and specific parameters are met.

> **WARNING**
>
> All CYPRES information covered in this manual is specific to Military CYPRES 2 models only. Military CYPRES 1 models are only in use by USMC Special Operations Command. Military CYPRES 1 and Expert CYPRES 1 rules and procedures have some discrepancies and should not be used in conjunction with Military CYPRES 2 information contained in this manual.

Note: Use of the Military CYPRES 2 in operational mode is recommended for all situations.

Note: It is essential that all personnel read this entire chapter before using and setting the Military CYPRES 2. The jumpmaster and parachutist must be familiar with all CYPRES model functions, procedures, and limitations.

Chapter 3

MILITARY CYPRES 2 PRINCIPLES OF OPERATION

3-2. When the parachutist arms the Military CYPRES 2, the reserve parachute deploys automatically if the parachutist reaches the preset altitude at the preset vertical velocity and meets other critical conditions. The Military CYPRES 2 deploys the reserve parachute by firing the release element and severing the reserve closing loop material. The reserve pilot chute is then free to launch and deploy the reserve parachute. If the parachutist reaches the preset altitude and does not meet the conditions to fire the release element (such as when the main parachute is fully deployed), the Military CYPRES 2 will not send the signal to fire the release element. In case the jump conditions change, the Military CYPRES 2 silently continues to monitor the parachutist's condition during canopy flight until the parachutist reaches 130 feet above the virtual drop zone (VDZ).

GENERAL OPERATION

3-3. The three Military CYPRES 2 models and the Expert CYPRES 2 model will only activate and fire the release element within the activation window. The Military CYPRES 2 will only fire the release element for parachute malfunctions that fall through the activation window and meet the vertical activation speed. All parachute malfunctions that fall faster than the vertical activation speed (such as pack closure, hard pull, bag lock, and horseshoe malfunctions with the canopy in the bag) and are within the activation window, will meet the conditions to fire the release element. For all other parachute malfunctions that cause the parachutist to fall slower than the vertical activation speed (such as single-riser separation, line over, pilot chute over the nose, line twists, closed end cells, broken control lines, and tension knots), the parachutist must activate the reserve manually. It must be understood that the Military CYPRES 2 will leave the activation window at 130 feet above the VDZ and will no longer operate.

ACTIVATION WINDOW

3-4. The activation window for default (training) and absolute (operational) modes is as follows:

- *Default (training) mode.* Once properly powered ON, the Military CYPRES 2 in default (training) mode arms itself 1,500 feet above the default activation altitude (750 feet above activation altitude for the Expert CYPRES 2). Once armed, the activation window will extend from the default activation elevation down to approximately 130 feet above the DAF elevation. For example, once powered ON prior to leaving the ground, the Military CYPRES 2 1500 35 A arms itself at 3,000 feet AGL (1,500 feet above the default setting of 1,500 feet AGL). The parachutist exiting the aircraft will need to fall approximately 1,000 feet to reach the vertical activation speed of 35 meters per second (78 mph/115 feet per second [fps]). If the parachutist enters the activation window (1,500 feet AGL to 130 feet AGL) and is falling faster than the vertical activation speed of 35 meters per second, the Military CYPRES 2 will fire the release element. The Military CYPRES 2 will go into a standby mode at 130 feet above the VDZ.

Note: The Military CYPRES 2 will go into an energy saving (standby) mode at 130 feet because this gives it an altitude buffer to make sure it goes into this mode. Without this buffer, the Military CYPRES 2 could possibly stay in the activation phase of the jump, which consumes the most battery power. This would greatly depreciate the battery life over time. Also, the Military CYPRES 2 would not save the jumper if it fired at 130 feet.

- *Absolute (operational) mode.* Once properly powered ON, the Military CYPRES 2 in absolute (operational) mode arms itself immediately. The activation window will extend from the default activation setting above the VDZ, set by the jumpmaster, down to approximately 130 feet above the VDZ. For example, once powered ON with a 5,000-foot VDZ setting, the Military CYPRES 2 1500 35 A is armed immediately, regardless of the location where it was powered ON. The parachutist exiting the aircraft will need to fall approximately 1,000 feet to reach the vertical activation speed of 35 meters per second (78 mph/115 fps). If the parachutist enters the activation window (1,500 feet above the VDZ to 130 feet above the VDZ) and is falling faster than the vertical activation speed of 35 meters per second, the Military CYPRES 2 will fire the release element.

MILITARY CYPRES 2 MODELS

3-5. The three Military CYPRES 2 models—1000 35 A, 1500 35 A, and 2500 29 A—are the models in present use for all tactical parachute systems (except for USMC Special Operations Command). All models have the same appearance, function, and theory of operation. The Military CYPRES 2 uses millibars absolute as the unit of measurement. The differences among the three models are their preset information, including the default activation altitude above the VDZ, the vertical activation speed, and the release unit configuration.

3-6. Different settings are required to tailor the Military CYPRES 2 to specific parachute equipment and mission applications. For quick identification and to help ensure proper settings, the three Military CYPRES 2 models have their presets displayed on the green (ON/OFF) button located on the control unit. The presets identify the model of a Military CYPRES that goes with a specific parachute system:

- The Military CYPRES 2 model 1000 35 A is used on the Military Tandem Tethered Bundle (MTTB) parachute and authorized nonstandard parachute systems.
- The Military CYPRES 2 model 1500 35 A is used with the MC-4, MJN-1, MJA-2, MT-2XX/SL, and SOV2-HH.
- The Military CYPRES 2 model 2500 29 A is used on the Sigma Vector–Military-Tandem Vector System (MTV-3) and the TORDS.

The preset information for each model can be found printed on the back of the control unit and on the front cover of the processing unit. Figure 3-1 shows the Military CYPRES 2.

DANGER

The jumpmaster must accurately identify the CYPRES model being used and understand the correct pressure setting method. Failure to identify the correct model for the parachute system and to properly set the CYPRES may result in the CYPRES not firing at the intended altitude, resulting in injury or DEATH to the parachutist.

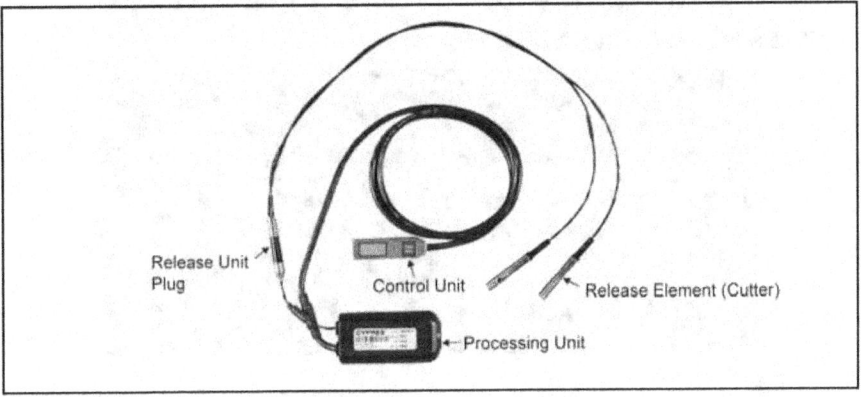

Figure 3-1. Military CYPRES 2 1500 35 A

Chapter 3

EXPERT CYPRES 2 MODEL

3-7. The Expert CYPRES 2 has the same look, function, maintenance, and theory of operation as the Military CYPRES 2 in training mode but is limited in use because it does not have an operational mode setting. While the Military CYPRES 2 uses millibars absolute as the unit for the setting, the Expert CYPRES 2 uses feet relative to the DAF. The Expert CYPRES 2 has its own preset activation altitude above the VDZ, activation speed, and release unit configuration. The Expert CYPRES 2 model is identified by the red (ON/OFF) button located on the control unit. The preset information is not printed anywhere on the outside of the Expert CYPRES 2. Figure 3-2 shows the Expert CYPRES 2.

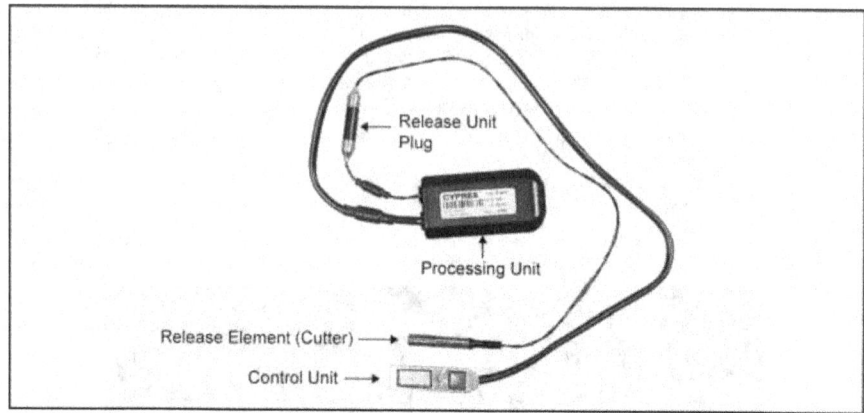

Figure 3-2. Expert CYPRES 2

3-8. The Expert CYPRES 2 model is used with authorized nonstandard parachute systems. The Expert CYPRES 2 model is set in feet and the display is graduated in feet. This setting is different from the three Military CYPRES 2 models which are set and displayed in millibars.

CYPRES MODEL IDENTIFICATION

DANGER

Failure to identify the correct Military CYPRES 2 model and setting method may result in configuring an improper setting, thus preventing the Military CYPRES 2 from firing when needed, resulting in injury or DEATH to the parachutist.

Failure to only use the Expert CYPRES 2 in authorized nonstandard parachute systems may result in the Expert CYPRES 2 not firing at the intended altitude, resulting in injury or DEATH to the parachutist.

3-9. The specific models of the CYPRES 2 are only authorized for use on the specified parachute systems listed in Table 3-1, page 3-5.

Table 3-1. CYPRES 2 model identification

Button Identification	Default Activation Altitude	Vertical Activation Speed	Pressure Setting Display Value	Authorized Parachute System
Military CYPRES 2 Green Button 1000 35 A	1,000 Feet	35 Meters per Second 78 mph 115 fps	Millibars Absolute	MTTB/Authorized Nonstandard Parachute Systems
Military CYPRES 2 Green Button 1500 35 A	1,500 Feet	35 Meters per Second 78 mph 115 fps	Millibars Absolute	MC-4, MJN-1, MJA-2, SOV2-HH, MT-2XX/SL, MMPS
Military CYPRES 2 Green Button 2500 29 A	2,500 Feet	29 Meters per Second 65 mph 95 fps	Millibars Absolute	MTV-3, TORDS
Expert CYPRES 2 Red Button (No Letters)	750 Feet	35 Meters per Second 78 mph 115 fps	+/- Feet Relative	Authorized Nonstandard Parachute Systems (Javelin)

3-10. The three Military CYPRES 2 models and the Expert CYPRES 2 model can be identified by the (ON/OFF) button on the control unit as described below:

- *Military CYPRES 2 Model 1000 35 A.* The control unit button is green, and the markings indicate that the Military CYPRES 2 is set to activate approximately 1,000 feet above the VDZ if the vertical speed is faster than approximately 35 meters per second (78 mph/115 fps). The A indicates that the pressure setting is in millibars absolute; the information in the control unit window display also reads in millibars (Figure 3-3). When the Military CYPRES 2 Model 1000 35 A is removed from the parachute system, the setting information can be read on the back of the control unit and on the front cover of the processing unit.

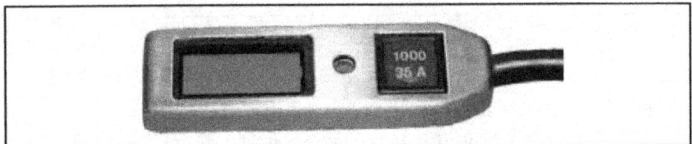

Figure 3-3. Military CYPRES 2 Model 1000 35 A control unit

- *Military CYPRES 2 Model 1500 35 A.* The control unit button is green, and the markings indicate that the Military CYPRES 2 is set to activate approximately 1,500 feet above the VDZ if the vertical speed is faster than approximately 35 meters per second (78 mph/115 fps). The A indicates that the pressure setting is in millibars absolute; the control unit window display also reads in millibars (Figure 3-4). When the Military CYPRES 2 Model 1500 35 A is removed from the parachute system, the setting information can be read on the back of the control unit and on the front cover of the processing unit.

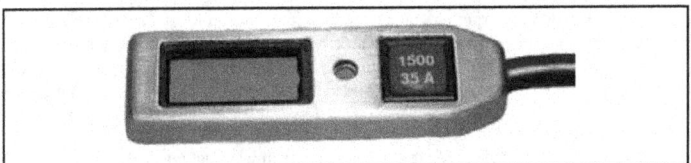

Figure 3-4. Military CYPRES 2 Model 1500 35 A control unit

- *Military CYPRES 2 Model 2500 29 A.* The control unit button is green, and the markings indicate that the Military CYPRES 2 is set to activate approximately 2,500 feet above the VDZ if the vertical speed is faster than approximately 29 meters per second (65 mph/95 fps). The A indicates the pressure setting is in millibars absolute; the control unit window display also reads in millibars (Figure 3-5). When the Military CYPRES 2 Model 2500 29 A is removed from the parachute system, the setting information can be read on the back of the control unit and on the front cover of the processing unit.

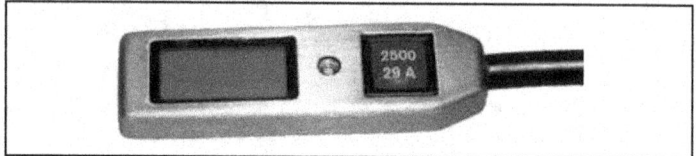

Figure 3-5. Military CYPRES 2 Model 2500 29 A control unit

- *Expert CYPRES 2 model.* The control unit button is red with no markings (Figure 3-6). If the vertical speed is faster than 35 meters per second (78 mph/115 fps), the Expert CYPRES 2 model is set to activate at approximately 750 feet above the DZ. The setting is graduated in feet and the control unit display is graduated in 30-foot increments.

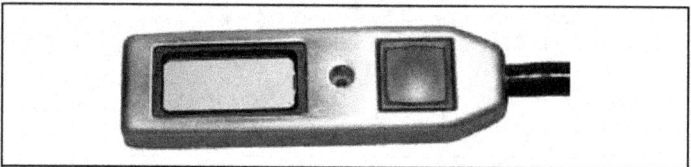

Figure 3-6. Expert CYPRES 2 control unit

COMPONENTS

3-11. The Military CYPRES 2 has only three components: the control unit, the processing unit with internal battery, and the release unit with one or two release elements.

CONTROL UNIT

3-12. The control unit houses a liquid crystal display (LCD) and a green ON/OFF button. It is attached to the processing unit by an electrical cable. The control unit provides the interface between the user and the processing unit. This allows the user to control functions such as powering ON and OFF, and setting the Military CYPRES 2 with the proper millibar setting for the absolute (operational) mode. During the power ON sequence, the LCD displays the power ON self-test information, the error codes, and the pressure setting. The user can see the Military CYPRES 2 is ON or OFF by observing the zero down arrow (0▼) setting for use in the default (training) mode, or by the proper millibar setting for use in the absolute (operational) mode. Once the power ON sequence and pressure setting is complete, the control unit is disengaged from the processing unit, and its only function is to power OFF the Military CYPRES 2. The pressure setting will remain displayed on the LCD. The control unit does not conduct any of the pressure readings or calculations performed by the Military CYPRES 2. The numerical information written on the control unit's single operating button lets the jumpmaster know which CYPRES 2 model is installed in the parachute system. Table 3-1, page 3-5, explains the numerical information. Figure 3-7, page 3-7, shows the control unit for the Military CYPRES 2 that has a default setting of 1,500 feet, fall rate setting of 35 meters per second (115 fps or 78 mph), and a pressure setting of absolute. When the Military CYPRES 2 is removed from the parachute system, the setting information can be read on the back of the control unit (Figure 3-8) and on the front cover of the processing unit (Figure 3-9).

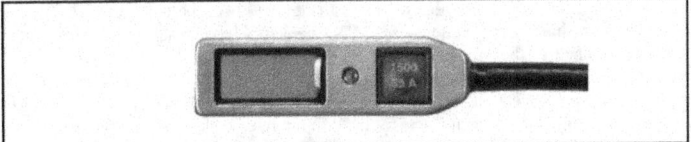

Figure 3-7. Military CYPRES 2 control unit

Figure 3-8. Back of Military CYPRES 2 control unit

Figure 3-9. Military CYPRES 2 processing unit

PROCESSING UNIT

3-13. The processing unit houses the microprocessor and the battery (Figure 3-9). The microprocessor conducts a self-test every time it is powered ON. The processing unit will stay ON and remain active for 14 hours from power ON, and then it will automatically power itself OFF. The microprocessor's software and sensors monitor the parachutist's altitude, vertical velocity, and other critical data points during free fall. It handles all critical calculations and functions to determine when a parachutist is in trouble so that the release elements can be fired. If the parachutist reaches the preset altitude at the preset vertical velocity and meets other critical conditions, the processing unit makes the decision to fire, sends an electrical charge to the release unit, fires the release elements, and severs the reserve closing loop material.

Note: The processing unit is protected from electromagnetic interference and static electricity, which means it is highly unlikely that radios and static electric shock will cause accidental discharge of the release elements.

Chapter 3

RELEASE UNIT

3-14. Release units are available for one-pin or two-pin reserve parachutes. The release unit contains a propellant actuated cutter called the release element (Figure 3-10). The number of release elements used depends on the parachute configuration. The MC-4 parachute system uses two release elements. The release unit is attached to the processing unit by the release unit plug. A used release element can be replaced during the reserve repack by unplugging the old release unit from the processing unit and plugging in the new release unit. In the event the parachutist meets all conditions to fire the release element, the processing unit sends an electrical input to the release element. A propellant inside the release element is electrically activated and, in turn, moves the release element knife approximately five millimeters to sever the reserve parachute closing loop(s) in order to open the reserve container. A CYPRES closing loop must be used to ensure proper operation. Once the release elements have fired, they must be replaced by the parachute rigger prior to repacking the reserve parachute. Release elements that have been fired are self-contained and remain pressurized—no attempt should be made to cut them open.

Note: The release element (cutter) is transportable on all military aircraft and does not require any special load planning or transportation considerations.

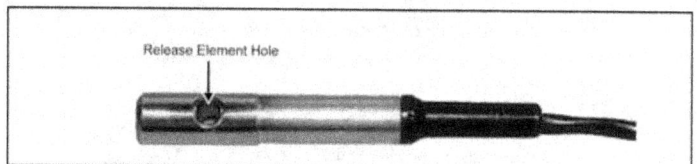

Figure 3-10. Military CYPRES 2 release unit

MAINTENANCE

3-15. Inspection, installation, maintenance, and storage of the Military CYPRES 2 shall be maintained by a ram-air pack-qualified parachute rigger.

BATTERY

3-16. The Military CYPRES 2 battery is replaced at 4 and 8 years from date of manufacture (plus or minus 6 months) during the periodic technical service performed by the manufacturer, SSK. The Military CYPRES 2 has a lifetime of 12.5 years from date of manufacture.

WATER LANDINGS

3-17. The Military CYPRES 2 is water resistant for 15 minutes at 5 meters. If a Military CYPRES 2 gets wet from a water landing, the rigger at the unit level is responsible for changing the filter.

SCHEDULED MAINTENANCE

3-18. The Military CYPRES 2 will be sent to the manufacturer for a periodic technical service at 4-year and 8-year intervals, plus or minus 6 months each, from the date of manufacture. The Military CYPRES 2 alerts the user when the scheduled maintenance is approaching. The maintenance due date and the unit's serial number are both easily retrievable (Figures 3-11 and 3-12, page 3-9). To access the serial number or the next maintenance due date without removing the unit from the parachute, the following steps are performed:
- Set the Military CYPRES 2 for use in the absolute (operational) mode.
- Enter a value outside of its operational range by selecting 0 for the first value and the numeral 1 (or 0) for the next three values.

The screen will momentarily go blank and then the serial number will appear in the display screen for approximately 5 seconds. The screen will go blank again and the next required maintenance date will appear on the screen.

Figure 3-11. Example of Military CYPRES 2 serial number

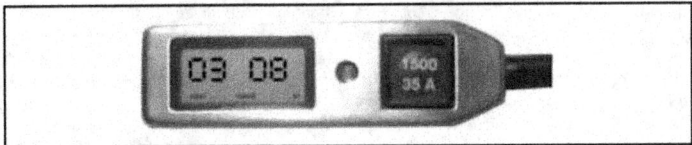

Figure 3-12. Example of next required maintenance date for Military CYPRES 2

GENERAL TERMS

3-19. The jumpmaster and parachutist should be familiar with the following terms as to how the Military CYPRES 2 performs its setting calculations:

- *Virtual drop zone.* The VDZ is defined as the virtual zero reference point established by the jumpmaster from which the Military CYPRES 2 makes its calculations. This reference point becomes the zero starting point, or the VDZ, for all Military CYPRES 2 calculations. There are two reasons for use of a VDZ in lieu of the actual DZ:
  - If a highest release point obstacle (HRPO) exists. This VDZ (highest elevation) must be used to provide the jumper with a safe distance above the obstacle for reserve deployment via the Military CYPRES 2.
  - To adjust the reserve parachute to a higher actuation altitude for a tactical HAHO option. This is SOP based only.
- *Highest release point obstacle.* For the Military CYPRES 2 setting calculations, a terrain feature near the release point is considered an obstacle if it is over 200 feet higher than the DZ. This is for HALO and HAHO jump operations. The obstacle is taken into consideration if it falls within the parameters below; the HRPO will become the VDZ on which all setting calculations are based:
  - 500-meter radius of the release point for operations up to 13,000 feet AGL.
  - 1,000-meter radius of the release point for operations above 13,000 feet AGL.

Note: Both HRPO radiuses are minimum distances and can be increased by the jumpmaster as needed.

- *Altimeter setting.* This setting is in inches of mercury (in/Hg) (QNH). The barometric pressure is corrected to MSL by taking the current station pressure and temperature, and adjusting it to MSL from the difference in elevation of where the reading was taken from. The pressure value of an aircraft altimeter scale is set so that it will indicate the altitude above MSL of an aircraft on the ground at the location for which the value was determined.
- *Unknown (combat) setting (29.92).* If the aircraft altimeter setting (in/Hg) is unknown, a value of 29.92 (in/Hg) is used to calculate the millibar setting of the Military CYPRES 2. The value of 29.92 (in/Hg) is the average pressure at 0 feet MSL, 59 degrees Fahrenheit, which is the around-the-world average.

Chapter 3

MODES OF OPERATION

3-20. The three Military CYPRES 2 models can be set in two modes: default (training) mode and absolute (operational) mode. The Expert CYPRES 2 can be set in two modes: default (training) mode and offset mode.

3-21. The primary jumpmaster for each jump is responsible for determining the proper mode and setting. The jumpmaster must properly identify the Military CYPRES 2 model to be used and fully understand its mode of operation to make the proper mode selection. The following paragraphs describe the Military CYPRES 2 modes of operation.

DEFAULT (TRAINING) MODE (MILITARY CYPRES 2 AND EXPERT CYPRES 2)

3-22. The default (training) mode is only used if the DZ and the DAF are the same location. Once powered ON in default (training) mode, the DAF/DZ elevation automatically becomes the zero reference point. The Military CYPRES 2 continuously samples atmospheric pressure changes and makes required adjustments to ensure that the zero reference point stays on the ground. For example, if all conditions are met, the Military CYPRES 2 Model 1500 35 A that is powered ON at the DAF/DZ will fire the release element at 1,500 feet above the DAF elevation. If all conditions are not met to fire the release element, the Military CYPRES 2 will remain active until it reaches 130 feet above the DAF elevation, at which time it will deactivate automatically. Once set in the default (training) mode, the Military CYPRES 2 must be powered OFF and powered back ON just prior to every lift at the DAF; this is a required safety measure for military use.

ABSOLUTE (OPERATIONAL) MODE (MILITARY CYPRES 2)

3-23. In absolute (operational) mode, the DZ or VDZ is calculated by the jumpmaster depending on operational requirements. By entering the desired millibar setting into the Military CYPRES 2, the jumpmaster tells the Military CYPRES 2 the absolute pressure of the location of the DZ or VDZ. The elevation corresponding to the pressure entered into the CYPRES calculator is now the zero reference point for the Military CYPRES 2. All calculations for the activation window and the activation altitude made by the Military CYPRES 2 are based off of this point. A VDZ may be programmed to any altitude within the device's operational range of -1600 feet to +36,000 feet MSL, which equates to 1075–0200 millibars.

3-24. Once set, the DZ/VDZ is locked into the millibar setting that corresponds to that altitude to start the Military CYPRES 2 calculations. The Military CYPRES 2 in absolute (operational) mode does not make adjustments for barometric pressure changes in weather. For example, if all conditions are met, the Military CYPRES 2 Model 1500 35 A in absolute (operational) mode with a VDZ set at 5,000 feet MSL (a mountain is at the release point) will fire at 6,500 feet MSL (VDZ MSL plus default activation equals Military CYPRES activation MSL). If all conditions are not met to fire the release element, the Military CYPRES 2 remains active until the parachutist reaches 130 feet above the VDZ (5,130 feet MSL) at which time it will deactivate automatically for the remainder of the canopy flight.

OFFSET MODE (EXPERT CYPRES 2)

3-25. The Expert CYPRES 2 does not have an absolute (operational) mode. However, the Expert CYPRES 2 can be set with an offset that will allow for a ±3,000-foot difference between the DAF and the DZ. This setting is derived by the jumpmaster using the same method employed to set the jumper's altimeter (altitude difference between the DAF MSL and DZ MSL equals offset). Rounded to the nearest 30-foot increment, the offset is entered into the Expert CYPRES 2. Once powered ON in offset mode, the VDZ (zero reference point) automatically becomes the DAF plus the amount of offset. For example, if all conditions are met, the Expert CYPRES 2 that is powered ON at the DAF with a default activation altitude of 750 feet and a +300-foot offset will fire the release element at 1,050 feet above the DAF elevation. If all conditions are not met to fire the release element, the Expert CYPRES 2 will remain active until it reaches 130 feet above the VDZ elevation, at which time it will deactivate automatically. Once set in the offset mode, the Expert CYPRES 2 must be powered OFF and powered back ON just prior to every lift at the DAF; this is a required safety measure for military use.

MODE DETERMINATION

3-26. The proper mode for the three Military CYPRES 2 models and the Expert CYPRES 2 must be determined by the primary jumpmaster before and during the jump operation. If the mission or conditions change, it may be necessary to adjust the setting or mode.

3-27. Absolute (operational) mode can be used in all situations and is the recommended mode to use for all military free-fall operations. To determine the use of the default (training) mode on the Military CYPRES 2, the following questions must be answered for every jump:

- Are the DAF and DZ at different locations?
- Is there a HRPO 200 feet or greater?
- Is this a HAHO jump?
- Is the desired Military CYPRES 2 activation different than the Military CYPRES 2 default values?
- Will the aircraft fly below the DAF/DZ elevation?
- Will the Military CYPRES 2 be powered ON in flight?
- Will the aircraft be pressurized?

3-28. If any of the questions were answered "YES," the Military CYPRES 2 must be used in absolute (operational) mode. If answered "NO" to every question, default (training) mode can be used. The jumpmaster will evaluate the operational parameters during the jump. If the parameters change, the Military CYPRES 2 may have to be reset. The ideal situation to use the three Military CYPRES 2 models and the Expert CYPRES 2 model in default (training) mode is when a DZ is selected with no terrain feature obstacles near the release point and the DAF and DZ are the same location and elevation MSL.

DEFAULT (TRAINING) MODE

3-29. The three Military CYPRES 2 models and the Expert CYPRES 2 model are set to operate in the default (training) mode with the same operational parameters. Both the Military and Expert CYPRES 2 models function in the same manner and use the same power ON sequence. While in default (training) mode, the parachutist's emergency procedures and pull altitudes do not change. The jumpmaster must ensure that all warnings, cautions, and rules are understood before selecting the default (training) mode of operation.

DANGER

Do not use the CYPRES 2 in default (training) mode unless the following conditions are met. Failure to adhere to these warnings may result in the CYPRES 2 not firing at the intended altitude and may result in injury or DEATH to the parachutist.

Prior to every lift while in default (training) mode, the CYPRES 2 will be powered OFF and powered back ON at the DAF. This ensures that all parachutists on the aircraft have the correct DZ setting. The CYPRES 2 will automatically reset the DZ under certain situations. Failure to power OFF and power back ON the CYPRES 2 just prior to every lift may result in the CYPRES 2 not firing at the intended altitude, which may result in injury or DEATH to the parachutist.

The jump aircraft must never fly below the DAF altitude. If the jump aircraft flies below the DAF altitude, the CYPRES 2 will reset the DZ, or zero reference point, to that lower altitude. Then the CYPRES 2 may fire at a lower-than-intended altitude, which may result in insufficient time for the reserve parachute to inflate, resulting in injury or DEATH to the parachutist.

Chapter 3

> **DANGER (Continued)**
>
> The cabin area of the jump aircraft must be depressurized during engine startup and takeoff to ensure that the cabin pressure will not build up above the air pressure on the ground. If the cabin pressure in the jump aircraft is allowed to build up (representing a lower altitude), the CYPRES 2 will automatically reset the VDZ, or ground reference, to the lower altitude and the CYPRES 2 may fire at a lower-than-intended altitude. This may result in insufficient time for the reserve parachute to inflate, resulting in injury or DEATH to the parachutist.
>
> The CYPRES 2 must not be powered ON in flight while in default (training) mode. Powering ON the CYPRES 2 while in flight will set the VDZ to the aircraft altitude or cabin pressure. This will result in the CYPRES 2 firing too high upon exit or not firing at all, depending on the exit altitude, resulting in injury or DEATH to the parachutist.

Conditions for Using All CYPRES 2 Models in Default (Training) Mode

3-30. While in default (training) mode, the three Military CYPRES 2 models and the Expert CYPRES 2 model may be used only under the conditions listed below. These conditions must be maintained throughout the operation with no exceptions. If any of these conditions are not met, the CYPRES 2 will not be used in the default (training) mode. The absolute (operational) mode will be used instead. The conditions include the following:

- Every CYPRES 2 on the lift must be powered OFF and powered back ON at the DAF while on the ground prior to every lift; this is a required safety measure for military use.
- Low-level flights en route to the DZ must not fly below the DAF MSL elevation.
- The aircraft cannot be pressurized.
- The DAF and DZ must be the same location.
- HAHO jumps cannot be done.
- The CYPRES default activation values must be used.
- Minimum vertical separation between reserve activation altitude and main deployment altitude is 2,000 feet for the 1500 model and 2,500 feet for the 2500 model. Minimum vertical separation between reserve activation altitude and main deployment altitude is 1,500 feet for the 1000 model, when used on nonstandard parachutes. When the 1000 model is used on the tandem bundle, the vertical separation is not applicable.
- There cannot be a HRPO. A terrain feature at the release point is an obstacle when that feature is 200 feet or higher than the DZ and is within—
    - A 500-meter radius of the release point for operations 13,000 feet AGL and below.
    - A 1,000-meter radius of the release point for operations above 13,000 feet AGL.

Jumpmaster and Pilot Considerations in Default (Training) Mode

3-31. When using the default (training) mode, the jumpmaster and pilot should consider the following:

- The three Military CYPRES 2 models arm themselves at an altitude of 1,500 feet above the default setting and the Expert CYPRES 2 model at an altitude of 750 feet above the default setting. For example, once powered ON prior to leaving the ground, the Military CYPRES 2 1500 35 A arms itself at 3,000 feet above the DAF.

- The aircraft should never descend to an altitude below the elevation of the DAF. The Military CYPRES 2 will automatically reset the VDZ, or ground reference, to the lower altitude and the Military CYPRES 2 may fire at a lower-than-intended altitude.
- If the aircraft can be pressurized, the pilot makes sure the cabin remains depressurized during engine startup and takeoff. He ensures that the cabin pressure does not build up above the air pressure on the ground. Jumpmasters onboard the aircraft may monitor this by observing their altimeters. If the cabin pressure in the aircraft is allowed to build up (representing a lower altitude), the Military CYPRES 2 will automatically reset the VDZ, or ground reference, to the lower altitude and the Military CYPRES 2 may fire at a lower-than-intended altitude.
- While descending, the aircraft should never exceed the vertical activation speed for the Military CYPRES 2 while in the activation window. Exceeding the vertical activation speed may cause the Military CYPRES 2 to fire the release element and deploy the reserve parachute. Aircraft may exceed the vertical activation speed of 6,900 feet per minute for the 1000 and 1500 35 A models or 5,700 feet per minute for the 2500 29 A model during descent while executing a tactical landing. The jumpmaster must brief the pilots not to exceed 5,000 feet per minute as this descent rate is easy to remember and covers all three Military CYPRES 2 models.

ABSOLUTE (OPERATIONAL) MODE

3-32. The absolute (operational) mode may be used under all conditions for HALO or HAHO jumps as long as the required absolute (operational) mode-setting parameters are followed. The absolute (operational) mode may be used for short or long flights. The Military CYPRES 2 will power itself OFF after 14 hours under any condition.

Operating Conditions for Absolute (Operational) Mode

3-33. The three Military CYPRES 2 models must be used in absolute (operational) mode for the following operating conditions:
- The Military CYPRES 2 is powered ON in flight.
- Low-level flights en route to the DZ are flying below the DAF MSL elevation.
- There is a HRPO of 200 feet or greater above the DZ.
- The Military CYPRES 2 activation altitude is different than the default activation altitude.
- DAF and DZ are at separate locations.
- The aircraft must be pressurized.

Rules for Using the Absolute (Operational) Mode

3-34. When the Military CYPRES 2 is used in absolute (operational) mode, all of the following apply to all operations and will be strictly followed:
- During in-flight power ON, the aircraft climb rate or descent rate will not exceed 1,000 feet per minute until all Military CYPRES 2 models on board are powered ON. The preferred method is to have the aircraft level off during Military CYPRES 2 setting.
- All parachutists on the same stick will have the same DZ/VDZ setting.
- The minimum VDZ setting for jump operations at 13,000 feet AGL and below is the height of the highest obstacle, if 200 feet or higher than the DZ, within 500 meters of the release point.
- The minimum VDZ setting for a jump operation at greater than 13,000 feet AGL is the height of the highest obstacle, if 200 feet or higher than the DZ, within 1,000 meters of the release point.
- Minimum vertical separation between reserve activation altitude and main deployment altitude is 2,000 feet for the 1500 model and 2,500 feet for the 2500 model. Minimum vertical separation between reserve activation altitude and main deployment altitude is 1,500 feet for the 1000 model, when used on nonstandard parachutes. When the 1000 model is used on the tandem bundle, the vertical separation is not applicable.

Chapter 3

- While in absolute (operational) mode, the Military CYPRES 2 may remain powered ON during multiple jumps as long as the operational parameters do not change. The altimeter setting for the DZ should be checked every hour during the operation, using the most accurate means available, and the Military CYPRES 2 setting will be recalculated. If the Military CYPRES 2 setting changes more than ±3 millibars or if the operational parameters change, the jumpmaster must recalculate and reset the Military CYPRES 2.

> **WARNING**
>
> The actual absolute pressure (QFE) at the VDZ must be entered into the Military CYPRES 2 in operational mode. This pressure can be determined by direct measurement with an instrument such as the Military CYPRES Portable Calibration Station. An actual Military CYPRES 2 can be calculated from the aircraft altimeter setting (QNH) and the MSL elevation of the VDZ using the approved CYPRES calculators; for example, Excel, circular calculator, or digital calculator.
>
> It is important to realize that flight service and weather stations normally report pressure as if it were sea level (aircraft altimeter setting [QNH]), and not the actual absolute pressure (QFE); thus, it is necessary to convert the aircraft altimeter setting (QNH) to the actual absolute pressure (QFE) for use with the Military CYPRES 2.

Calculations for the Drop Zone in Absolute (Operational) Mode

3-35. When using the Military CYPRES 2 in operational mode, the jumpmaster will calculate the millibar setting by obtaining the following information:
- Actual absolute air pressure at the DZ.
- Both the current aircraft altimeter setting (QNH) in/Hg and DZ elevation MSL.

> **WARNING**
>
> Pressure readings should be as current as possible, preferably updated every hour by the drop zone safety officer (DZSO) and recorded from the nearest source to the DZ.
>
> Jumpmasters must use their best judgment when obtaining the aircraft altimeter setting off-site from the DZ. Within ± 20 miles is a good reference. Depending on the geographic location of the DZ or, in many cases (HAHOs), the HARP, in reference to atmospheric conditions, pressure values could be significantly different from one valley to another that are only separated by a single ridgeline. Jumpmasters should make note of pressure differences and weather conditions in relation to the location they obtained the aircraft altimeter setting from and to the location of the DZ. The distances away from the DZ/VDZ for obtaining the pressure can greatly increase if meteorological conditions are favorable.

Cybernetic Parachute Release System

> **WARNING (Continued)**
>
> The DZ elevation used for calculating the millibar setting is the highest point of elevation (MSL) given on the Air Force (AF) IMT Form 3823 (Drop Zone Survey).

Calculations for the Virtual Drop Zone in Absolute (Operational) Mode

3-36. In some scenarios, the jumpmaster must use a VDZ, described as a virtual line in the sky, which is a higher elevation MSL than that of the actual DZ. These scenarios include the following:

- HALO jumps or HAHO jumps with a low reserve activation altitude that have a HRPO of 200 feet or greater than the DZ elevation within a—
  - 500-meter radius of the release point for operations up to 13,000 feet AGL.
  - 1,000-meter radius of the release point for operations above 13,000 feet AGL.
- HAHO with a high reserve activation altitude setting for a tactical operation based on SOP.

3-37. The jumpmaster will calculate the millibar setting for a VDZ by obtaining the following information:

- Current aircraft altimeter setting (QNH) in/Hg.
- The higher VDZ elevation MSL.

> **WARNING**
>
> When using a VDZ or higher elevation than the DZ, the jumpmaster must also change the pull altitude of his jumpers. Jumpers must maintain 2,000 feet (1500 35 A model) of vertical separation between pull altitude and reserve activation altitude (2,500 feet of vertical separation for the 2500 29 A model CYPRES). When used with nonstandard parachutes, 1,500 feet of vertical separation is required for the 1000 35 A model CYPRES.

Calculations for Unknown Setting (29.92) in Absolute (Operational) Mode

3-38. When using the Military CYPRES 2 in operational mode, if the jumpmaster cannot obtain a current aircraft altimeter setting for a precise measurement of station pressure, he must use 29.92 in/Hg for his millibar calculation. Because the pressure may not actually be 29.92 in/Hg, the jumpmaster must plan for the reserve activation altitude being possibly higher or lower than planned for, as follows:

- If the actual pressure is higher, the Military CYPRES 2 will activate on the high side.
- If the actual pressure is lower, the Military CYPRES 2 will activate on the low side.

3-39. The jumpmaster must add a safety factor into his calculations in order for the reserve to have enough altitude to fully inflate and save his jumpers' lives. To prevent activation on the low side, the jumpmaster will add 500 feet to the DZ or HRPO VDZ for a safety factor. This new elevation is the elevation the jumpmaster will use to calculate the millibar. The jumpmaster does not need to add a 500-foot safety factor to the VDZ if the jump profile is a HAHO with a high reserve activation altitude as there is unlimited altitude for the reserve to fully deploy if it fires on the low side (with no obstacles).

3-40. The jumpmaster will calculate the millibar setting by using and obtaining the following information:

- Unknown aircraft altimeter setting (QNH) = 29.92 in/Hg.
- DZ or VDZ elevation MSL + 500 feet for safety factor = VDZ elevation MSL.

Chapter 3

> **DANGER**
>
> When using unknown setting of 29.92 in/Hg, the jumpmaster must also add 1,000 feet to the jumpers' pull altitude as a safety factor in case the Military CYPRES 2 fires on the high side. Failure to do so could cause a dual canopy deployment, resulting in injury or DEATH to the parachutist.

Note: Unknown setting of 29.92 in/Hg may be used for tactical training jumps as long as all calculations are done using the appropriate safety factors.

Jumpmaster and Pilot Considerations for Using the Absolute (Operational) Mode

3-41. When using the absolute (operational) mode, the jumpmaster and pilot must consider the following:

- While operating in absolute (operational) mode, the three Military CYPRES 2 models arm themselves as soon as they are powered ON.
- When in the absolute (operational) mode, the Military CYPRES 2 can be set in both a pressurized and a depressurized aircraft while in flight. During an in-flight power ON for an unpressurized cabin, the aircraft climb rate or descent rate will not exceed 1,000 feet per minute or a steady pressurized rate within 1,000 feet per minute for a pressurized cabin until all Military CYPRES 2 models are powered ON. Leveling off is preferred.
- While descending, the aircraft should never exceed the vertical activation speed for the Military CYPRES 2 while in the activation window. Exceeding the vertical activation speed may cause the Military CYPRES 2 to fire the release element and deploy the reserve parachute. Aircraft may exceed the vertical activation speed of 6,900 feet per minute for the 1000 and 1500 35 A models or 5,700 feet per minute for the 2500 29 A model during descent while executing a tactical landing that will cause the Military CYPRES 2 to activate in the aircraft. The jumpmaster must brief the pilots not to exceed 5,000 feet per minute as this descent rate is easy to remember and covers all three Military CYPRES 2 models.
- Descent to an altitude below the elevation of the DAF will not affect the Military CYPRES 2 in the absolute (operational) mode.
- Once the aircraft descends through the VDZ altitude, the Military CYPRES 2 will deactivate itself and will not fire the release element. Therefore, if the jump altitude is lowered below the VDZ, all Military CYPRES 2 models on the aircraft must be reset.

OPERATING PROCEDURES

3-42. There are two modes for the three Military CYPRES 2 models and two modes for the Expert CYPRES 2 model:

- Default (training) mode used for both the Military CYPRES 2 models and the Expert CYPRES 2 model.
- Absolute (operational) mode used for the three Military CYPRES 2 models.
- Offset mode used for the Expert CYPRES 2 model.

3-43. The three Military CYPRES 2 models and the Expert CYPRES 2 model operate and power ON in the same way. The only difference among the models is the activation speed and altitude. The button on the control unit is the only means the user has to control the CYPRES 2. The parachutist performs two actions: powering ON and powering OFF the CYPRES 2.

Cybernetic Parachute Release System

POWER ON PROCEDURES FOR MILITARY CYPRES 2 AND EXPERT CYPRES 2 IN DEFAULT (TRAINING) MODE

> **CAUTION**
>
> For the default (training) mode only, the Military CYPRES 2 must be powered ON at the DAF while on the ground; it must not be powered ON inside a flying aircraft. The user must initialize on the ground at ground level to be accurate.

3-44. During the power ON sequence, the CYPRES 2 conducts a self-test. The jumper must watch the display during the entire power ON self-test. Table 3-2 explains the error codes.

Table 3-2. CYPRES 2 power ON self-test error codes in default (training) mode

Code	Meaning
1111 2222	One or both of the attached release units are not correctly electrically connected to the unit. The reason may be a cable break, the cutter plug could be disconnected, or the release unit(s) may have activated.
3333	Excessive variations in ambient air pressure have been measured during the self-test period. The unit is unable to obtain consistent values for the ambient air pressure at ground level. Possible reasons could be an attempt to switch on the CYPRES 2 in the training mode in an airborne aircraft while exceeding a climb rate or descent rate of more than 1,500 feet per minute.
7777	Low battery. The battery capacity is large enough to cover most of the usage profile, but in extreme situations a low-battery indication may show up. In this case, Airtec or SSK should be contacted before the next use.

> **CAUTION**
>
> The jumper should press the control unit button with the tip of a finger. He should not use a fingernail or a sharp object. Prolonged use of a fingernail or a sharp object will wear the letters off of the button and possibly wear a hole in the button material, thus rendering the CYPRES 2 unserviceable.

Note: If a button click is missed or a button is pressed too soon before the light comes on, the CYPRES 2 will not power ON. If the CYPRES 2 fails to power ON, the sequence should be started over again in default (training) mode.

Note: The four-click initiation cycle is designed to avoid accidental activation.

3-45. The jumper starts the power ON process for the Military CYPRES 2 in the default (training) mode by pressing the button on the control unit four times as follows:
- Press the button on the control unit with the tip of a finger.
- When the light-emitting diode (LED) illuminates, press the button again while the light is on.
- Repeat above step two more times for a total of four times.

When the power ON steps are successful, the display will come on and the self-test will start to count down, which should last for 10 seconds (Figure 3-13, page 3-18).

Chapter 3

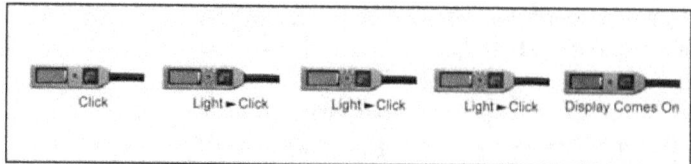

Figure 3-13. Power ON sequence for Military CYPRES 2
in default (training) mode

3-46. The jumper must watch the display during the self-test period. The display will start with 10 (Figure 3-14) and then show a rapid countdown to zero with the arrow pointing down (0▼), referred to as zero down. If the self-test is successful, the Military CYPRES 2 will remain powered ON.

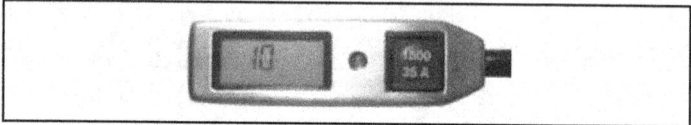

Figure 3-14. Beginning of Military CYPRES 2 self-test countdown
in default (training) mode

3-47. During the countdown, the display pauses between one and zero to display the current barometric pressure in millibars (Figure 3-15). The jumpmaster should always take note of this number.

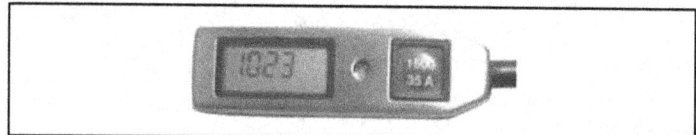

Figure 3-15. Military CYPRES 2 displaying current barometric pressure
in millibars

Note: The Military CYPRES 2 will not display battery voltage because the user cannot replace the battery. If battery voltage is low, the Military CYPRES 2 will not power ON.

3-48. The Military CYPRES 2 will continue the countdown to a reading of 0▼ (Figure 3-16). Once the 0▼ reading is displayed, the Military CYPRES 2 has passed the self-test and is powered ON, ready to monitor a jump. While in the default mode, the Military CYPRES 2 will arm itself 1,500 feet above the default activation altitude.

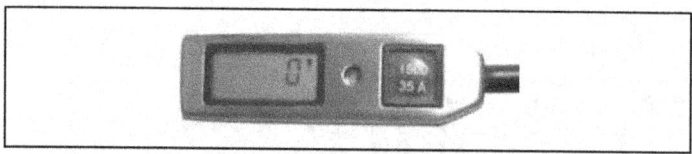

Figure 3-16. Military CYPRES 2 set in default (training) mode

3-49. If a functional deficiency in the Military CYPRES 2 is detected during the power ON self-test, the Military CYPRES 2 will display an error code and power OFF. Prior to the Military CYPRES 2 powering off, the jumper should note the error code in the display (Figure 3-17).

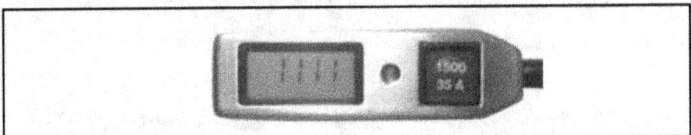

Figure 3-17. Example of Military CYPRES 2 error code

3-50. The Military CYPRES 2 automatically powers OFF after 14 hours. During the 14 hours, the Military CYPRES 2 settings will adjust for barometric pressure changes.

POWER ON PROCEDURES FOR MILITARY CYPRES 2 IN ABSOLUTE (OPERATIONAL) MODE

3-51. During the power ON sequence, the Military CYPRES 2 conducts a self-test. The jumper must watch the display during the entire power ON self-test. Table 3-3 explains the error codes.

Table 3-3. CYPRES 2 power ON self-test error codes in absolute (operational) mode

Code	Meaning
1111 2222	One or both of the attached release units are not correctly electrically connected to the unit. The reason may be a cable break, the cutter plug could be disconnected, or the release unit(s) may have activated.
3333	Excessive variations in ambient air pressure have been measured during the self-test period. The unit is unable to obtain consistent values for the ambient air pressure at ground level. Possible reasons could be an attempt to switch on the CYPRES 2 in the training mode in an airborne aircraft while exceeding a climb rate or descent rate of more than 1,500 feet per minute.
7777	Low battery. The battery capacity is large enough to cover most of the usage profile, but in extreme situations a low-battery indication may show up. In this case, Airtec or SSK should be contacted before the next use.

CAUTION

The jumper should press the control unit button with the tip of a finger. He should not use a fingernail or a sharp object. Prolonged use of a fingernail or a sharp object will wear the letters off of the button and possibly wear a hole in the button material, thus rendering the CYPRES 2 unserviceable.

Note: If a button click is missed or a button is pressed too soon before the light comes on, the Military CYPRES 2 will not power ON. If the Military CYPRES 2 fails to power ON, the sequence should be started over again in absolute (operational) mode.

Note: The four-click initiation cycle is designed to avoid accidental activation.

3-52. The jumper starts the power ON process for the Military CYPRES 2 in the absolute (operational) mode by pressing the button on the control unit four times as follows:
- Press the button on the control unit with the tip of a finger.
- When the LED illuminates, press the button again while the light is on.

Chapter 3

- Repeat above step two more times for a total of four times.
- Hold the button down on the fourth press.

When the power ON steps are successful, the display will come on and the self-test will start to count down, which should last for 10 seconds (Figure 3-18).

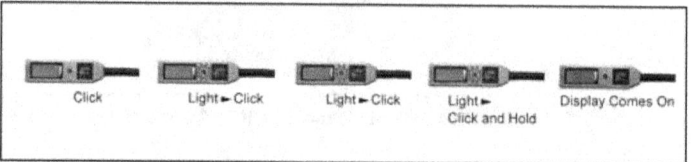

Figure 3-18. Power ON sequence for Military CYPRES 2 in absolute (operational) mode

3-53. The jumper must watch the display during the self-test period. The display will start with 10 (Figure 3-19) and then show a rapid countdown to 0 ▼.

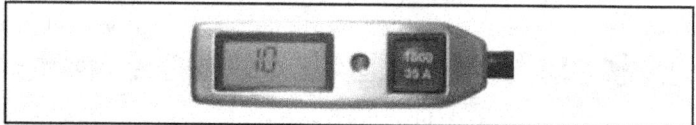

Figure 3-19. Beginning of Military CYPRES 2 self-test countdown in absolute (operational) mode

3-54. The self-test cycle takes 10 seconds to complete. There is a brief pause between 1 and 0 where the current barometric pressure is displayed in millibars (Figure 3-20).

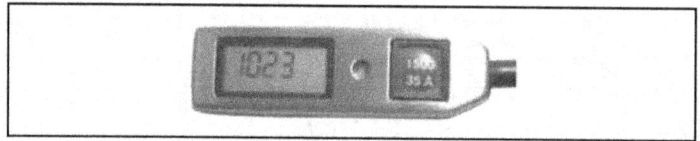

Figure 3-20. Military CYPRES 2 displaying current barometric pressure in millibars

3-55. Once the 0 ▼ reading is displayed (Figure 3-21), the Military CYPRES 2 has passed the self-test.

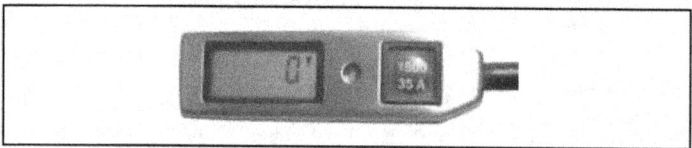

Figure 3-21. Military CYPRES 2 set in absolute (operational) mode

3-56. If a functional deficiency in the Military CYPRES 2 is detected during the power ON self-test, the Military CYPRES 2 will display an error code and power OFF. Prior to the Military CYPRES 2 powering off, the jumper should note the error code in the display (Figure 3-22, page 3-21).

Cybernetic Parachute Release System

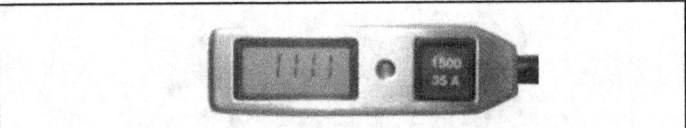

Figure 3-22. Example of Military CYPRES 2 error code

3-57. Upon completion of the self-test, the Military CYPRES 2 will display the millibar setting of 1000. To set the appropriate millibar setting, the jumper performs the following steps:

- The 1 will alternate with 0. Release the button to choose 0 or 1. The chosen value remains on the display (Figure 3-23).

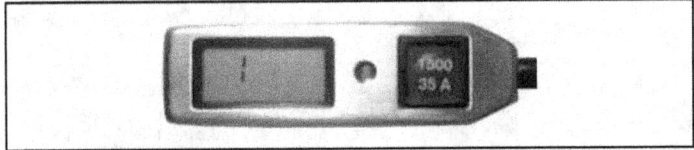

Figure 3-23. First value of 1 chosen for millibar setting

- Press and hold the button again. The second digit counts from 0 through 9. To select the second value, release the button when the desired value appears. This value remains on the display (Figure 3-24).

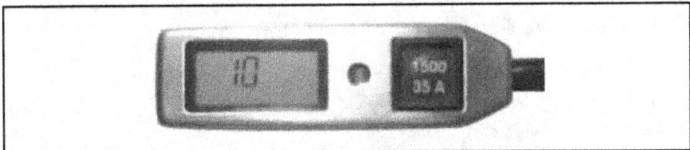

Figure 3-24. Second value of 0 chosen for millibar setting

- Press and hold the button again. The third digit counts from 0 through 9. To select the third value, release the button when the desired value appears. This value remains on the display (Figure 3-25).

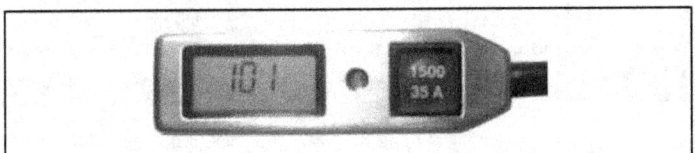

Figure 3-25. Third value of 1 chosen for millibar setting

- Press and hold the button again. The fourth digit counts from 0 through 9. To select the final value, release the button when the desired value appears. This value remains on the display (Figure 3-26, page 3-22).

Chapter 3

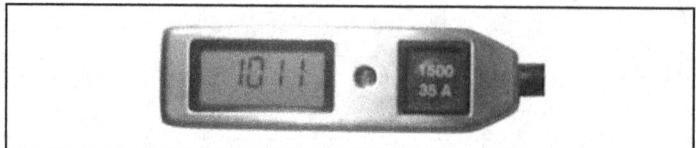

Figure 3-26. Final value chosen and Military CYPRES 2 set

3-58. To change an incorrect entered number, the jumper presses the button until the value shows up again. (After 9, the display restarts automatically with 0.) In order to start over completely because of an incorrect input in a previous value, without powering off, the jumper—
- Inputs a number into all four values.
- Before the light goes out after the fourth value input, presses and holds the button again. The display will start over at the first value again.

Note: If the user tries to enter a pressure of less than 200 millibars (approximately 39,000 feet above sea level) or more than 1,075 millibars (approximately 1,600 feet below sea level), the Military CYPRES 2 switches itself off. The blank display indicates that the desired adjustment is outside the specified parameters.

3-59. The pressure adjustment and the display indication remain until the unit is switched off. To change the setting, the jumper switches the Military CYPRES 2 off and on again.

3-60. The Military CYPRES 2 automatically turns off after 14 hours. During the 14 hours, the Military CYPRES 2 settings will not adjust for barometric pressure changes.

POWER ON PROCEDURES FOR EXPERT CYPRES 2 IN OFFSET MODE

3-61. The Expert CYPRES 2 can be set in an offset mode that allows the default altitude to be changed by ±1,500 feet (±3,000 feet with units produced after 2006) from the DAF. This allows the default setting to be adjusted if the DZ elevation is higher or lower than the DAF or if the default activation level needs to be raised. In the offset mode, the Expert CYPRES 2 powers ON the same way as the Military CYPRES 2 in the absolute (operational) mode except that the button cannot be released while making the setting. As soon as the button is released, the setting is locked into the Expert CYPRES 2. If the desired offset is missed, the Expert CYPRES 2 must be powered OFF and powered back ON in offset mode to adjust the setting. During the power ON sequence, the Expert CYPRES 2 conducts a self-test. The jumper must watch the display during the entire power ON self-test. Table 3-4 explains the error codes.

Table 3-4. Expert CYPRES 2 power ON self-test error codes in offset mode

Code	Meaning
1111 2222	One or both of the attached release units are not correctly electrically connected to the unit. The reason may be a cable break, the cutter plug could be disconnected, or the release unit(s) may have activated.
3333	Excessive variations in ambient air pressure have been measured during the self-test period. The unit is unable to obtain consistent values for the ambient air pressure at ground level. Possible reasons could be an attempt to switch on the CYPRES 2 in the training mode in a car, driving uphill or downhill, or in an airborne aircraft.
7777	Low battery. The battery capacity is large enough to cover most of the usage profile, but in extreme situations a low-battery indication may show up. In this case, Airtec or SSK should be contacted before the next use.

CAUTION

The jumper should press the control unit button with the tip of a finger. He should not use a fingernail or a sharp object. Prolonged use of a fingernail or a sharp object will wear the letters off of the button and possibly wear a hole in the button material, thus rendering the Expert CYPRES 2 unserviceable.

Note: If a button click is missed or a button is pressed too soon before the light comes on, the Expert CYPRES 2 will not power ON. If the Expert CYPRES 2 fails to power ON, the sequence should be started over again in offset mode.

Note: The four-click initiation cycle is designed to avoid accidental activation.

3-62. The jumper starts the power ON process for the Expert CYPRES 2 in the offset mode by pressing the button on the control unit four times as follows:

- Press the button on the control unit with the tip of a finger.
- When the LED illuminates, press the button again while the light is on.
- Repeat above step two more times.
- Hold the button down on the fourth press. Do not let up finger pressure on the button.

When the power ON steps are successful, the display will come on and the self-test will start to count down, which should last for 10 seconds (Figure 3-27).

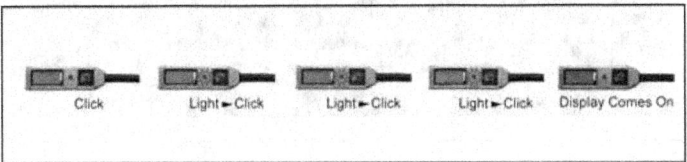

Figure 3-27. Power ON sequence for Expert CYPRES 2 in offset mode

3-63. The jumper must watch the display during the self-test period. The display will start with 10 and then show a rapid countdown to zero with the arrow pointing down (0▼), referred to as zero down. The self-test cycle takes 10 seconds to complete (Figure 3-28).

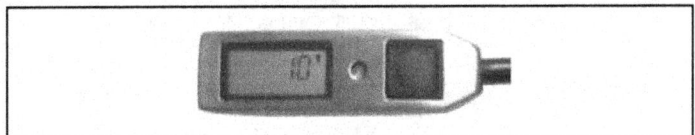

Figure 3-28. Expert CYPRES 2 displaying countdown

3-64. If a functional deficiency in the Expert CYPRES 2 is detected during the power ON self-test, the Expert CYPRES 2 will display an error code and power OFF. Prior to the Expert CYPRES 2 powering off, the jumper should note the error code in the display (Figure 3-29, page 3-24).

Chapter 3

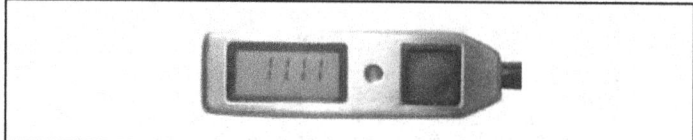

Figure 3-29. Example of Expert CYPRES 2 error code

3-65. Once the 0▼ reading is displayed (Figure 3-30), the Expert CYPRES 2 has passed the self-test in offset mode.

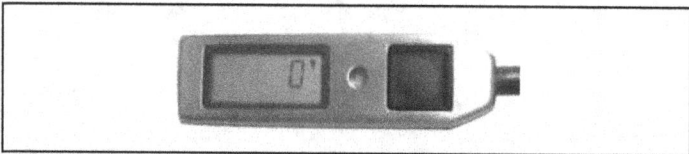

Figure 3-30. Expert CYPRES 2 control unit displaying countdown at zero down in offset mode

DANGER

In the offset mode, an arrow up means the offset is higher than the default altitude of 750 feet; an arrow down means the offset is lower than the default altitude of 750 feet. Failure to ensure that the arrow is pointing in the correct direction could result in the Expert CYPRES 2 not firing at the intended altitude, resulting in injury or DEATH to the parachutist.

Note: The offset mode is good for one jump only. The jumper must reset the offset mode every time the same jump profile is repeated.

3-66. After reaching 0, the numbers on the display screen will advance in 30-foot increments up to 1,500 feet with the arrow changing from up to down at each increment. The direction of the arrow indicates whether the designated landing area is higher or lower than the DAF (the jumper's current position). Once the desired offset is reached, the jumper releases the button when the arrow is pointing in the desired direction. The desired setting will stay on the display screen. If the desired setting is missed, the jumper powers OFF the Expert CYPRES 2 and repeats the whole process. Figure 3-31 shows the Expert CYPRES 2 set at 120 feet above the default altitude of 750 feet. The Expert CYPRES 2 is now set to activate at 870 feet above the DAF (the jumper's current position).

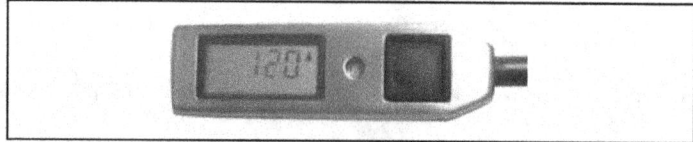

Figure 3-31. Expert CYPRES 2 set at 120-foot offset

3-67. The Expert CYPRES 2 automatically powers off after 14 hours. During the 14 hours, the Expert CYPRES 2 settings will adjust for barometric pressure changes.

POWER OFF PROCEDURES FOR CYPRES 2

3-68. The power OFF procedures are the same for all Military and Expert CYPRES 2 models and in every mode. It is the reverse of the power ON process (Figure 3-32).

> **CAUTION**
>
> The jumper should press the control unit button with the tip of a finger. He should not use a fingernail or a sharp object. Prolonged use of a fingernail or a sharp object will wear the letters off of the button and possibly wear a hole in the button material, thus rendering the CYPRES 2 unserviceable.

3-69. The jumper starts the power OFF process by pressing the button on the control unit four times as follows:

- Press the button on the control unit with the tip of a finger.
- When the LED illuminates, press the button again while the light is on.
- Repeat above step two more times.

When the power OFF steps are successful, the display will shut off. If the CYPRES 2 does not power OFF, the jumper should repeat above three steps.

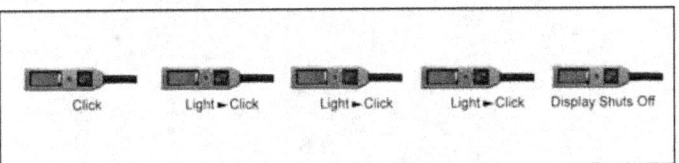

Figure 3-32. Power OFF sequence for CYPRES 2

USING MILITARY CYPRES 2 CALCULATORS

3-70. In situations where the Military CYPRES 2 absolute (operational) mode must be used, two sets of information are required to calculate the millibar setting: the altimeter setting for the intended DZ and the MSL elevation of the DZ or VDZ. The jumpmaster obtains the DZ altimeter setting from the pilot or from a weather station within as close a range as possible to the DZ. The jumpmaster does not use the actual barometric pressure of the DZ, but instead uses the altimeter setting for the DZ. If the current DZ altimeter setting information is unavailable from the pilot or weather station, the unknown altimeter setting of 29.92 in/Hg, which is 1013 millibars at 0 foot MSL, should be used. Once the jumpmaster obtains the current aircraft altimeter setting and DZ/VDZ MSL elevation, he will calculate the millibar setting using an approved Military CYPRES 2 calculator.

3-71. Tools authorized to use for the millibar setting calculation include the following:

- Military CYPRES Absolute Adjust Circular Calculator (Whiz Wheel) (Figure 3-33, page 3-26).
- Personal Digital Assistant (PDA) computer/Military CYPRES Absolute Model Calculator (Figure 3-34, page 3-27).
- Personal computer (PC)-based Excel spreadsheet calculator.

Chapter 3

- Military CYPRES Web site: http://www.ssk.us/military_calc.asp.
- Green digital Military CYPRES Calculator from SSK (Figure 3-35, page 3-27).

> **WARNING**
>
> Use of any other device to get the altimeter setting is unauthorized; for example, Suunto, Kestrel, or other personal device.

3-72. When on a mission with limited weather information, the aircrew can provide the altimeter setting for the DZ en route to the drop area. The altimeter (pressure) setting is given in inches of mercury (in/Hg) to the nearest one-hundredth of an inch. The altimeter setting will always be for the intended DZ. Once the altimeter setting of the intended DZ has been determined, the primary jumpmaster will use an approved CYPRES calculator to determine the setting of each Military CYPRES 2 model on the mission. Once the primary jumpmaster has determined the settings, the assistant jumpmaster will independently determine the settings. If any discrepancy is found in the results, the primary and assistant jumpmasters will work together to determine the correct settings.

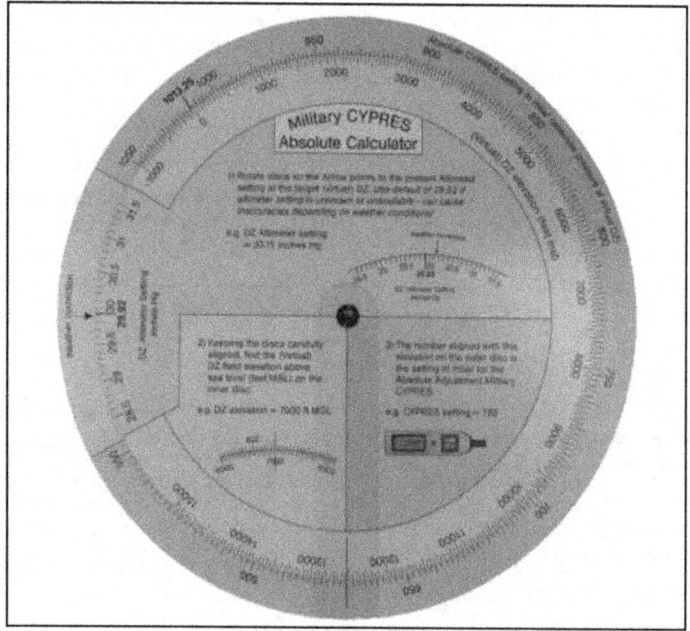

Figure 3-33. Military CYPRES Absolute Adjust Circular Calculator (Whiz Wheel)

Cybernetic Parachute Release System

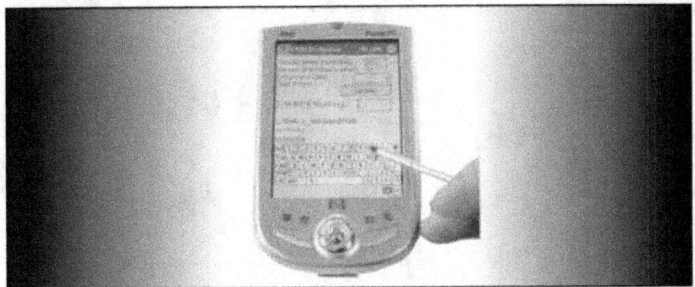

Figure 3-34. PDA computer with Military CYPRES Absolute Model Calculator software download

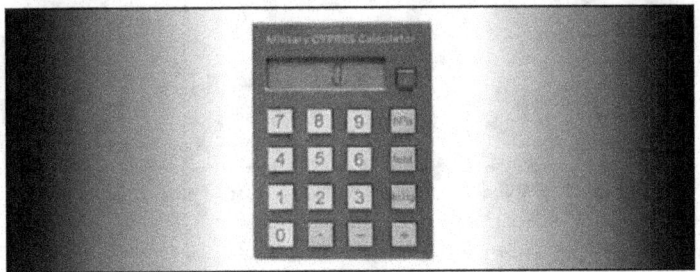

Figure 3-35. Military CYPRES Calculator

MILITARY CYPRES ABSOLUTE ADJUST CIRCULAR CALCULATOR

3-73. The jumpmaster obtains the forecasted aircraft altimeter setting for the DZ. If flying a mission with limited weather information, the aircrew can provide the altimeter setting en route to the drop area. The altimeter (pressure) setting will be given in inches of mercury (in/Hg). The jumpmaster obtains the setting to the nearest one-hundredth of an inch. Using the Military Absolute Adjust Circular Calculator (Figure 3-36A, page 3-28), the jumpmaster determines the absolute adjust millibar setting by—

- Rotating the discs so the weather correction (QNH) arrow points to the present aircraft altimeter setting at the target (virtual) DZ. A default of 29.92 is used if the altimeter setting is unknown or unavailable.

Note: This setting can cause inaccuracies depending on weather conditions; for example, DZ altimeter setting = 30.15 in/Hg (Figure 3-36B).

- Keeping the discs carefully aligned, finding the VDZ field elevation above sea level (feet MSL) on the inner disc, and placing the "clock hand" black indicator line on the ground elevation of the desired (virtual) DZ (for example, DZ elevation = 7,100 feet) (Figure 3-36C). The number aligned with this elevation on the outer disc is the setting in millibars for the absolute adjustment for the Military CYPRES (example 787 millibars) (Figure 3-36C).

Chapter 3

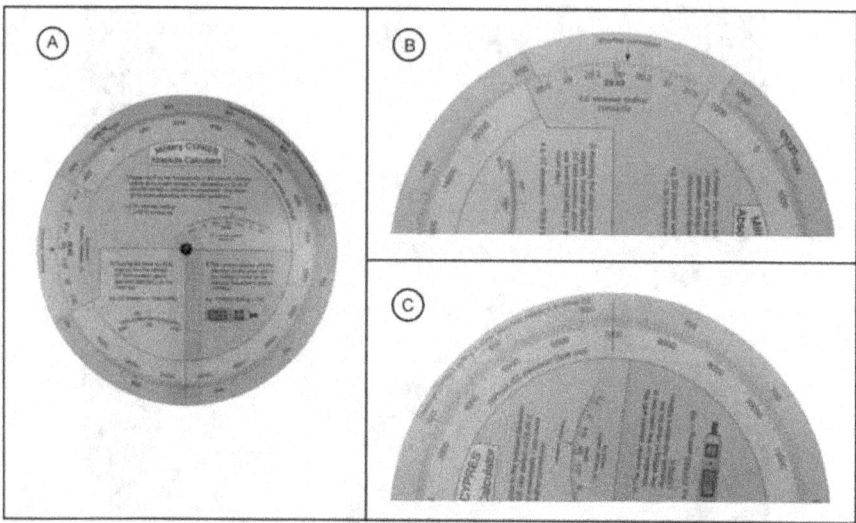

Figure 3-36. Military CYPRES Absolute Adjust Circular Calculator

MILITARY CYPRES ABSOLUTE ADJUST MODEL CALCULATOR

3-74. All downloads (software programs) for the Military CYPRES Absolute Adjust Model Calculator can be located at www.ssk.us. To use the online Military CYPRES Absolute Adjust Model Calculator, jumpmasters may go to www.ssk.us. There is no one brand of calculator that must be used for this application. The military does carry one iPAQ Pocket PC, National Stock Number (NSN): 5180-09-000-3952. The only requirement is that the PDA or Pocket PC have the proper software (example: Microsoft-based program, Excel) to accept downloading of one of the following:

- A zipped version of the Military Calculator for Microsoft Excel.
- A nonzipped version of the Military Calculator for Microsoft Excel.
- The PDA versions of Military Calculator (Microsoft OS-VB, Win CE, and Pocket PC).

Note: To download the files, user goes to www.ssk.us and places his mouse over the link on the Web page, right-clicks, then chooses "Save Target As" option for Internet Explorer or "Save Link Target As" option for Netscape Navigator.

USING THE PDA/POCKET PC

3-75. If the atmospheric (absolute) air pressure values to perform the altitude adjustment are not known, it is possible to do the altitude adjustment using the iPAQ Military CYPRES Absolute Calculator. This calculator can be ordered separately. To use the PDA/Pocket PC, parachutists do the following:

- Power ON the PDA.
- Enter the altimeter setting (Figure 3-37, page 3-29) of the location (in either Hg or millibars).
- Enter the elevation at the VDZ (Figure 3-37) (in either feet or meters).
- Select from the drop-down box the elevation scale (either feet or meters).

- Click the 'Calculate!' button.
- The value for the Military CYPRES setting (Figure 3-37) is displayed in the box.

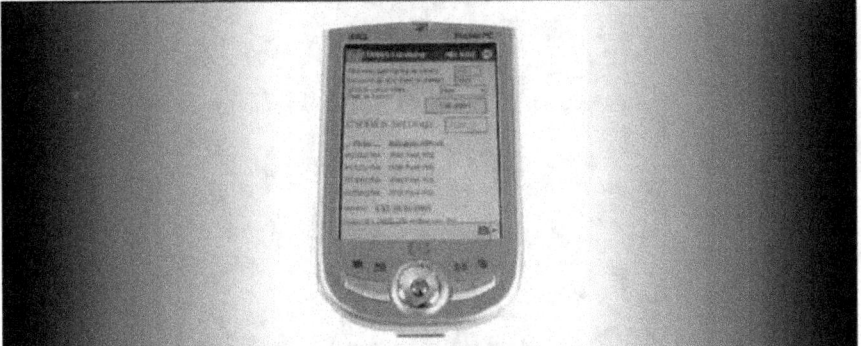

Figure 3-37. Value for the Military CYPRES setting is displayed in the CYPRES setting box

USING THE MILITARY CYPRES CALCULATOR

3-76. If the atmospheric (absolute) air pressure values to perform the altitude adjustment are not known, it is possible to do the altitude adjustment using the Military CYPRES Calculator feet/hPa/in Hg developed by Airtec. This calculator can be ordered separately (meter scale also available).

Note: If jumpmasters want to set a Military CYPRES 2 in operational mode and nobody is able to tell them the air pressure of their target, then they should use the Military CYPRES calculator, or go to the Military CYPRES User's Guide at www.ssk.us.

Figure 3-38, page 3-30, shows the usage instructions located on the back of the Military CYPRES calculator.

USING THE ON–LINE MILITARY CYPRES ABSOLUTE ADJUST MODEL CALCULATOR

3-77. Parachutists may log on to www.ssk.us and click to use the online "Military CYPRES Absolute Adjust Model Calculator" on the SSK Military Industries, Inc. home page. Figures 3-39 to 3-43, pages 3-30 through 3-32, show examples of the online calculator.

Note: Parachutists must use this calculator with "Absolute Adjustment" Military CYPRES units only ("Abs. Adj." nomenclature on control unit). Detailed procedures are in the Absolute Adjustment Military CYPRES User's Guide, as well as additional information on how to utilize all of the CYPRES capabilities.

Chapter 3

If you jump to another elevation, always program the actual air pressure in hPa (equivalent millibar) of your Target Drop Zone into your Military CYPRES 2.
You can execute the programming prior to take off on the ground or during flight or even while being in an active pressure cabin.

In case you cannot acquire the air pressure info, then take help from this calculator:

If your target is where your take off location is but at a higher altitude:
* enter your ambient ground pressure
  (Military CYPRES 2 tells you that in the selftest)
* press "hPa"
* press "+"
* enter the amount of feet between yourself and your Target Drop Zone
* press "feet"
Display shows the air pressure in hPa of your Target Drop Zone.

If your target is at another location and air pressure at target is unknown:
* find out how many feet your Target Drop Zone is above sea level
* enter that number (followed by "-", if target should be below sea level)
* press "feet"
* press "hPa"
Display shows the air pressure in hPa of your Target Drop Zone.
Maybe a bit imprecise, as it doesn't respect the weather at your target.

If target's QNH (pressure at msl, in US called altimeter setting) is available:
* enter the QNH
* press "hPa" or "In.Hg", whatever your scale is
* press "+" or "-"
* enter the height or depth in feet of your DZ above or below sea level
* press "feet"
Display shows the air pressure in hPa of your Target Drop Zone.

Automatic power off 3 minutes after last touch.

Figure 3-38. Usage instructions for the Military CYPRES calculator

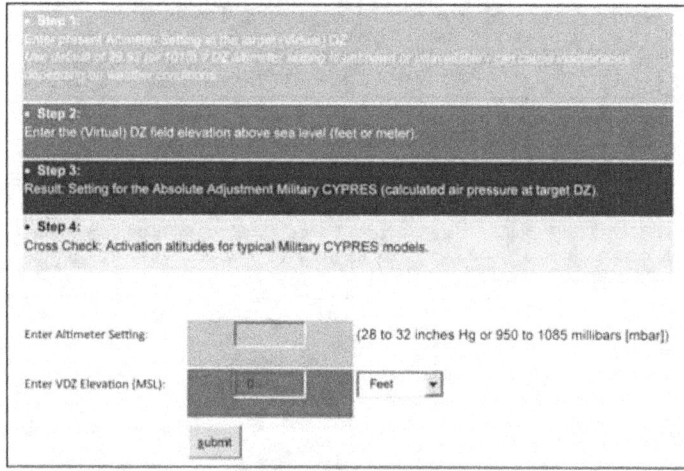

Figure 3-39. Instructions and first page of online "Military CYPRES Absolute Adjust Model Calculator"

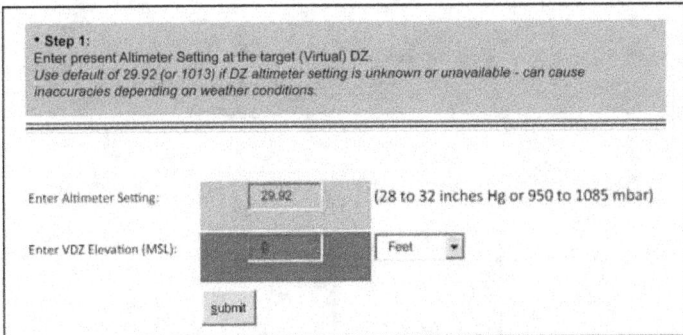

Figure 3-40. Step 1: Military CYPRES Absolute Adjust Model Calculator

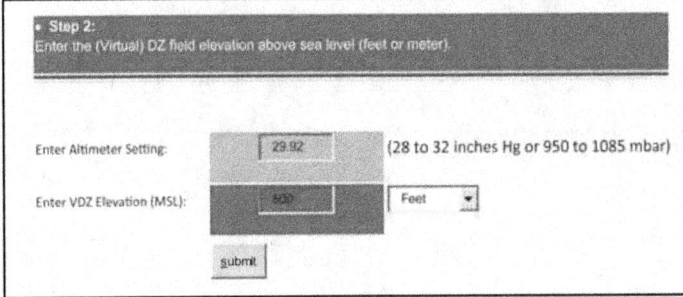

Figure 3-41. Step 2: Military CYPRES Absolute Adjust Model Calculator

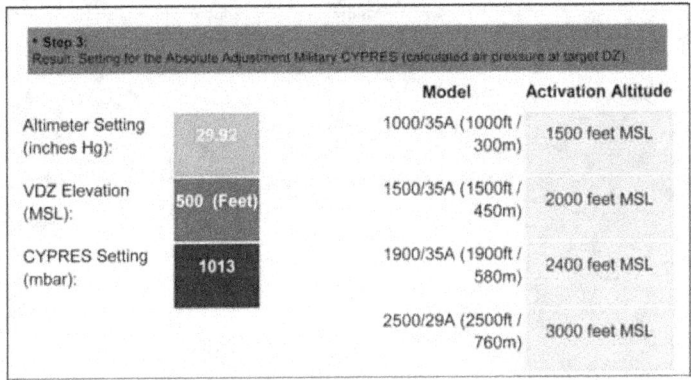

Figure 3-42. Step 3: Military CYPRES Absolute Adjust Model Calculator

Chapter 3

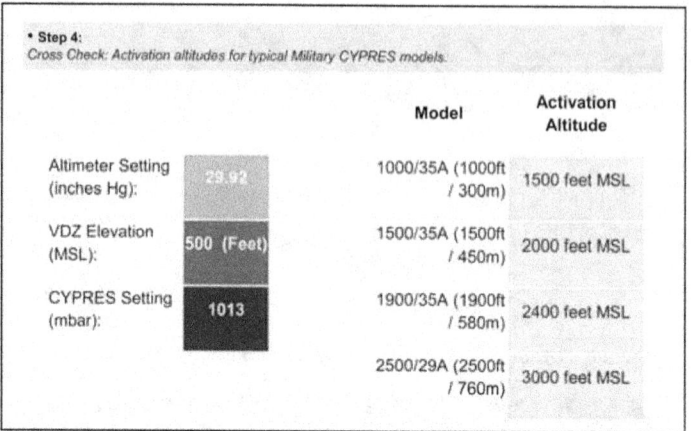

Figure 3-43. Step 4: Military CYPRES Absolute Adjust Model Calculator

MILITARY CYPRES COMMERCIAL AIR TRAVEL

3-78. A CYPRES-equipped rig may be transported in freight and passenger airplanes without restrictions. All of its components (for example, electronics, power supply, loop cutter, control unit, plugs, cables, and casing), as well as the complete system, contain parts and materials that are approved by the U.S. Department of Transportation (DOT) and other competent agencies worldwide, and are not subject to any transport regulations. Because of the size of a rig, it is recommended to check it in as normal luggage and not take it on board as hand luggage. In case of questions or objections from the security personnel, parachutists should use the card in Figure 3-44, page 3-33. The card shows an X-ray of a complete rig with the Military CYPRES 2. Depending on type and design of the rig, the X-ray on the security's screen may vary. Presently, the Parachute Industry Association and the United States Parachute Association are working with the Transportation Security Agency concerning traveling with parachutes.

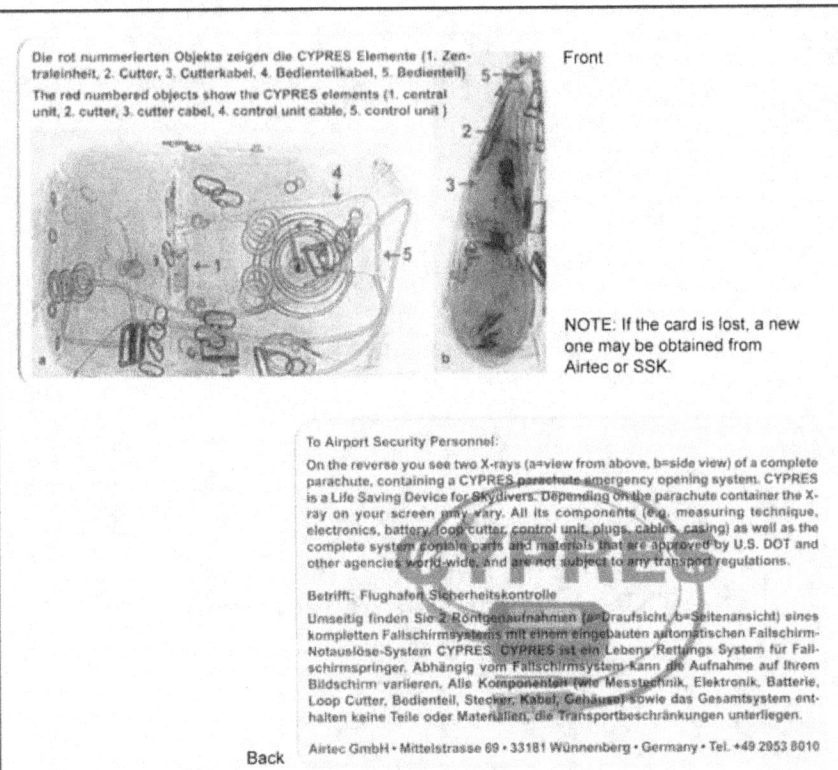

Figure 3-44. Military CYPRES 2 air travel card

This page intentionally left blank.

Chapter 4

Use of Oxygen in Support of Military Free-Fall Operations

MFF parachuting is physically demanding. It exposes the parachutist to temperature extremes, rapid pressure changes, and long exposures at altitudes requiring supplemental oxygen. To prepare for this environment, the MFF parachutist must be thoroughly familiar with the physiological effects of oxygen, oxygen use, and the operation of oxygen equipment. All personnel participating in MFF operations must meet the physiological training requirements outlined in Appendix B, regardless of altitude and type of aircraft used.

OXYGEN HANDLING AND SAFETY

4-1. Because of the limited contact with oxygen and its handling, personnel may not fully appreciate the danger involved. Improper use and handling can result in property damage, serious injury, and death. Personnel handling oxygen should always adhere to the warnings provided below.

DANGER

Keep oil and grease away from oxygen. Do not handle oxygen equipment with greasy hands or clothing.

No substance will be in the mouth of any jumper while on life-support equipment (oxygen) at any time, to include smokeless tobacco (dip), chewing gum, or food.

Keep equipment clean and free from petroleum-based products, lubricants, hydraulic fluid, and dirt. A drop of oil or lubricant coming in contact with pure oxygen under certain circumstances can cause an explosion.

Keep oxygen away from any source of ignition or fire/flame. Small fires rapidly become large fires in the presence of oxygen supplies. Never permit smoking near oxygen equipment, while handling oxygen supplies, or when using oxygen life-support equipment.

Handle cylinders and valves with extreme caution. Before opening cylinder valves, ensure cylinder is firmly supported. Never drop or tip over an oxygen cylinder. Dropping a cylinder can damage or break the valve, allowing gas to escape under pressure, with the potential for propelling the cylinder a great distance and with great force. Only open and close oxygen valves by hand and never strike the valve with any tool or object to loosen it. If the parachutist or technician cannot open and close the oxygen valve by hand, the cylinder must be returned to the depot for repair.

Chapter 4

PHYSIOLOGICAL EFFECTS OF HIGH-ALTITUDE MILITARY FREE-FALL OPERATIONS

4-2. Most physiological effects of high-altitude MFF operations fall into the category of pressure-change hazards. These hazards usually include various physiological symptoms. Based on Class C physiological mishaps since 1984, the most common types have been sinus blocks and ear blocks, hypoxia, decompression sickness (DCS), and hyperventilation. Each of these symptoms is discussed in the following paragraphs. Procedures for physiological and oxygen equipment-related emergencies are also discussed.

SINUS BLOCKS AND EAR BLOCKS

4-3. Sinus blocks and ear blocks normally occur when an MFF parachutist jumps with a head cold or some other type of upper respiratory illness. Sinus blocks and ear blocks usually occur during free-fall descent or during aircraft pressurization. Performing a Valsalva maneuver as the parachutist feels his ears getting "full" can clear most ear blocks. A Valsalva maneuver may clear a sinus block but may require additional medical attention. Use of nasal sprays may alleviate the symptoms associated with sinus and ear blocks.

Note: Chewing gum will not be used to clear blocked ears when wearing the oxygen mask.

HYPOXIA

4-4. Hypoxia is a condition caused by lack of oxygen. A reduction in the partial pressure of oxygen in the atmosphere occurs as the parachutist ascends. When the parachutist inhales, he receives fewer oxygen molecules. The reduction of the partial pressure inhibits the body's ability to transfer oxygen to the tissues. The most common symptoms of hypoxia are blurred or tunnel vision, color blindness, dizziness, headache, nausea, numbness, tingling, euphoria, belligerence, loss of coordination, and lack of good judgment. Corrective action for a parachutist who becomes hypoxic is to place him on 100-percent oxygen and inform the jumpmaster and/or physiological technician. In extreme cases, it may be necessary to descend the aircraft and evacuate the parachutist to the nearest medical facility. If hypoxia goes unrecognized and uncorrected, it can result in seizures, unconsciousness, or even death.

DECOMPRESSION SICKNESS

4-5. DCS is a condition caused by the release of nitrogen from body tissues. DCS usually occurs during unpressurized flights above 18,000 feet MSL, but can occur at lower altitudes. Many factors contribute to DCS. Facial hair can cause an insufficient seal of the oxygen mask to the parachutist's face, rendering prebreathing ineffective. Poor physical conditioning and fatigue will make the individual more susceptible to DCS. Alcohol use dehydrates the body, constricting the capillaries and decreasing the efficiency of the cardiovascular system. Nicotine from tobacco use hardens arteries and restricts blood flow to the capillaries, reducing the efficiency of the cardiovascular system. Smoking also reduces the efficiency of the lungs. Parachutists should be aware of the symptoms of DCS and constantly monitor themselves on board the aircraft and after return to the ground. Some parachutists may have symptoms of DCS during flight that are not readily noticeable. Minor symptoms may be confused with discomfort from the parachute and equipment. Other individuals may choose not to report what may be considered to be minor problems. Although these symptoms usually resolve themselves upon the jumper's return to ground, some personnel may continue to have symptoms. These individuals require prompt medical evaluation since their illness is more severe.

Use of Oxygen in Support of Military Free-Fall Operations

> **WARNING**
>
> If untreated, decompression sickness may result in debilitating and/or permanent medical disorders.

4-6. There are four types of DCS: the bends, chokes, neurological (central nervous system) hits, and skin manifestations. Each of these is discussed in the following paragraphs.

The Bends

4-7. The bends are the most common type of DCS. The most frequent symptom is a deep, dull, and penetrating pain in major movable joints that can increase to agonizing intensity. This pain may be significant enough to make the parachutist feel as if he cannot move the joint. The affected parachutist might also go into shock. Corrective action for a parachutist who experiences the bends is to—

- Place him on 100-percent oxygen.
- Inform the jumpmaster and/or physiological technician.
- Descend the aircraft and pressurize the cabin to as close to sea level as possible.
- Evacuate to the nearest medical facility with a recompression chamber. A flight surgeon or aeromedical examiner will determine if compression therapy is required.

The Chokes

4-8. The chokes are a rare but potentially life-threatening form of DCS. They are similar to the bends, but occur in the smaller blood vessels of the lungs, resulting in poor gas exchange and oxygenation of the blood. The most common symptoms are a deep, sharp pain near the breastbone; a dry, nonproductive cough; the inability to take a normal breath; a feeling of suffocation and apprehension; and possible shock symptoms, such as sweating, fainting, and cyanosis. Corrective action for a parachutist who experiences the chokes is the same as that stated for the bends in paragraph 4-7.

Neurological Hits

4-9. Neurological hits occur in extreme cases of DCS when the central nervous system becomes affected. The affected parachutist may experience vision disturbances, headaches, partial paralysis, loss of orientation, delirium, and vertigo. Corrective action for a parachutist who experiences neurological hits is the same as that stated for the bends in paragraph 4-7.

Skin Manifestations or Paresthesia

4-10. Skin manifestations or paresthesia is caused by nitrogen bubbles forming at the subcutaneous layer of the skin. The most common symptoms are itching, hot and cold flashes, a creepy feeling or gritty sensation, mottled reddish or purplish rash, and a tingling feeling of the affected area. Corrective action for a parachutist who experiences any of these symptoms is to—

- Place him on 100-percent oxygen.
- Keep him from scratching or exercising the affected area.
- Inform the jumpmaster and/or physiological technician.

4-11. Normally, the condition will dissipate upon descent. However, if the parachutist is incapacitated due to the condition, further corrective action is to—

- Descend the aircraft and pressurize the cabin to as close to sea level as possible.
- Evacuate to a medical facility with a recompression chamber. A flight surgeon or aeromedical examiner will determine if compression therapy is required.

Chapter 4

HYPERVENTILATION

4-12. Hyperventilation is a condition characterized by abnormal shallow and rapid breathing. Fear, anxiety, stress, intense concentration, or pain normally causes hyperventilation. Symptoms are similar to hypoxia and include lightheadedness, visual impairment, dizziness, numbness and tingling of the extremities, and loss of coordination and judgment. Personnel should conduct the following corrective actions:

- Calm the parachutist and have him talk, which will make him reduce his rate and depth of breathing. The goal is to achieve a breathing rate of 12 to 16 breaths per minute.
- Because of the similarity to hypoxia, continue or place him on 100-percent oxygen.
- Inform the jumpmaster and/or physiological technician.
- Reevaluate the parachutist's conscious state. If he is not responsive, treat the situation as an in-flight emergency and evacuate the parachutist to the nearest medical facility.

PHYSIOLOGICAL AND OXYGEN EQUIPMENT-RELATED EMERGENCIES

4-13. Procedures for physiological and oxygen equipment-related emergencies are discussed below. Personnel should—

- For in-flight emergencies, make sure the jumpmaster, oxygen safety technician, and aircraft commander (also USAF physiological technician if flight is above 20,000 feet MSL) are made aware of the problem.
- Ensure that the parachutist is receiving 100-percent oxygen from the console, the walk-around bottle, or an onboard aircraft regulator.
- Attempt to establish communications with the parachutist. Identify the problem and take corrective actions, to include immobilizing the affected areas, if possible.
- If the problem becomes progressive or severe, inform the aircraft commander of the nature of the problem and declare an in-flight emergency.
- Descend the aircraft and pressurize the cabin to as close to sea level as possible.
- Evacuate to a medical facility with a recompression chamber. A flight surgeon or aeromedical examiner will determine if compression therapy is required.

4-14. Parachutists should be aware of the symptoms of DCS and monitor themselves on return to the ground. Some parachutists may have symptoms of DCS during flight that they do not notice due to discomfort from the parachute and equipment worn or that they do not report. Although these symptoms usually resolve themselves upon returning to ground, some personnel may continue to have symptoms. These personnel require prompt medical evaluation since their illness is more severe.

OXYGEN FORMS

4-15. Oxygen is an odorless, colorless, tasteless gas that makes up 21 percent of the atmosphere. The remaining atmosphere consists of 78-percent nitrogen and 1 percent of other trace gases. There are four types of oxygen in use today—aviation, medical, welding, and research. Aviation oxygen is the only one suitable for MFF operations. The following paragraphs discuss the various forms of aviator's oxygen and their associated containers.

GASEOUS OXYGEN

4-16. Gaseous aviator's breathing oxygen is designated Grade A, Type I, Military Specification MIL-0-27210E. No other manufactured oxygen is acceptable. The difference between aviator's and medical or technical (welder's) oxygen is the absence of water vapor. The purity requirement for aviator's oxygen is 99.5 percent by volume. It may not contain more than 0.005 milligram of water vapor per liter at 760 millimeters of mercury at 68 degrees Fahrenheit. It must be odorless and free from contaminants, including drying agents. The other types of oxygen may be adequate for breathing, but they usually contain excessive water vapor that, with the temperature drop encountered at altitude, could freeze and restrict the flow of oxygen through the oxygen system the parachutist uses. The two types of gaseous aviator's breathing oxygen are as follows:

- *Gaseous—low-pressure.* Low-pressure aviator's breathing oxygen is stored in yellow, lightweight, shatterproof cylinders. These cylinders are filled to a maximum pressure of 450 pounds per square inch (psi); however, they are normally filled in the range of 400 to 450 psi. They are considered empty when they reach 100 psi. If a cylinder is stored at a pressure less than 50 psi for more than 2 hours, it must be purged because of the water condensation that forms.
- *Gaseous—high-pressure.* High-pressure aviator's breathing oxygen is stored in lime green, heavyweight, shatterproof bottles stenciled with AVIATOR'S BREATHING OXYGEN. These bottles can be filled to a maximum pressure of 2,200 psi; however, they are normally filled in the range of 1,800 to 2,200 psi.

LIQUID OXYGEN

4-17. Liquid aviator's breathing oxygen is designated Grade B, Type II, Military Specification MIL-0-27210E. The most common use of liquid oxygen is in storage facilities and for aircraft oxygen supplies because a large quantity can be carried in a small space.

OXYGEN REQUIREMENTS

4-18. The lower density of oxygen at high altitude causes many physiological problems. For this reason, MFF parachutists and aircrews need additional oxygen. Table 4-1, page 4-6, contains USAF-established requirements for supplemental oxygen for the MFF parachutist during unpressurized flight. Air Force Instruction (AFI) 11-409, *High-Altitude Airdrop Mission Support Program*, outlines these requirements. The following briefly describe the requirements:

- All personnel will prebreathe 100-percent oxygen at or below 16,000 feet MSL pressure or cabin altitude below 16,000 feet MSL pressure on any mission scheduled for a drop at or above 20,000 feet MSL.
- The required prebreathing time will be completed before the 20-minute warning and before the cabin altitude ascends through 16,000 feet MSL.
- Any break in prebreathing requires restarting the prebreathing period or removing the individuals whose prebreathing was interrupted from the mission.
- Prebreathing requires the presence of sufficient USAF physiological technician support onboard the aircraft.
- All personnel onboard during unpressurized operations above 10,000 feet MSL and higher will use oxygen. (Exception: Parachutists may operate without supplemental oxygen during unpressurized flights up to 13,000 feet MSL provided the time above 10,000 feet MSL does not exceed 30 minutes each sortie.)

Note: Portable oxygen bottles or locally procured oxygen systems may not be used for prebreathing; the quick-don/smoke mask is emergency equipment and is not approved for prebreathing or other parachute operations conducted at or above 13,000 feet MSL.

4-19. MFF parachuting is physically demanding. The higher jump altitudes associated with MFF operations expose the body to rapid pressure changes that require the use of supplemental oxygen. As a result, the MFF parachutist must—

- Conduct no more than three prebreather sorties in a 24-hour period.
- Not conduct MFF operations within 24 hours of making a nonoxygen dive.
- Wear a clear face shield/goggles on MFF operations that require prebreathing.

Note: The jumpmaster and the oxygen safety technician must be able to see the eyes of the jumpers to determine if they are having any physiological problems.

Table 4-1. Supplemental oxygen requirements for MFF parachutists

Altitude	Oxygen Requirement	Prebreathe Time*	Maximum Exposure Time per Sortie**
Below 10,000 Feet (ft) MSL	N/A	N/A	N/A
10,000 ft to 12,999 ft MSL	Supplemental	N/A	Supplemental Oxygen Required Only When Time Exceeds 30 Minutes
13,000 ft to 19,999 ft MSL	Supplemental	N/A	Unlimited
20,000 ft to 24,999 ft MSL	100-Percent Oxygen	30 Minutes	110 Minutes
25,000 ft to 29,999 ft MSL	100-Percent Oxygen	30 Minutes	60 Minutes
30,000 ft to 34,999 ft MSL ***	100-Percent Oxygen	45 Minutes	30 Minutes
35,000 ft MSL or Above	100-Percent Oxygen	75 Minutes	30 Minutes

NOTES:
* No more than 3 prebreather sorties in a 24-hour period unless otherwise restricted.
** Maximum exposure time per sortie is when cabin altitude reaches maximum planned altitude; extended or delayed ascent times expose everyone onboard to greater DCS risk. Missions that require staggered altitude drops will use accumulative times per sortie information for mission planning.
   Example: Mission-planned drops at 35,000 ft MSL, 29,999 ft MSL, and 24,999 ft MSL: 30 minutes upon reaching 35,000 ft MSL, descend to 29,999 ft MSL—spend only 30 minutes (60 accumulative); descend to 24,999 ft MSL—spend only 50 minutes (110 minutes accumulative).
*** No personnel will be exposed to unpressurized flight above 30,000 ft MSL more than 3 times each 7 days and must have a minimum of 24 hours between exposures.

OXYGEN LIFE-SUPPORT EQUIPMENT

4-20. Life-support equipment consists of the oxygen mask (MBU-12/P), the portable bailout oxygen system (PBOS) with the AIROX VIII assembly, the Parachutist Oxygen Mask (POM) with the portable bailout Parachute Oxygen System, the six-man prebreather portable oxygen system, the MA-1 portable oxygen assembly, and the prebreather attachment. This equipment is discussed in the paragraphs below.

OXYGEN MASK

4-21. The oxygen mask is designed to be worn with parachutist helmets that have bayonet lug receivers for the mask's harness assembly or by utilizing the IOH (referred to as "skull cap") to attach the mask when wearing the ACH. Oxygen enters the face piece through the valve located at the front of the mask. Exhaled air passes out through the same valve. The construction of the valve's exhalation port allows a pressure of only 1 millimeter of mercury greater than the pressure of the oxygen being supplied by the regulator to force open the valve and allow exhaled air to pass to the atmosphere. A 17.5-inch-long convoluted silicone hose with a 3/4-inch internal diameter attaches to the mask. Inside the hose is an antistretch cord that prevents extreme stretching and hose separation during free fall. The mask has an integral microphone that adapts to the aircraft's communication system.

> **WARNING**
>
> No type of petroleum, oils, and lubricant products (commercial sunblock, camouflage paint, and lip balm) will be used by MFF parachutists while on oxygen life-support equipment.

Note: No substance will be in the mouth of any jumper while on oxygen life-support equipment, to include smokeless tobacco (dip), chewing gum, or food.

4-22. There are several types of oxygen masks currently in use by different Services. The most common of these masks are described below.

MBU-12/P

4-23. The MBU-12/P pressure-demand oxygen mask is a replacement for the MBU-5/P mask (Figure 4-1, page 4-8). It has a soft, supple silicone rubber face piece integrally bonded to a plastic hard shell. It seals firmly during pressure breathing. It comes in four sizes to provide proper fit and superior comfort during extended wear. The lower profile design and four-point suspension are more stable than the MBU-5/P mask during free fall. Antiroll webs at the nose seal prevent downward roll-off. The integral face piece and hard shell design permit good downward vision and increased head mobility. The MBU-12/P oxygen mask was a system originally designed and produced as an aviator's oxygen mask. Its oxygen hose can be mistaken for a rip cord, which is hazardous. The POM system will replace the MBU-12/P system.

Parachutist Oxygen Mask

4-24. The POM breath-demand oxygen mask is an in-flight oxygen breathing device used either with a flight helmet and microphone or with a quick-don suspension assembly called the IOH. These masks are manufactured by Gentex Corporation and Carleton Technologies. The POM mask is available in four sizes, indicated by the marking on the outer edge of the soft shell. The four sizes are small narrow, medium narrow, medium wide, and large wide. The POM connects to the portable prebreather assembly and the 106-cubic-inch portable oxygen system through use of the hydraflow HS-57 oxygen breathing hose to ensure unrestricted oxygen. The PHANTOM mask and regulator assembly combines the latest Gentex parachutist oxygen mask (PM HALO/HAHO) with a high-performance Carleton miniature oxygen regulator. The PHANTOM system is much more comfortable and provides significantly improved performance over existing systems. The Gentex mask includes a high-flow, noncompensated exhalation valve along with an integral antisuffocation valve to protect the user in the event of oxygen supply depletion. The Carleton miniature regulator installs directly into the standard inlet port of the Gentex mask. The universal compatibility of Carleton's equipment eliminates the cost and logistics problems that would typically come with the development of a custom mask. The regulator can be installed or removed in under 30 seconds, which lends itself to improved maintainability and reliability.

Chapter 4

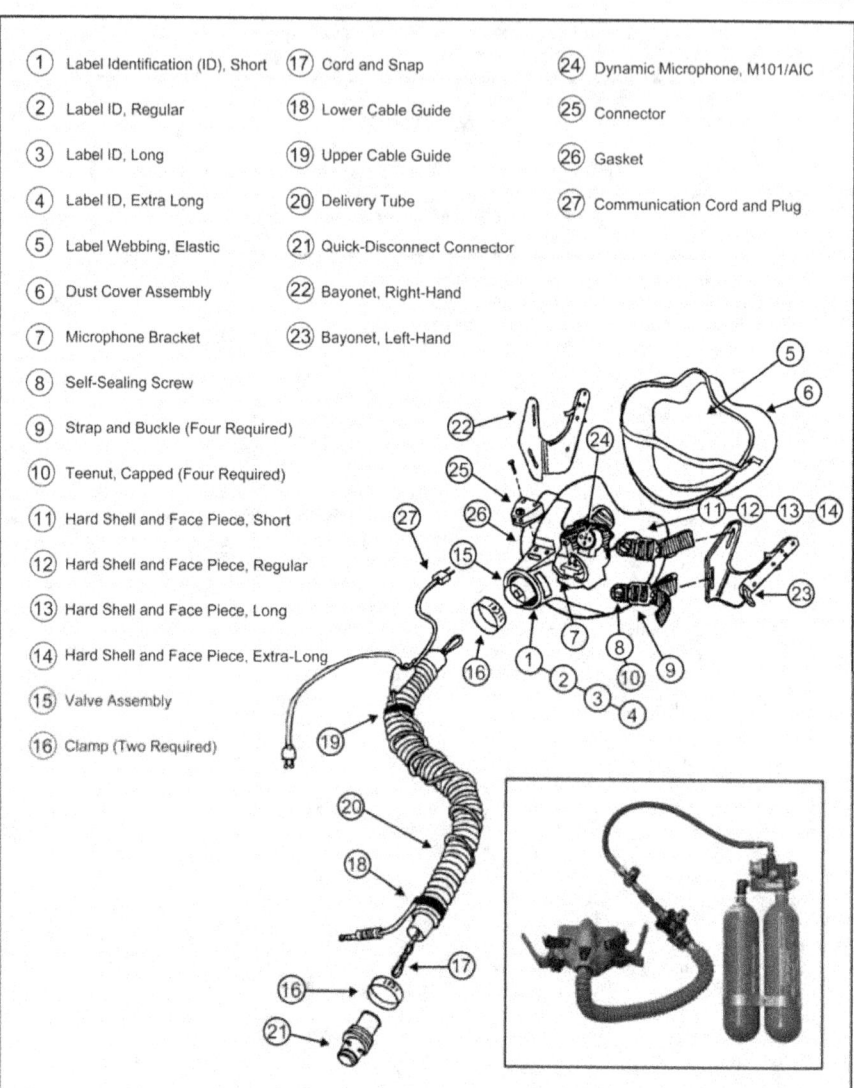

Figure 4-1. MBU-12/P pressure-demand oxygen mask components

4-25. The POM provides the parachutist with a high oxygen-flow capacity to improve breathing comfort during long missions with an automatic dilution shut-off feature when connected to a prebreathing console. Figure 4-2, page 4-9, shows the POM.

Figure 4-2. Parachutist Oxygen Mask

4-26. The POM has a high-flow, noncompensated exhalation valve and an integral antisuffocation valve to protect the parachutist in the event of oxygen supply depletion. There is a diffuser over the inhalation valve of the regulator to provide improved mixing of the oxygen in the mask. The mask allows the jumpers to pinch their nostrils and perform a Valsalva maneuver in order to release sinus pressure. The configuration of the screws and nuts used to fix the mask attachment straps are designed in a way to reduce rotational friction and ease maintenance.

Improved Oxygen Harness (Skull Cap)

4-27. The IOH was designed specifically for use with the POM and ACH during MFF operations. The IOH (Figure 4-3) is fabricated from absorbent cotton polyester fabric, tubular nylon straps, hook-pile tape (HPT) (Velcro), and Fastex buckles. The IOH is worn directly on the parachutist's head underneath the ACH. After donning the ACH over the IOH, the parachutist mounts the POM by attaching the four half-inch Fastex buckles attached at each end of the 1/2-inch tubular nylon straps. The straps with Fastex buckles are fully adjustable to accommodate a wide range of face and head shapes. The IOH also contains sizing channels made of 2-inch-wide pile tape sewn down the center of the cap. When worn with communication equipment (Figure 4-4, page 4-10) (for example, the MICH), the headset will sit between the ACH and IOH (skull cap).

Figure 4-3. The improved oxygen harness

Chapter 4

Figure 4-4. Complete Parachutist Oxygen Mask with MICH

Advanced Combat Helmet Accessory Rail Connector

4-28. The advanced combat helmet accessory rail connector (ACH-ARC) provides accessory direct mounting for ACH-style ballistic helmets. Rails and accessories are secured during MFF, combat movement, or tactical use. The ACH-ARC's low-profile, lightweight, snag-free design mount addition does not impede mobility and incorporates a dynamic breakaway feature to help prevent head and neck injury when exposed to extreme torque with no permanent damage to the unit. Tough, fiber-reinforced rails bolt directly to the helmet, holding a slide-and-lock Picatinny adapter for mounting of the POM with the additional plug-in oxygen mask receptacles (Figure 4-5, page 4-11). The straps with Fastex buckles are fully adjustable to accommodate a wide range of face and head shapes for proper fit of the POM. The rail connector also contains Picatinny adapter locks and sizing channels that are slot-triggered for ease of tightening, loosening, and fitting the oxygen mask. The ACH-ARC—

- Does not require drilling; it uses existing chinstrap mounting holes.
- Fits ACH, MICH, TC 2000, and MICH 2002 gunfighter helmets in sizes medium to extra large.
- Does not fit the MICH 2001 (high ear-cut) helmet or the enhanced combat helmet.
- Includes the oxygen single-strap kit or the oxygen double-strap kit (Figure 4-5) for mounting the POM and the MBU-12/P oxygen mask.

4-29. IAW HQ, U.S. Army Developmental Test Command memorandum, subject: Safety Confirmation for the Advanced Combat Helmet (ACH) in Support of Materiel Release and Fielding, dated 21 June 2011, the following precautions should be followed for use of the ARC, the oxygen single-strap kit, and the oxygen double-strap kit with the ACH during MFF operations:

- MFF parachutists should ensure they are fitted with the correct size helmet and follow fit and wear instructions IAW TM 10-8470-204-10, *Operator's Manual for Advanced Combat Helmet (ACH)*.
- Preventive maintenance checks and services (PMCS) procedures found in the ACH TM should be used for the ACH and the ARC. Users must check for loose or missing screws during PMCS. Missing screws should be replaced and loose screws should be tightened. If screws remain loose, they should be secured with thread-locking compound, NSN 8030-01-104-5392.
- Prior to rigging for MFF, each parachutist and the MFF jumpmaster should—
  - Inspect the strap kit for frayed or cut webbing, cracked or damaged plastic components, and inoperable head-lock tabs. All damaged items should be replaced and the strap kit reinspected before use.
  - Inspect the strap kit and ensure the swivel clips are securely fastened to the webbing, the swivel clip securely locks into the accessory rail connector tab, accessory rail connector tab securely locks into the accessory rail, and the rear-strap buckle and/or front-pull release buckle are operable. Parachutist should remount and reinspect all incorrectly or loosely mounted items and/or replace broken or defective components.

Use of Oxygen in Support of Military Free-Fall Operations

- Ensure the rear-strap buckle and/or front-pull release buckle are correctly inserted and the corresponding strap is not twisted. If the strap is incorrectly inserted and/or twisted, the parachutist should remove the buckle, rotate 180 degrees, reinsert, and then reinspect the buckle and strap assembly.
- Inspect the ARC for cracked or damaged plastic components, inoperable swivel clips, and inoperable ARC tab adapters. Parachutist should replace all damaged items that fail to lock in place.
- Solvents and steel or metal bristle brushes may damage the ARC and strap kit. Parachutist should only use a medium bristle brush and/or mild detergent to clean soil and debris from the rail system and its components.

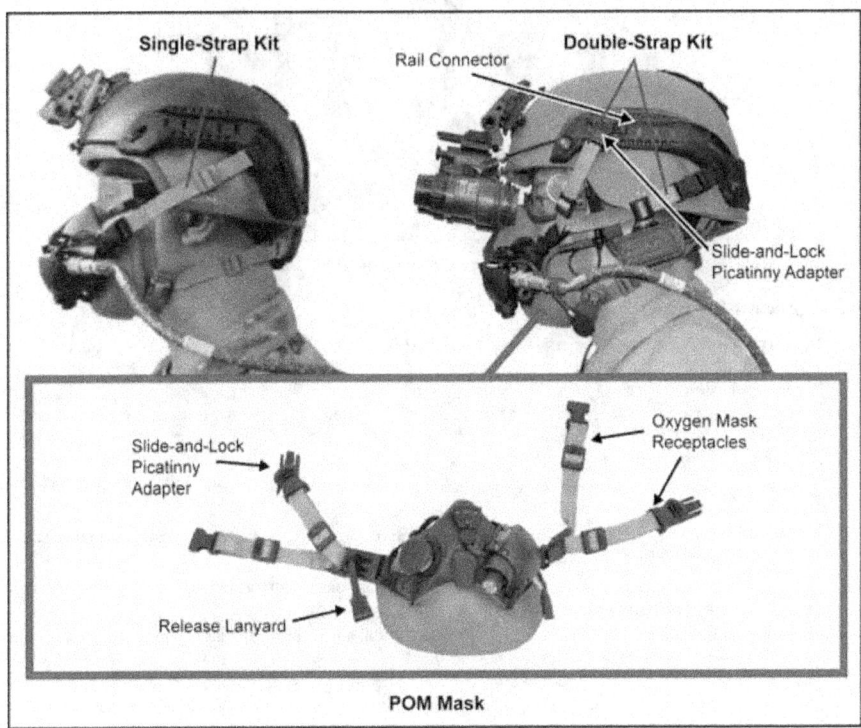

Figure 4-5. Advanced combat helmet accessory rail connector with oxygen single-strap and double-strap kits

Oxygen Supply Hose Assembly

4-30. The oxygen supply hose assembly consists of a four-pin male coupling assembly with an attached end cap, a 98- or 240-inch hose assembly, a flow indicator, and a low-pressure hose assembly. The oxygen supply hose assembly is used primarily by military parachutists for HALO and HAHO missions. The hose interfaces with the six-person portable oxygen console to provide a supplemental 100-percent oxygen source prior to jumping from the aircraft. The oxygen supply hose assembly also incorporates a breathing regulator, which reduces the flow from the oxygen console to regulate the mask's breathing pressure.

4-31. The POM uses a Hydraflow HS-57 crush-proof oxygen breathing hose to ensure unrestricted oxygen flow to the parachutist and is easily attached and detached with the quick-disconnect fitting (Figure 4-6). The POM can fully integrate with the existing Parachutist High-Altitude Oxygen System (PHAOS) equipment with no modification required. When the POM is used with the American Safety Flight System (ASFS) bail-out storage and delivery system, only minor modifications to the auxiliary equipment is required: the outlet pressure of the prebreathing console must be adjusted, and an adapter must be incorporated into the outlet of the existing bail-out system to permit the connection of the oxygen hose to the console.

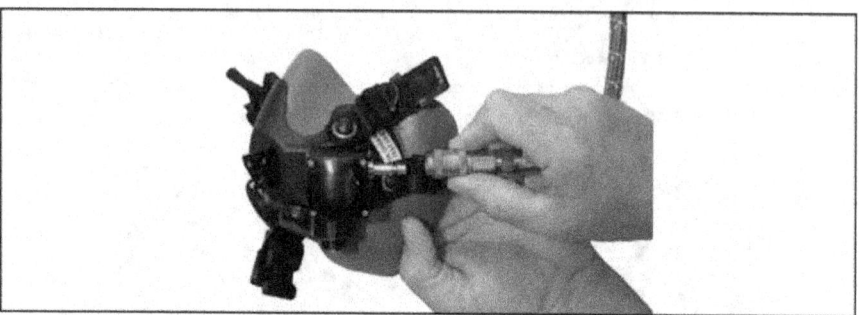

Figure 4-6. Parachutist Oxygen Mask and HS-57 quick disconnect

Fitting the MBU-12/P Pressure-Demand Oxygen Mask

4-32. Trained personnel must supervise mask fitting (Figure 4-7, page 4-13). When the mask fits properly, it should create a leak-tight seal around the sealing flange throughout the range of pressure-breathing forces administered by regulators. The mask has a four-point suspension harness with offset bayonet connectors that the parachutist attaches to the receivers mounted on his helmet to fit the mask. For safety, and to make sure of proper fit, the MFF parachutist should be issued the same mask and helmet for each operation. To fit the oxygen mask, the parachutist—

- Loosens the adjustment screws on the receivers on the helmet (depending on the type of helmet and bayonet receivers).
- Places the mask over his face and inserts each bayonet lug into its bayonet receiver to the second locking position (Figure 4-7A).
- Adjusts the mask straps until the mask is comfortable and snug, but not so snug that the mask hinders his vision (Figure 4-7B). He also secures any excess straps.
- To test for a proper seal, pulls the two pins of the antisuffocation valve toward the chrome ring, closing the antisuffocation valve, and inhales (Figure 4-7C). If the mask leaks around the face portion, he readjusts the four straps and once again checks for a proper seal. If any other portion of the mask leaks, the mask must be replaced. If a seal cannot be made at the face portion, he exchanges the mask for the next size and repeats the fitting process.
- Tightens the receiver adjustment screws and secures the excess straps if a proper seal is achieved (depending on the type of helmet and bayonet receivers).

Use of Oxygen in Support of Military Free-Fall Operations

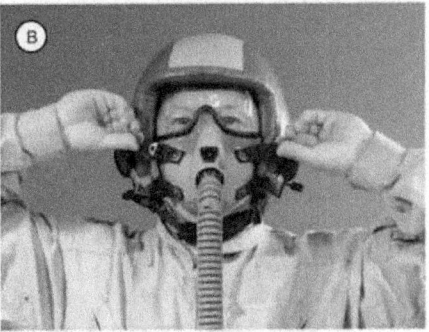

Figure 4-7. Fitting the MBU-12/P oxygen mask

Fitting the New Parachutist Oxygen Mask With Improved Oxygen Harness

4-33. Trained personnel must supervise mask fitting. When the mask fits properly, it should create a leak-tight seal around the sealing flange throughout the range of pressure-breathing forces administered by regulators. The mask has a four-point suspension harness strap and buckle assembly that attaches to the IOH (skull cap) connectors that the parachutist wears on his head to fit the mask. For safety and to ensure proper fit, the MFF parachutist should be issued the same mask and skull cap for each operation. To fit the oxygen mask, the parachutist should perform the following procedures:

- *Step 1*: Center IOH in a comfortable position on the head.
- *Step 2*: Place mask over face with left hand and have jumpmaster secure all four Fastex buckles to the IOH.
- *Step 3*: While jumper is holding mask with left hand, lightly tighten right top strap; switch hands and lightly tighten left top strap to hold mask into place. Grasp both top straps and tighten two top straps equally so mask stays centered on face and is tight enough to create a good seal. The POM is equipped with an integral antisuffocation valve to protect the parachutist in the event of

Chapter 4

oxygen supply depletion or for when fitting mask without oxygen being attached. If parachutist cannot exhale, mask must be considered unserviceable.
- *Step 4*: Tighten two bottom straps equally, keeping mask centered.
- *Step 5*: Move head back and forth two or three times. Alternately tighten straps until a good seal is achieved.
- *Step 6*: Once this procedure is completed, have the mask inspected by an oxygen technician. Figure 4-8 shows a completely and properly fitted mask.
- *Step 7*: After oxygen technician checks mask, secure any excess straps by rolling them inboard and securing rolls with masking tape.
- *Step 8*: Breathe through mask to determine if proper fit has been achieved. Parachutist should be able to breathe through ambient air port on mask; however, no air should enter around nose, cheeks, or chin. If mask is leaking air, readjust it for a proper fit. If a proper fit cannot be established, try a different size mask.

Figure 4-8. Properly fitted mask

Fitting the New Parachutist Oxygen Mask with Bayonet Connectors

4-34. To fit the new POM with bayonet connectors, the parachutist should perform the following procedures:
- *Step 1*: Place mask over face and insert bayonet connectors into the second locking position in bayonet receiver. Two clicks will be heard on each side indicating the second locking position. Setting mask in this position will allow adjustments to be made.
- *Step 2*: Ensure helmet is in a comfortable position on head. Tighten two top straps equally so mask stays centered on face and is tight enough to create a good seal. The POM is equipped with an integral antisuffocation valve to protect the parachutist in the event of oxygen supply depletion or for when fitting mask without oxygen being attached. If parachutist cannot exhale, mask must be considered unserviceable.
- *Step 3*: Tighten two bottom straps equally, keeping mask centered.
- *Step 4*: Move head back and forth two or three times. Alternately tighten straps until a good seal is achieved.
- *Step 5*: Once this procedure is completed, have the mask inspected by an oxygen technician.
- *Step 6*: After oxygen technician checks mask, secure any excess straps by rolling them inboard and securing rolls with masking tape (Figure 4-9, page 4-15).
- *Step 7*: Breathe through mask to determine if proper fit has been achieved. Parachutist should be able to breathe through ambient air port on mask; however, no air should enter around nose, cheeks, or chin. If mask is leaking air, readjust it for a proper fit. If a proper fit cannot be established, try a different size mask.

Use of Oxygen in Support of Military Free-Fall Operations

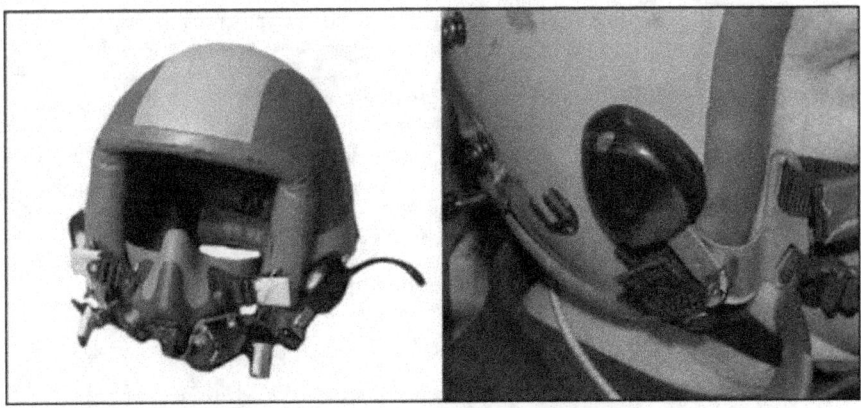

Figure 4-9. Parachutist Oxygen Mask with bayonet connectors and taped straps

Fitting the New Parachutist Oxygen Mask With Advanced Combat Helmet Accessory Rail Connector With Oxygen Single-Strap and Double-Strap Kits

4-35. To fit the new POM with double-strap kits to the ACH-ARC, the parachutist should perform the following procedures:

- *Step 1*: Place mask over face and insert slide-and-lock Picatinny adapter swivel clips into accessory rail connector. Take slack out of all straps by pulling down on them with equal pressure. Setting mask in this position will allow adjustments to be made.
- *Step 2*: Ensure helmet is in a comfortable position on head. Tighten two top straps equally so mask stays centered on face and is tight enough to create a good seal. The POM is equipped with an integral antisuffocation valve to protect the parachutist in the event of oxygen supply depletion or for when fitting mask without oxygen being attached. If parachutist cannot exhale, mask must be considered unserviceable.
- *Step 3*: Tighten two bottom straps equally, keeping mask centered.
- *Step 4*: Move head back and forth two or three times. Alternately tighten straps by sliding strap retainers until a good seal is achieved.
- *Step 5*: Once this procedure is completed, have the mask inspected by a jumpmaster/oxygen technician.
- *Step 6*: Breathe through mask to determine if proper fit has been achieved. Parachutist should be able to breathe through ambient air port on mask; however, no air should enter around nose, cheeks, or chin. If mask is leaking air, readjust it for a proper fit. If a proper fit cannot be established, try a different size mask.

Note: To fit the new POM with single-strap kit to the ACH-ARC, the parachutist should perform steps 1, 2, 4, 5, and 6 above.

Cleaning the Oxygen Mask

4-36. The parachutist cleans his oxygen mask after each use IAW TM 55-1660-247-12, *Operation, Fitting, Inspection and Maintenance Instructions With Illustrated Parts Breakdown for MBU-12/P Pressure-Demand Oxygen Mask*. He carefully wipes all surfaces with gauze pads or a similar lint-free material dampened with 70-percent isopropyl alcohol (rubbing alcohol). If isopropyl alcohol is not available, a solution of warm water and a mild liquid dishwashing detergent, such as Ivory, Joy, or Lux, is used. To rinse, the parachutist wipes the mask with swabs soaked in clean water, taking care not to wet the electronic parts. He

Chapter 4

allows the mask to air dry and stores it in a dust-free environment, away from heat and sunlight. If the mask needs more extensive cleaning, the parachutist turns it in to the supporting life-support facility.

100-CUBIC-INCH AND 106-CUBIC-INCH PORTABLE BAILOUT OXYGEN SYSTEM WITH PARACHUTIST OXYGEN SYSTEM ASSEMBLY

4-37. When used with the MBU-12/P oxygen mask, the PBOS is connected to the AIROX VIII, which is a constant-flow oxygen metering system consisting of a pressure reducer and an oxygen and air controller with an integrated prebreather adapter. These components increase oxygen duration and permit comfortable exhalation with standard military pressure-demand masks and associated connectors (Figure 4-4, page 4-10). This system requires minimum maintenance and—

- Has been approved for use from 0 to 35,000 feet MSL.
- Has an 8.2 to 9.3 liters-per-minute nominal oxygen flow to the parachutist.
- Has an oxygen reducer.
- Interfaces with the MBU-12/P oxygen mask.
- Has an oxygen and air controller that mates with the CRU-60/P or MC-3A connectors.
- Has a charging valve.
- Has a 20-micron oxygen/60 mesh air inlet filter.
- Contains two 2.6-inch siphon tubes that protect the oxygen reducer from foreign matter in the cylinders.
- Has a toggle-type ON/OFF control.
- Has an oxygen relief valve.
- Reduces exhalation difficulty associated with constant-flow oxygen systems.
- Uses two 50-cubic-inch or 53-cubic-inch high-pressure cylinders.
- Weighs approximately 10.5 pounds.

4-38. When used with the POM (Figure 4-10, page 4-17), the Twin-53 (106-cubic-inch) portable oxygen system consists of two 53-cubic-inch oxygen cylinders rated at 1,800 psi and is connected to a pressure reducer for oxygen delivery at a nominal pressure of 50 psi. With a nominal inlet pressure of 50 psi, the oxygen regulating system is capable of delivering 8.2 to 9.3 liters per minute of oxygen to the parachutist. The Twin-53 system is interconnected via an oxygen hose (Figure 4-10) with its oxygen output controlled by a mask-mounted regulator. A complete breathing system is formed when the Twin-53 system is connected to an oxygen mask. When used with a 100-percent oxygen regulator, this system has a maximum operating altitude equal to that of the aircraft service ceiling. The Twin-50 bottle system has two 50-cubic-inch oxygen cylinders rated at 2,100 psi, with the same outlet pressure as the Twin-53 system. As the connection point at the manifold of the Twin-50 oxygen system is identical to that of the Twin-53, the POM is compatible with the Twin-50 bottle system.

4-39. The AIROX VIII assembly (Figure 4-11, page 4-17) provides the MFF parachutist with a standoff parachuting capability up to 35,000 feet MSL. It extends the duration of two 53-cubic-inch oxygen cylinders and permits the use of any pressure-demand mask and associated oxygen connectors. The AIROX VIII assembly also eliminates the back pressure associated with constant-flow oxygen systems and requires almost no maintenance. TM 1-1680-377-13&P-5, *Operator's, Unit, and Direct Support Maintenance Manual Including Repair Parts and Special Tools List for Helicopter Oxygen Systems*, provides further information.

4-40. The parachutist cannot overbreathe the system. When inhaling more volume than the unit delivers, an ambient air valve opens up negating the breathing starvation sensation felt with other constant-flow systems as cylinder pressure decreases.

4-41. The AIROX VIII assembly has a special prebreather adapter that allows simultaneous hookup of the prebreather unit and the bailout system to the AIROX unit. The parachutist makes only one disconnection upon standing up. The connection from the prebreather connects to the ambient air port on the AIROX unit, thus preventing any ambient air from entering the parachutist's system while prebreathing. When preparing to exit the aircraft, the parachutist stands up, turns on the bailout system, disconnects from the prebreather, and jumps.

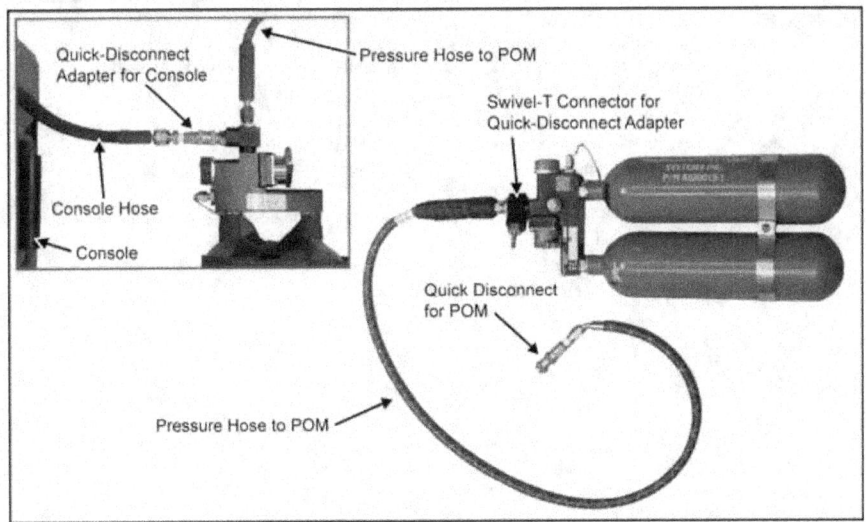

Figure 4-10. The 106-cubic-inch PBOS with the quick-disconnect oxygen hose

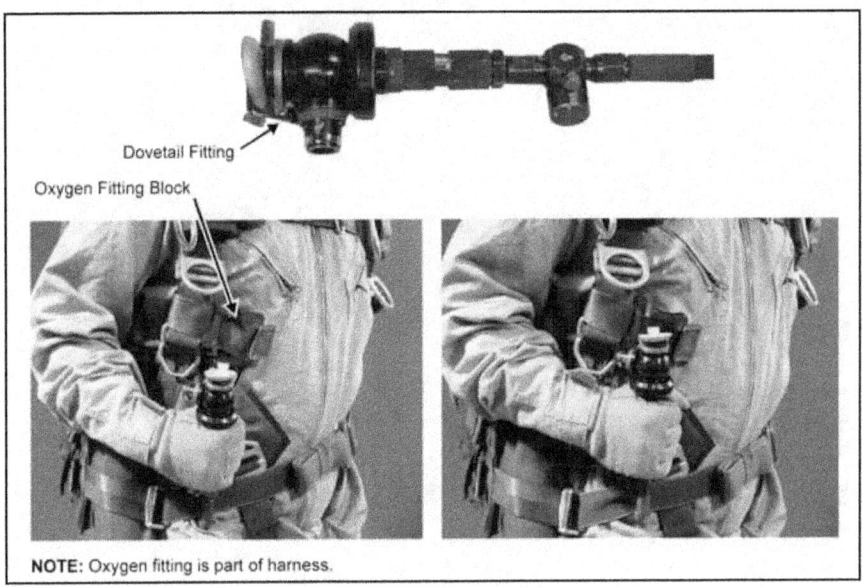

Figure 4-11. AIROX VIII assembly

Rigging the Portable Bailout Oxygen System with the AIROX VIII Assembly to the Ram-Air Personnel Parachute System

4-42. To rig the PBOS with the AIROX VIII assembly to the RAPPS (Figure 4-12, page 4-19), the parachutist—

- Places the oxygen cylinders into the detachable pouch with the ON/OFF valve to his front. He secures it with the hook-pile straps. He threads the waistband through the center keepers on the detachable pouch (Figure 4-12A).
- Fastens the waistband (Figure 4-12B).
- Tightens the right wing flap over the oxygen bottles (Figure 4-12C).
- Routes the oxygen hose between his body and the right main lift web and under the waistband on his right side (Figure 4-12D and E).
- Routes the oxygen hose over the waistband and secures the dovetail fitting in the oxygen fitting block (Figure 4-12F).
- Tightens the waistband.
- Ensures the center keepers point toward the body.

Use of Oxygen in Support of Military Free-Fall Operations

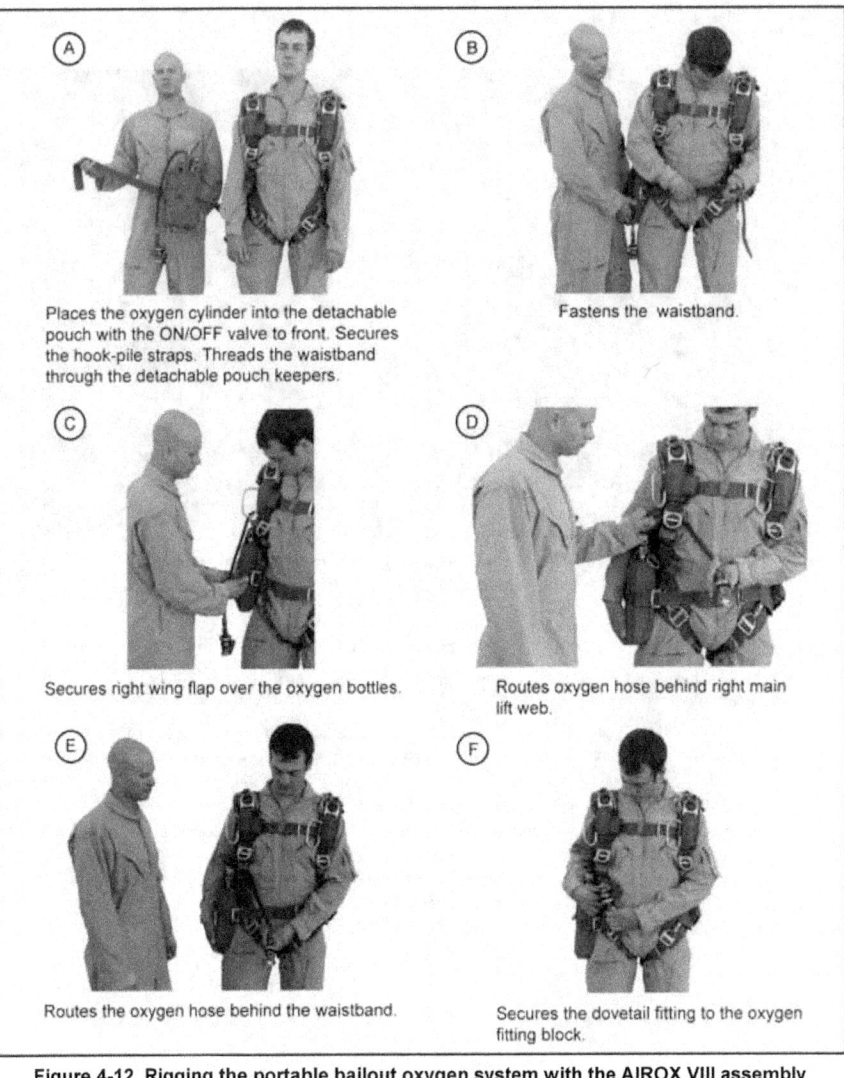

Figure 4-12. Rigging the portable bailout oxygen system with the AIROX VIII assembly to the RAPPS

Chapter 4

4-43. Figure 4-13 shows the completed rigging of the PBOS with the AIROX VIII assembly to the RAPPS.

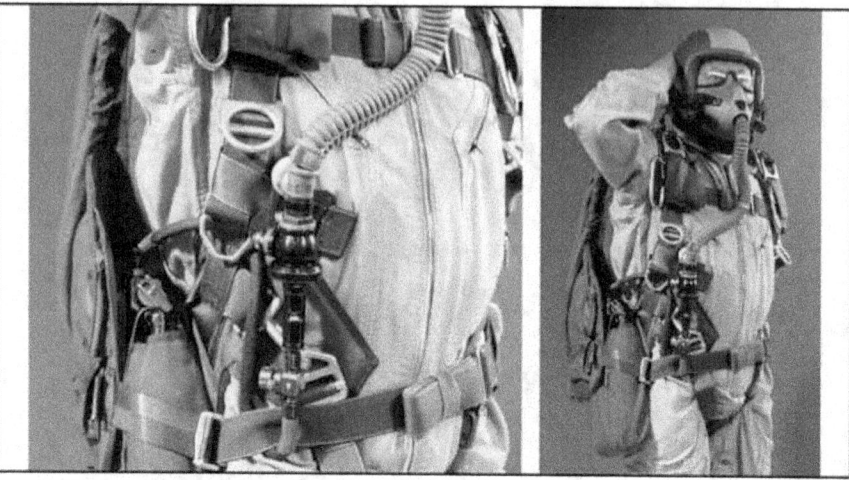

Figure 4-13. Completed rigging of the portable bailout oxygen system with the AIROX VIII assembly to the RAPPS

Rigging the Parachutist Oxygen Mask/American Safety Flight System to the Parachutist

4-44. To rig the complete POM assembly with the PBOS to the RAPPS (Figure 4-14, page 4-21), the parachutist—

- Places the oxygen cylinders into the detachable pouch with the ON/OFF valve to his front. He secures it with the hook-pile straps (Figure 4-14A) rerouted away from the on/off switch. He threads the waistband through the center keepers on the detachable pouch (Figure 4-14A).
- Secures right wing flap over the oxygen bottles (Figure 4-14B).
- Fastens the waistband and routes the oxygen supply under the right riser protective flap (Figure 4-14C).
- Routes the rigid oxygen supply hose assembly and quick disconnect between his back (Figure 4-14D) and the parachute container.
- Pulls the oxygen supply rigid hose assembly and quick disconnect up behind the neck and over the left shoulder (Figure 4-14E).
- Attaches the quick disconnect of the HS-57 hose to the POM (Figure 4-14F).

Use of Oxygen in Support of Military Free-Fall Operations

Places the oxygen cylinder into the detachable pouch with the ON/OFF valve to front. Secures the hook-pile straps. Threads the waistband through the detachable pouch keepers.

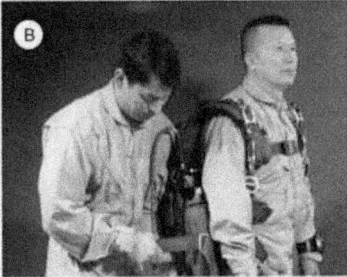
Secures right wing flap over the oxygen bottles.

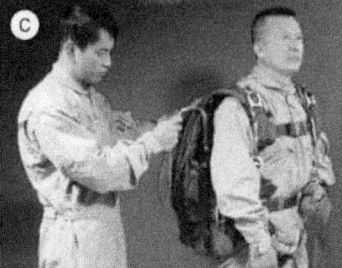
Fastens the waistband and routes oxygen supply under the right riser protective flap.

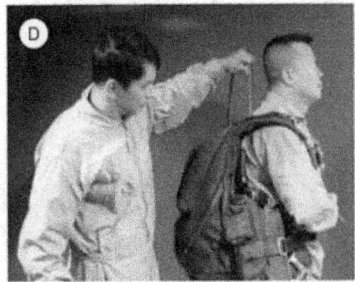
Routes oxygen supply hose between jumper's back and parachute container.

Routes oxygen supply hose behind neck and over left shoulder.

Attaches POM to IOH and attaches the quick-disconnect hose to the POM.

Figure 4-14. Completed rigging of the POM/ASFS with the Parachutist Oxygen System assembly to the RAPPS

SIX-MAN PREBREATHER PORTABLE OXYGEN SYSTEM

4-45. The six-man prebreather portable oxygen system allows six parachutists to prebreathe 100-percent oxygen from a 1,292-cubic-inch supply that can be used during flight on fixed-wing aircraft and helicopters as

Chapter 4

a supplemental or emergency oxygen source that was designed as a self-contained, easy-to-operate, small, lightweight, and nearly maintenance-free oxygen system (Figure 4-15). TM 1-1680-377-13&P-5 provides further information. The six oxygen-supply hose assemblies interconnect the oxygen console with the breathers' oxygen masks. Each of the hoses incorporates a low-pressure regulator which reduces the flow from the oxygen console to breathing pressure. Oxygen is stored under 1,800 psi or 2,100 psi in two 646-cubic-inch tandem-connected storage cylinders which are charged through the filler valve assembly connected to the oxygen charging assembly (P/N T80-30007-9), which is in turn attached to a high-pressure oxygen source. Oxygen duration is based on altitude and individual consumption requirements.

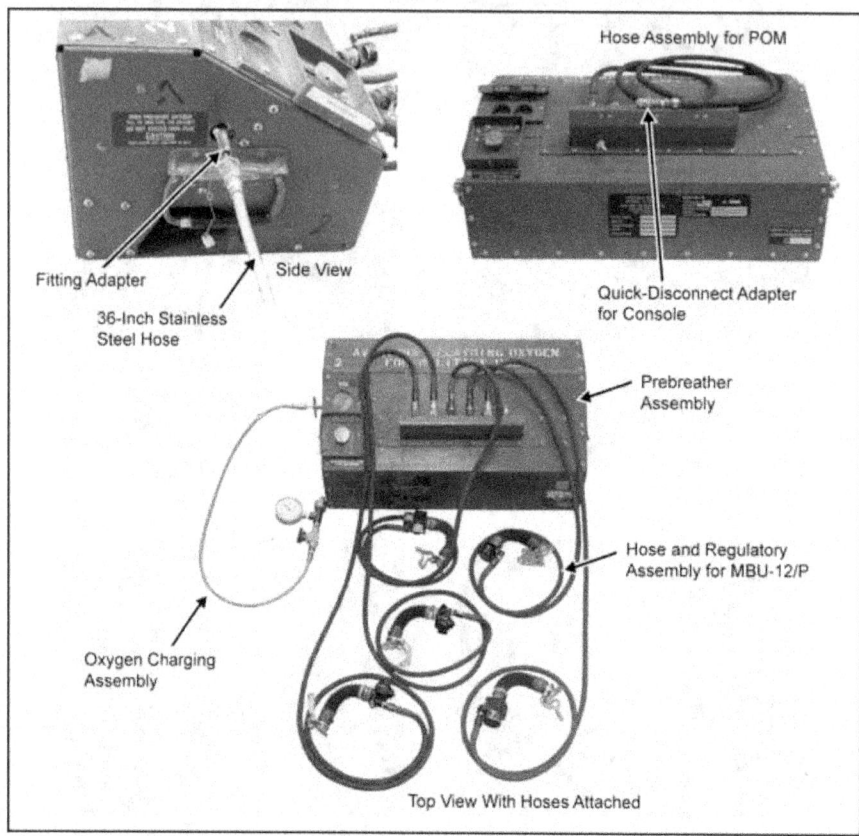

Figure 4-15. Six-Man Prebreather Portable Oxygen System

4-46. The six-man prebreather portable oxygen system is secured to the existing floor fittings on the C-17 aircraft. On the C-130 aircraft, the 5,000-pound tie-downs are used to secure the console. On the V-22 Osprey, the consoles are restrained on the floor on the starboard side using ratchet straps IAW Navy requirements for a 20G load. The outer housing consists of 4130 aircraft sheet steel, and recesses or steel guards protect the system's critical components. The oxygen console dimensions are: length—27 5/16 inches, width—13 6/16 inches, and height—11 inches. The weight is 91 pounds (empty) and 103 pounds (charged). The hoses are two each of three lengths: 72 inches, 90 inches, and 98 inches. Color-coding identifies certain parts, such as hoses and their mating parts, to prevent their misconnection.

4-47. The six-man console system has 100-percent oxygen capability for six individuals for approximately one hour at 10,000 to 35,000 feet MSL. The console is primarily intended for use by MFF parachutists during HAHO and HALO operations.

> *Note*: With the CRU-79/P regulator, the system has an operational ceiling of 50,000 feet MSL.

4-48. Other system features include the following:
- Weighs 106 pounds when filled.
- Measures 27.3 inches wide, 13.37 inches deep, and 10.99 inches high.
- Can provide oxygen for one to six parachutists.
- Has modular components.
- Is constructed to survive an 8G (gravitational force) forward crash load.
- Has a recessed refilling point.
- Has an easily gripped and guarded ON/OFF knob.
- Has color-coded and -indexed oxygen connectors to help ensure proper hose connections, and includes optional hose lengths to fit parachutist seating requirements.
- Has a steel guard around oxygen hose connectors.
- Interfaces with any pressure-demand mask and associated connectors.
- Can be refilled while being used.

OXCON OXYGEN CONSOLE

4-49. The OXCON is a portable, self-contained, deck-mounted oxygen supply system designed to deliver 100-percent aviator's breathing oxygen to as many as six parachutists from ground level up to 35,000 feet MSL for approximately 1.5 hours. The OXCON supplies 100-percent oxygen to the parachutist's mask by means of a quick-disconnect low-pressure delivery hose (65 to 75 psi) connected to the bailout bottle worn on the right side of the parachute.

4-50. Up to six delivery hoses can be attached to the OXCON—three on the front of the unit and three on the back. Some hoses are equipped with a flow indicator for a visual confirmation that oxygen is flowing to the parachutist. To ensure the parachutist's oxygen supply is not interrupted when it is time to disconnect from the OXCON, the connection of the hose cannot be made without the bailout bottle's ON/OFF lever in the ON position. The 70 psi coming from the OXCON's delivery hose overrides the bailout bottle's operation, ensuring that the parachutist gets 100-percent oxygen to the mask and does not consume any of the oxygen from the bailout bottle. The OXCON is supplied with six delivery hoses and a jumpmaster hose extension. Because of its configuration, three hoses on the left side and three on the right, the OXCON must be centerline mounted in the aircraft.

> *Note*: Figure 4-16, page 4-24, displays one example of rigging the OXCON and K-bottle inside the aircraft. The load master has final approval authority for securing equipment inside the aircraft.

> *Note*: Some aircrews may require no metal-to-metal contact between the OXCON and the deck of the aircraft.

4-51. To extend the duration of the OXCON, a K-bottle of aviator's breathing oxygen may be attached and cascaded into the console when it is rigged on the aircraft. CGU-1B 5,000-pound cargo straps are used to secure the OXCON and the K-bottles to the deck of the aircraft. To ensure that no metal-to-metal contact occurs, a piece of plywood can be placed under the OXCON, but cradles must be used to hold the K-bottles in place.

Chapter 4

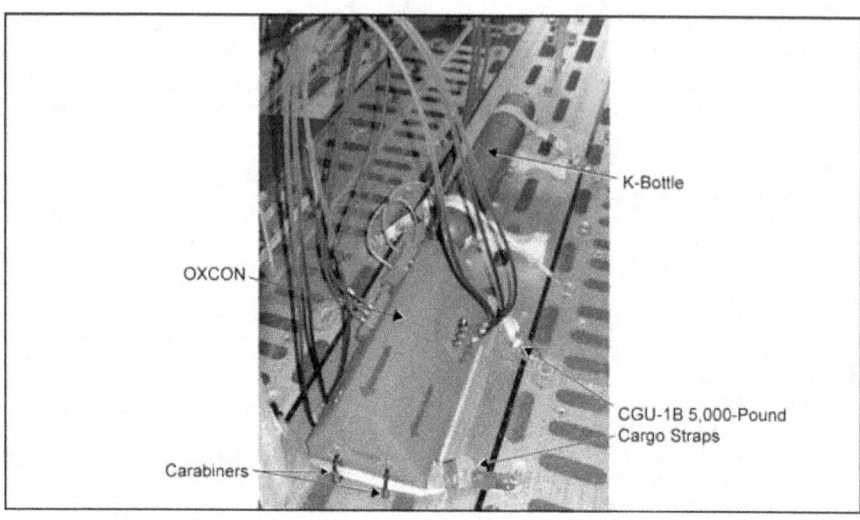

Figure 4-16. OXCON rigged in C-130 aircraft

4-52. To rig the OXCON on the deck of approved aircraft, the following procedures must be performed during pre-stage:

- Determine how many OXCONs are going to be used during the mission.
- Position the OXCONs with the arrow pointing in the direction of flight. Ensure all gear is positioned properly before securing.
- Look for the location of tie-down rings and the accessibility of hoses to the jumpers.

Note: Some squadrons may require no metal-to-metal contact between the OXCON and the deck of the aircraft.

- If supplemental K-bottles are used, ensure wooden 4x4-inch cradles are available for each bottle. Place cradles in the proper direction and within reach of the charging assemblies.

Note: Use Stubai 85 carabiners or equivalent with at least 5,000-pound breaking strength to tie down because the ends of the tie-down straps do not fit through the tie-downs on the OXCON.

Note: The locking barrels on the carabiners need to face out and close downward. They should be taped so that vibration of the aircraft does not unlock the carabiner barrels.

- Fasten both sides of the cradle to the deck using CGU1-B 5,000-pound cargo tie-downs with the direction of pull toward each other and tape the excess straps.

WARNING

OXCON hoses are under high pressure. Accidentally unhooking the hose from the OXCON unit may result in injury to a parachutist.

Use of Oxygen in Support of Military Free-Fall Operations

- Tightly stretch a length of 1/2-inch tubular nylon between the centerline stanchions in the aircraft to hold the hoses up and out of the way.
- If supplemental K-bottles are used, keep the charging assembly between the K-bottle and OXCON securely looped and taped so it cannot get snagged (Figure 4-17).
- Wrap the cargo strap around the K-bottle once before tightening (Figure 4-18).

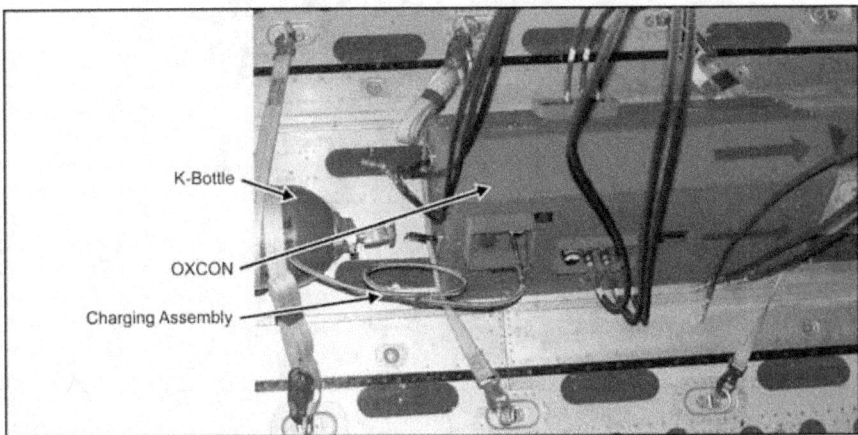

Figure 4-17. Charging assembly looped and taped out of the way of parachutists

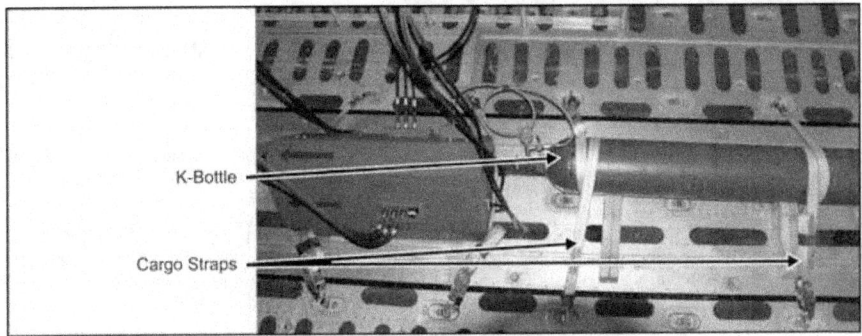

Figure 4-18. Side and top view of strap on K-bottle

MA-1 PORTABLE OXYGEN ASSEMBLY

4-53. The MA-1 portable oxygen assembly is a low-pressure system capable of supplying the parachutist with breathing oxygen for normal or emergency use (Figure 4-19, page 4-26). It is commonly called the walk-around bottle. The MA-1 is filled from the aircraft's oxygen supply. Pressure is indicated on the cylinder pressure gauge. The cylinder is considered full at 300 psi and empty at 100 psi. The MA-1 is operated by placing the selector knob at one of the four settings (NORM [normal], 30M, 42M, and EMER [emergency]) and breathing directly through the connector regulator unit (CRU) connector receiver port or an attached oxygen mask.

Note: The hose connectors for the POM are not compatible with the portable MA-1 portable oxygen bottles on USAF aircraft.

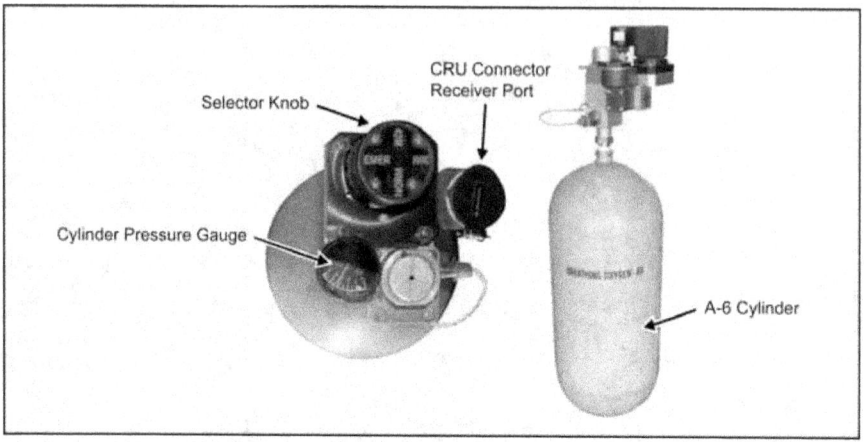

Figure 4-19. MA-1 Portable Oxygen Assembly

PREBREATHER ATTACHMENT

4-54. The prebreather oxygen assembly is normally located under the troop seats, and the oxygen supply hoses are routed up and behind the seats. The prebreather may also be positioned centerline in the aircraft using 10,000-pound tie-down fittings (C-17), 5,000-pound tie-down fittings (C-130), or securing straps.

4-55. When using 10,000-pound tie-down fittings, the parachutist places the two large holes in the base plate of the prebreather over existing 10,000-pound tie-down fitting holes in the floor of the aircraft. Through the openings in the side of the prebreather, he places two 10,000-pound fittings (one through each end) into the mating receptacle now visible through the prebreather's base plate. He then locks the fittings in place. These fittings will provide all the security necessary to hold the prebreather in place.

4-56. When using the oxygen console tie-down assembly, the parachutist places the two large holes in the prebreather's base plate over the attached 5,000-pound ringed tie-down fittings. Next, he places the securing adapters over the exposed rings and pushes the pins through the holes in the adapters until they lock. These fittings will provide all the security necessary to hold the prebreather in place (Figure 4-20, page 4-27).

4-57. Cargo straps are not necessary for added security when using the 10,000-pound tie-down fittings or oxygen console tie-down assembly. If cargo straps are used in place of the tie-down fittings, the parachutist places the straps through the securing access holes at each end of the prebreather and cinches tightly to existing fittings.

Note: The prebreather carrying handles are not stressed for use as securing points.

Use of Oxygen in Support of Military Free-Fall Operations

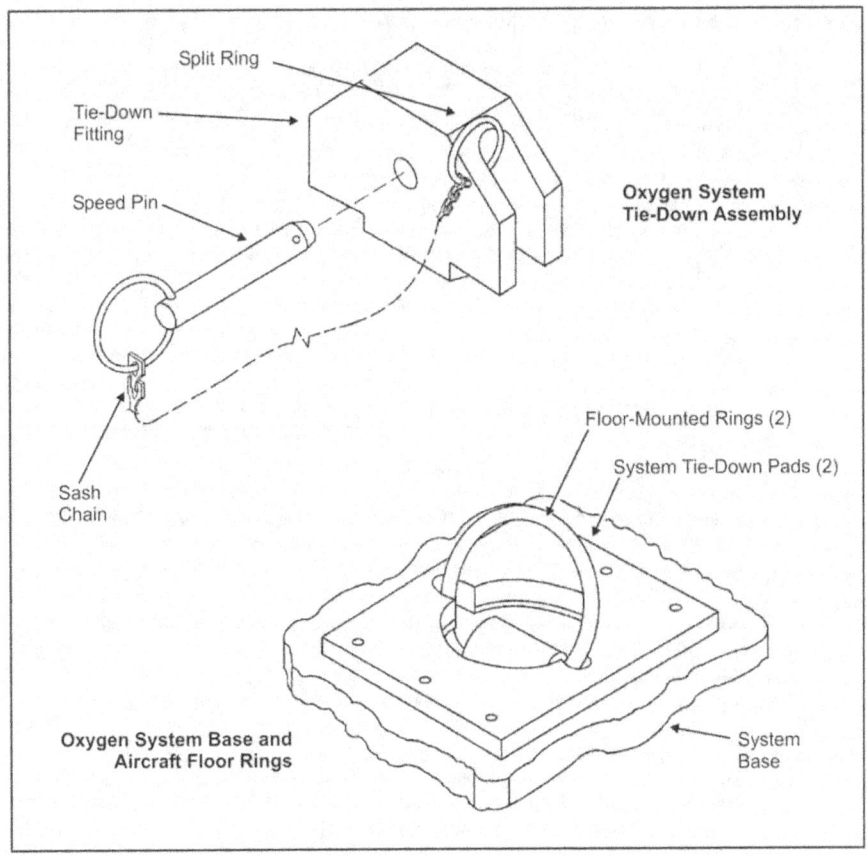

Figure 4-20. Tie-down assembly and installation

THE "PRICE" CHECK

4-58. Each letter of the acronym PRICE represents an area of or a specific item of oxygen equipment that the parachutist must check. The PRICE check makes no provision for inspecting the mask or protective helmet. The parachutist must check each of the following:

- *P—Pressure*: Checks for full pressure on the particular system in use.
- *R—Regulator*: Checks everything on the particular regulator in use. He checks for dents, cracks, broken gauges, grease or oil, and movement of dials and levers. He checks the entire oxygen delivery system for leaks.
- *I—Indicator*: Checks to ensure the flow indicator shows that gas is flowing through the regulator from the storage system.
- *C—Connections*: Checks all hose connections.
- *E—Emergency equipment*: Performs a complete check on any emergency oxygen equipment and the complete bailout system.

Chapter 4

OXYGEN SAFETY PERSONNEL AND PREFLIGHT CHECKS

4-59. Oxygen safety personnel (rigger or another free-fall jumpmaster with experience on the equipment) must be onboard each aircraft during MFF operations using supplemental oxygen. They must have received physiological training and unit-level technical training on the oxygen systems being used. For jumps from 20,000 feet or above, one USAF physiological technician per 16 jumpers will be requested with the aircraft and will be onboard for the jump. The oxygen safety personnel or the USAF physiological technician will—

- Plan for all oxygen equipment required for the mission. He will provide one additional mask of each size and one additional complete bailout system per six parachutists, and plan for one additional open oxygen station per every six parachutists in the event of a hose or regulator failure.
- Conduct preflight inspection and preflight operational checks of all oxygen equipment (Figures 4-21 and 4-22, pages 4-29 through 4-31).
- Supervise the transportation of and installation onboard the aircraft of prebreathers and oxygen cylinders.
- Issue oxygen supply hoses to each parachutist and supervise hose connection.

WARNING

Never partially close the shutoff valve during oxygen use. Closing the valve (even partially) will result in a restriction of oxygen flow to the parachutist, possibly incapacitating and/or causing serious injury to personnel.

- Prior to the aircraft procedure signal MASK being given, fully open shutoff valves on prebreathers.
- After the aircraft procedure signal MASK is given, ensure the parachutists don their masks properly and receive oxygen.
- Periodically check oxygen pressure and equipment function during use (every 10 minutes).
- Monitor each parachutist for signs of hypoxia, the bends, the chokes, neurological hits, and skin manifestations or paresthesia.
- Assist the parachutist with the activation of the bailout systems and inspect all bailout systems to make sure they were activated.
- Check the parachutist's hose connections on the AIROX VIII or OXCON oxygen console. If the parachutist still indicates a problem, the technician activates the bailout system, moves the parachutist to an open station, and the technician deactivates the bailout system.

WARNING

It is essential that problems associated with faulty bailout bottles are identified early in the prebreathing cycle. If the affected parachutist has not been breathing 100-percent oxygen, the prebreathing clock starts over once a new bailout bottle is properly installed. This could delay the jump or disqualify the affected parachutist from jumping.

Use of Oxygen in Support of Military Free-Fall Operations

Preflight Inspection of Portable Bailout Oxygen System

- ☐ Cylinders are lime green and stenciled in white with the words AVIATOR'S BREATHING OXYGEN.
- ☐ No cracks, dents, or gouges are in the cylinders.
- ☐ Cylinder clamp and roller are secured and on the bottom one-third of the cylinders.
- ☐ Cylinders are tight into the pressure reducer body.
- ☐ Reducer body is not cracked or damaged.
- ☐ Filler valve, pressure gauge, and relief valve are tight into the pressure reducer body.
- ☐ Cap on the filler valve is secure, and the filler cap lanyard is secured to both the cylinder and filler valve.
- ☐ Pressure gauge face is not damaged, and the dial indicator is not sticking.
- ☐ ON/OFF control valve is secured to the pressure reducer body with four Allen screws.
- ☐ Guide rails of the ON/OFF control valve are undamaged. Operating lever operates properly, and the detent will hold the valve in the ON and the OFF positions.
- ☐ Union elbow is secured tightly to the top of the pressure reducer, and the elbow directs the hose over the pressure gauge.
- ☐ Hose assembly is not frayed or crushed, and the cloth covering is not worn and is free of oil and other contaminants.
- ☐ Hose assembly is securely attached to the union elbow and flow indicator.
- ☐ There is no obvious damage to the flow indicator body, the arrow points toward the AIROX, and the flow indicator is securely attached to the AIROX.
- ☐ View glass is clear, indicating a no-flow condition, and the white sleeve, yellow sleeve, and spring are present.
- ☐ Blue tamper-proof dot is present directly below the ambient air port.
- ☐ Equalization port is free of foreign objects or debris.
- ☐ Brass set screw and brown tamper-proof dot are present.
- ☐ Body of the AIROX is not damaged or cracked.
- ☐ Ambient air port is securely attached to the AIROX and not damaged, and the safety lock wire and screw are intact.
- ☐ Chrome ring is present and rotates freely.
- ☐ Gasket is present, clean, and free of nicks or tears.
- ☐ Inlet orifice is free of foreign objects or debris, and the screen is present and not damaged.
- ☐ Cover of the outlet orifice is spring-loaded and seats properly.
- ☐ Outlet orifice is free of foreign objects or debris, and the screen is present and not damaged.
- ☐ Dovetail mounting plate is securely attached to the bracket.
- ☐ There is no damage to the dovetail mounting plate.
- ☐ Locking lever is spring-loaded and functions properly.

Preflight Operational Function Check Procedures

- ☐ Ensure the system is fully charged at 70 degrees Fahrenheit.
- ☐ Connect a mask to the outlet orifice and ensure that it is secure and that excessive force is not required to connect and disconnect.
- ☐ Turn the system on and seal the mask to the face.
- ☐ Inhale—yellow sleeve (on flow indicator) rises.
- ☐ Exhale—yellow sleeve falls. Inhalation should be normal with no undue exertion.
- ☐ Ensure there is no oxygen flow from the relief valve.
- ☐ Turn the system off, reseal the mask to the face, and ensure parachutist can breathe through the ambient air port.
- ☐ Connect a hose and regulator assembly to the ambient air port; ensure that it is secure and that excessive force is not required to connect and disconnect.

Figure 4-21. Portable Bailout Oxygen System preflight inspection and operational checklist

Chapter 4

Preflight Inspection of 6-Man Prebreather
Unit has no obvious damage.
Gauge faces are not broken.
Dial indicators are not sticking.
All screws are present and not coming loose.
Handles are not separating from unit.
Filler cap is present and tied down to unit.
All female disconnect plugs are present and tied down to disconnect.
Female disconnects are not distorted, and the pins of the male connectors of hose assemblies will engage with the collar of the female disconnect.
Female disconnects are safety-wired to the adjacent female disconnect.
Connector manifold guard does not interfere with the operation of the female disconnects or male connectors of the hose and regulator assembly.
Both sets of screws in the ON/OFF knob are present and not backing out.
ON/OFF valve stem is not bent.
Container is not cut, damaged severely, or corroded.
Unit is fully charged to 1,800 psi at 70 degrees Fahrenheit.
Preflight Inspection of the Hose and Regulator Assembly
Each male connector has the proper amount of pins (red: 2 pins, yellow: 3 pins, gray: 4 pins), and the mating probe is not distorted.
Male connector is tight into hose assembly.
Wire wrapping is not frayed, and hose is not crushed.
Cloth covering is free of oil and other contaminants.
Red male connector is connected to 72-inch hose, yellow connector to 90-inch hose, and gray connector to 98-inch hose.
Hose is tightly connected to regulator.
Regulator is not cut or cracked.
No foreign object or debris is in equalization port.
Hose and check-valve assembly is clamped to regulator, and clamp is safety wired.
Cover is spring-loaded and seats evenly over check valve.
Check valve is spring-loaded.
Preflight Operational Function Check Procedures
Turn the shutoff valve counterclockwise to the fully opened position (about 5 1/2 turns) (Figure 4-23, page 4-31).
Ensure the reducer pressure gauge indicates 40 to 60 psi (Figure 4-23).
Remove each disconnect plug, depress the poppet of each disconnect (Figure 4-24, page 4-32), and ensure oxygen flows from each disconnect.
Close shutoff valve and ensure reducer pressure remains steady (40 to 60 psi).
Bleed off the pressure through the disconnect manifold.
Install all hose and regulator assemblies to their appropriate disconnect (Figure 4-24). (Be sure to bleed manifold pressure before attaching hose and regulator assemblies.)
Connect an MBU-12/P or POM mask to each hose and regulator assembly.
Open shutoff valve (about 5 1/2 turns).
Listen for and feel the oxygen flow from each mask. Disconnect all but one mask and note the reducer pressure for 3 to 5 seconds. The reducer pressure should not drop below 40 psi.

Figure 4-22. Sample prebreather preflight inspection and operational function checklist

Use of Oxygen in Support of Military Free-Fall Operations

☐ Hold the mask to the face and inhale. Inhalation shall be normal with no undue exertion to breathe oxygen. Remove mask from hose and regulator assembly; ensure check valve closes and that there is no flow from the hose and regulator assembly. Repeat the above step for each hose and regulator assembly.

☐ Close shutoff valve and bleed manifold pressure through one or more check valves until reducer pressure indicates zero.

☐ Monitor reducer pressure for 15 minutes. Ensure gauge indicator remains at zero.

Figure 4-22. Sample prebreather preflight inspection and operational function checklist (continued)

CAUTION

Failure to properly connect the hose and regulator assemblies to the prebreather using the above procedures could possibly damage the diaphragm of the CRU-79/P regulator and render the equipment inoperative.

WARNING

Personnel must NEVER partially close the shutoff valve during oxygen use; it will result in a restriction of oxygen flow to the parachutist.

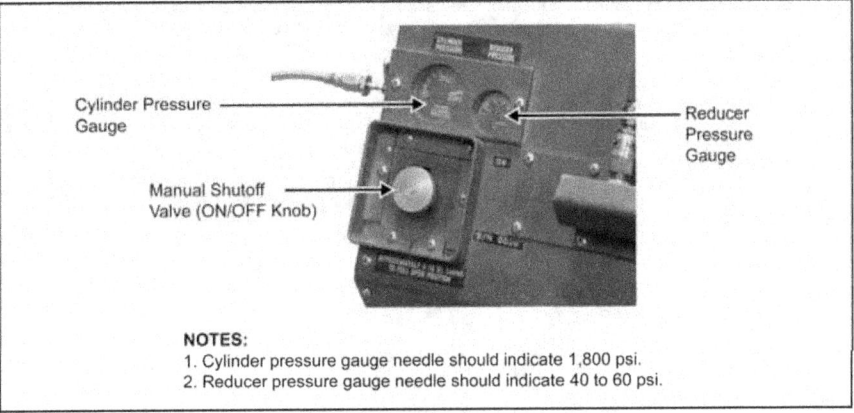

NOTES:
1. Cylinder pressure gauge needle should indicate 1,800 psi.
2. Reducer pressure gauge needle should indicate 40 to 60 psi.

Figure 4-23. Pressure gauge and manual shutoff valve

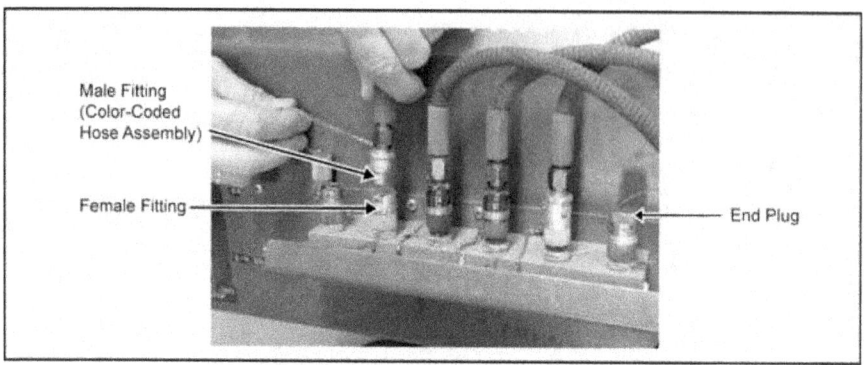

Figure 4-24. Removing end plugs and depressing poppets

OXYGEN HANDLING AND SAFETY

4-60. Because of limited contact with oxygen and its handling, personnel may not fully appreciate the danger involved. Improper use and handling can result in property damage, serious injury, and death. Personnel handling oxygen must—

- *Keep oil and grease away from oxygen.* They must not handle oxygen equipment with greasy hands or clothing. They do not let fittings, hoses, or any other oxygen equipment get smeared with petroleum-based products, lubricants, hydraulic fluid, or dirt. A drop of oil or lubricant in the wrong place can cause an explosion.
- *Keep oxygen away from fires.* Small fires rapidly become large fires in the presence of oxygen supplies. Personnel handling oxygen must never permit smoking near oxygen equipment, while handling oxygen supplies, or when using oxygen life-support equipment.
- *Handle cylinders and valves carefully.* Before opening cylinder valves, they make sure the cylinder is firmly supported. They never let a cylinder drop or tip over. Dropping a cylinder can damage or break the valve, allowing the gas to escape and to propel the cylinder a great distance, which is an obvious hazard. Personnel open and close the valves only by hand. If they cannot open and close them by hand, they must return the cylinder to the depot for repair.

Chapter 5

Equipment and Weapon Rigging Procedures

Free-fall parachutists will normally operate with individual equipment that includes clothing and equipment in keeping with the climatic conditions, food, and survival items. In addition, each parachutist will have a weapon, free-fall parachutist's jump helmet, goggles, and altimeter. Free-fall parachutists jump and carry all detachment equipment and supplies as individual loads. If selected items must be dropped as accompanying supplies, they pack these supplies in appropriate aerial delivery containers.

EQUIPMENT AND WEAPON PACKING CONSIDERATIONS

5-1. The parachutist can attach or wear his individual equipment and weapon in several configurations (weapons, for example, exposed, placed in containers, or a mix of the two). Unit SOPs specify ways to pack equipment that are consistent with safety requirements. As a rule, units pack hard, bulky, or irregularly shaped (nonaerodynamic) items in containers. Parachutists can use rucksack rigging systems approved by their Service test board.

5-2. The parachutist packs his individual equipment in a container, kit bag, parachutist drop bag (PDB), or the medium or large combat pack. He then attaches it to the equipment rings on the parachute's main lift web. He may front- or rear-mount the combat pack using the harness, single point release (HSPR); improved equipment attaching sling (Spider Harness); or the H-harness (modified). He may attach both a front- and rear-mounted rucksack and equipment as long as he is under the 360-pound "all-up" total weight (to include personnel, gear, and weight of canopy suspended below the parachute). He should lower combat packs or any equipment that weighs more than 35 pounds.

5-3. The parachutist pads fragile items, such as weapon sights. He does not place crushable items, such as the protective mask, directly under the attaching harnesses. Exposed weapons or equipment, snap hooks, and projections are potential safety hazards that the parachutist tapes.

PARACHUTIST AND PARACHUTE LOAD LIMITATIONS

5-4. Commanders must not overload the parachutist with equipment. The variety and weight of equipment and weapons that can be attached to a parachutist (Tables 5-1 through 5-4, pages 5-1 through 5-3) may exceed the safe design limits of the MC-4 RAPPS. Overloading can result in parachute damage, unsafe descent rates, and injury to the parachutist. Also, the parachutist's actions and the time available to release the tie-down straps and to lower the equipment may interfere with his control of the parachute close to the ground.

Table 5-1. Container weight limits

Description	Maximum Container Load (Pounds [lb])	Maximum Rigged Weight (lb)*
Medium Combat Pack	50	55.56
Large Combat Pack	70	75.96
PDB, Medium	140	145.96 (Not to Exceed All-up Weight of Canopy)

Chapter 5

Table 5-1. Container weight limits (continued)

Description	Maximum Container Load (lb)	Maximum Rigged Weight (lb)*
PDB, Large	Not to Exceed All-up Weight of Canopy	Not to Exceed All-up Weight of Canopy

*Weight of H-harness attaching sling.

Table 5-2. Parachute load limits

Description	Weight (lb)	Reference	Remarks
Maximum Load-Bearing Capacity of MC-4 RAPPS on Deployment	360	Natick Research and Development Command	Increased weight will reduce canopy service life or destroy canopy (for example, blown cells).
Air Movement Planning Weight of Combat-Equipped Free-Fall Parachutist	305	None	Parachutist with one equipment container and weapon.

Table 5-3. Weight of parachutist with two equipment loads

Container Type	Container Maximum Internal Weight	Weight of Container	Suspended Weight of MC-4 RAPPS With Oxygen	Fatigue Uniform, Helmet, Mask, and Boots	Soldier Weight	M4 Rifle With Magazine	Total Suspended Weight*
Kit Bag	50	3	43.15	15	205	6.9	323.75
Medium Combat Pack	50	5.56	43.15	15	205	6.9	326.31
Large Combat Pack	70	5.96	43.15	15	205	6.9	346.71

*Weight of parachutist in pounds.

Table 5-4. Weight of parachutist with two equipment loads and basic load

Weapon Load Type	Weapon Load With Ammunition (1)	Weight of Large Combat Pack	Soldier Weight	Fatigue Uniform, Helmet, Mask, and Boots (2)	Body Armor With 70 Ounces of Water (Camel-Back Bladder)	Suspended Weight of MC-4 RAPPS With Oxygen	Remaining Weight of MC-4 RAPPS With Oxygen	Total Suspended Weight*
M4 Rifleman	31	5.96	205	15	24.5	43.15	35.39	360

Table 5-4. Weight of parachutist with two equipment loads and basic load (continued)

Weapon Load Type	Weapon Load With Ammunition (1)	Weight of Large Combat Pack	Soldier Weight	Fatigue Uniform, Helmet, Mask, and Boots (2)	Body Armor With 70 Ounces of Water (Camel-Back Bladder)	Suspended Weight of MC-4 RAPPS With Oxygen	Remaining Weight of MC-4 RAPPS With Oxygen	Total Suspended Weight*
M203 Gunner	40	5.96	205	15	24.5	43.15	26.39	360
M60 Machine Gunner	54.4	5.96	205	15	24.5	43.15	11.99	360

*Weight of parachutist in pounds.
(1) Includes basic load of ammunition, grenades, claymore, bayonet, and cleaning kit.
(2) Weight of uniform does not include winter gear (for example, parka, liners, underwear).

HOOK-PILE TAPE (VELCRO) LOWERING LINE ASSEMBLY

5-5. Figure 5-1, page 5-4, shows the steps (A through D) for stowing an HPT lowering line assembly. The current HPT lowering line assembly (National Stock Number [NSN] 1670-01-067-6838) consists of—

- An 8- or 15-foot lowering line (the 8-foot lowering line is recommended for most equipment) made of 1-inch-wide tubular nylon.
- A 9- by 7-inch nylon duck retainer (stow pocket) sewn to the upper end. The flaps have HPT sewn to the edges.
- A metal (parachute harness) ejector snap with a yellow safety release.

Note: Maintenance Advisory Message (MAM) Number 90-16 authorizes the modification of the 15-foot HPT lowering line to an 8-foot HPT lowering line. Additionally, it provides procedures for altering the HPT lowering line. Military Occupational Specialty (MOS) 92R are the only personnel authorized to alter the HPT lowering line IAW MAM Number 90-16.

Note: The yellow release lanyard may be removed or, if it remains attached to the HPT lowering line, it should be taped with two wraps of 1-inch masking tape around the end approximately 1 inch from the bottom of the lanyard, securing it to the lowering line and leaving 1 to 2 inches exposed at the top of the lanyard.

Note: To help prevent inadvertent, premature release of the lowering line, the parachutist places a medium-weight double-looped retainer band around the middle of the stowed lowering line retainer pocket before attaching it to the combat pack. Also, the parachutist places a heavy-weight double-looped retainer band around the quick-ejector snap.

5-6. The steps for stowing an HPT lowering line assembly, as shown in Figure 5-1, are as follows:
- Starting with the looped end to the left, neatly S-fold the tubular nylon until the quick-ejector snap is coming out the right side (Figure 5-1A).
- Mate the Velcro on the excess tubular nylon to the Velcro on the bottom closing flap (Figure 5-1B).
- Close the top flap, encasing the folded tubular nylon (Figure 5-1C).

Chapter 5

- Close off both running ends by realigning the Velcro (Figure 5-1D).
- Remove the yellow release lanyard or tape it with one complete wrap of masking tape. Place a doubled heavyweight retainer band (Figure 5-1E) around the center of the folded lowering line and a tripled heavyweight retainer band around the quick-ejector snap.

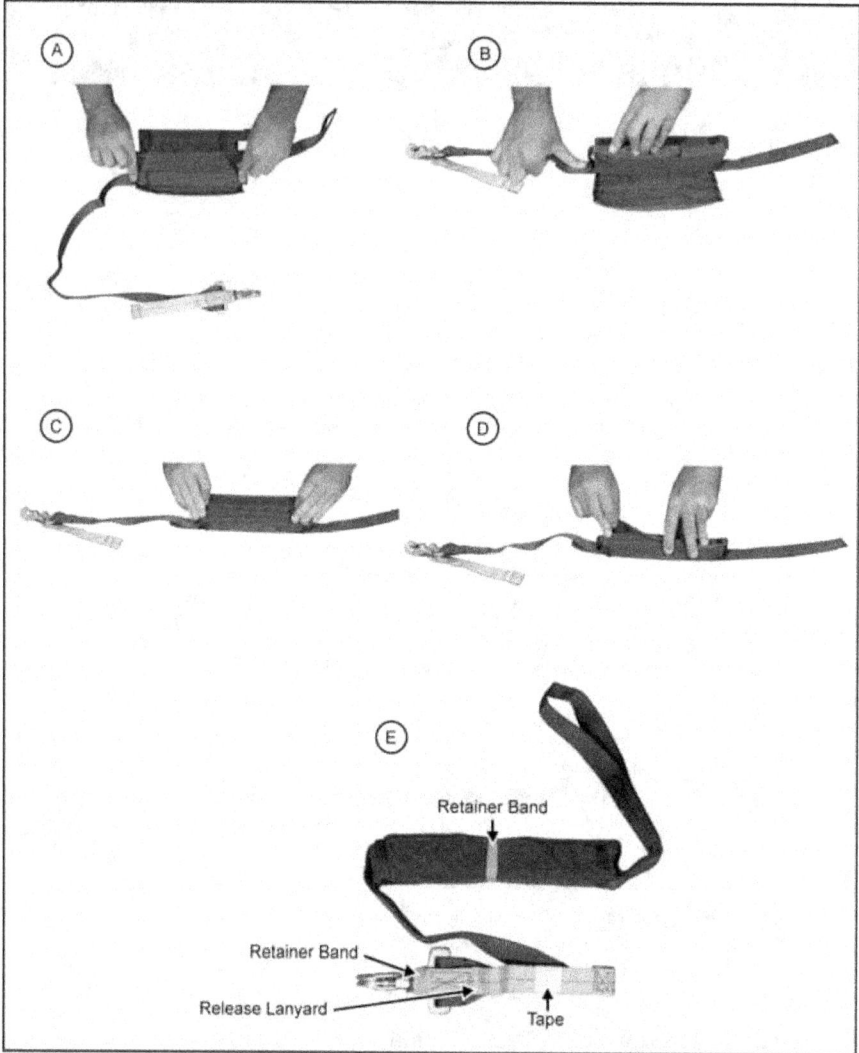

Figure 5-1. Stowing the HPT lowering line assembly

COMBAT PACKS AND OTHER EQUIPMENT CONTAINERS

5-7. The following paragraphs discuss the use of harnesses, equipment attachment slings, and lowering lines in preparing and rigging kit bags and different packs.

H-Harness (Modified)

5-8. The modified H-harness consists of two 84-inch nylon straps held together by two 11-inch straps (Figure 5-2). One end of each strap has two friction adapters attached 3 inches apart. Two 24-inch or 36-inch equipment attachment straps with adjustable lugs and two quick-release ejector snap hooks are part of the assembly. The H-harness is used to rig the kit bag and combat packs to the parachute harness.

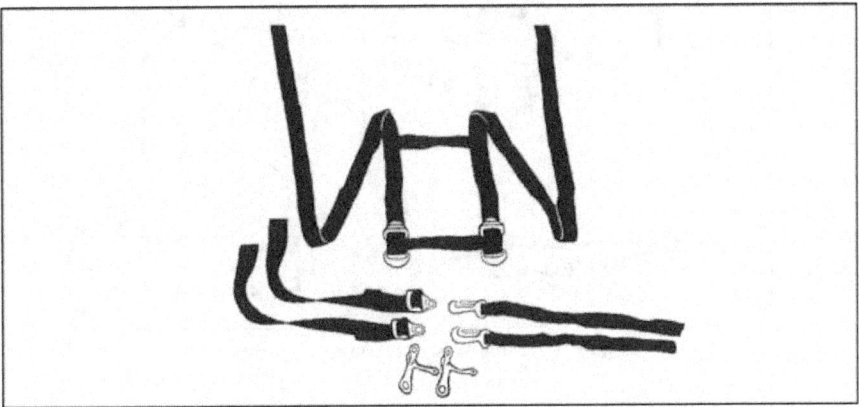

Figure 5-2. H-harness with attaching straps

Aviator's Kit Bag/MC-4 Kit Bag

5-9. The parachutist uses the canvas aviator's kit bag or the MC-4 kit bag to jump individual equipment, such as the load-carrying equipment or properly padded machine gun groups.

Preparing the Bag

5-10. The parachutist packs the equipment IAW the unit SOP. He carefully places sharp-edged objects in the bag so that they are not against his body when he attaches the bag to the parachute harness. He unfastens the snaps, undoes the slide fastener, and folds down the top of the kit bag (about one half its filled bulk) to pack the equipment. When packed, the parachutist zips the bag and fastens the snaps. He gathers up the excess bag material and folds it on top so as to expose the handles.

Attaching the H-Harness to the Kit Bag

5-11. The parachutist takes the two end web adapters and lays out the harness (with the adapters nearest the body and the second two adapters on top). He connects the equipment attachment straps as outlined below. The parachutist—

- With the adjustable lug nearest the body, threads the attachment strap's end under the attaching bar of the second friction adapter and back over the top of the bar.
- Tightens the strap, leaving about 3 inches between the nap and the bar, and repeats this step for the remaining strap.
- Places one quick-release snap hook on each adjustable lug.

Chapter 5

- Lays out the H-harness with the attachment straps down and the snap hook openings up.
- Attaches the H-harness to the kit bag by centering the bag on the harness 6 inches from the snap hooks.
- Places the H-harness straps around the kit bag and threads them through the friction adapters to form a quick release.
- Threads the snap hooks on the attaching straps through the handles of the kit bag. He rolls and tapes any excess straps (Figure 5-3).

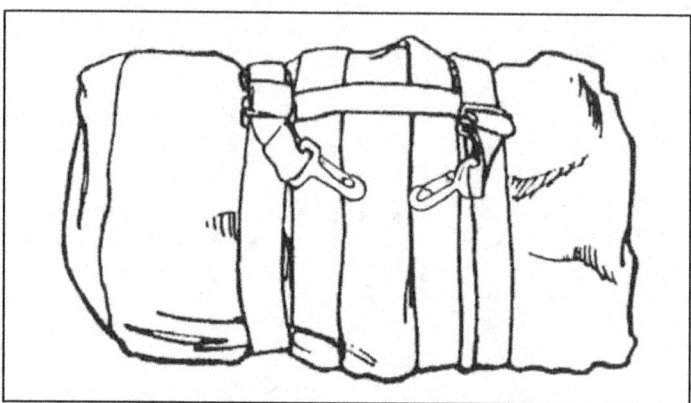

Figure 5-3. H-harness attached to the kit bag

Attaching the Kit Bag to the Parachutist

5-12. When completely rigged, the parachutist attaches the H-harness to himself. He runs the attachment straps through the handles of the kit bag and then attaches them to the equipment attachment rings on the parachute harness. If wearing a front-mounted aviator's kit bag and a rear-mounted combat pack, the parachutist hooks up the kit bag quick-release snap hooks to the equipment attachment rings first. He then hooks up the combat pack quick-release snap hooks to the outside of the kit bag's snap hooks.

COMBAT PACKS

5-13. The parachutist attaches medium and large combat packs by using the modified H-harness or the improved equipment attachment sling. Combat packs can be either front- or rear-mounted.

Packing the Combat Pack

5-14. The parachutist—

- Places equipment in the combat pack and places padding between the load and the front portion of the pack.
- Fills the outside pockets with nonfragile items (full pockets help to position the H-harness and attachment sling).
- Closes the combat pack by engaging the drawstrings and tie-down straps.
- Routes the running ends of the waist straps behind the frame and secures them by tying or taping.

Equipment and Weapon Rigging Procedures

Rigging the Medium Combat Pack Without the Pack Frame

5-15. The parachutist—
- Turns the pack upside down.
- Places the H-harness on his pack so that the cross straps are in front of the pack and the friction adapters are touching the bottom of the pack.
- Runs the harness straps over the top of the pack and crosses the straps at the center of the back of the pack.
- Runs the straps through the friction adapters.
- Threads the equipment attaching straps through the intermediate friction adapters.
- Attaches the quick-release snap hooks to the adjustable lugs.

Rigging the Medium and Large Combat Packs With the Pack Frame, Modified H-Harness, and Lowering Line

5-16. The parachutist—
- Positions the modified H-harness on the floor or ground with the friction adapters down. He places the pack, frame up, over the harness making sure that the cross straps are to the top of the pack and the friction adapters are touching (or near) the bottom of the frame (Figure 5-4).
- Runs the harness straps over the top of the pack and then under the top portion of the frame.
- Runs the harness straps under the horizontal bar of the frame and crosses them at the center of the back of the pack. He continues to run the straps under the frame and secures them to the friction adapters.
- Routes the loop end of the lowering line under the crossed diagonal straps. He passes the running end of the lowering line through its own loop and tightens it, making sure he centers the lowering line at the intersection of the straps.
- Secures the lowering line stow pocket to the pack frame with retainer bands. He leaves the portion with the quick-ejector snap free for attachment to the parachute harness.
- Threads the equipment attaching straps through the intermediate friction adapters, attaches a quick-release snap hook to each adjustable lug, and rolls and tapes any excess straps.

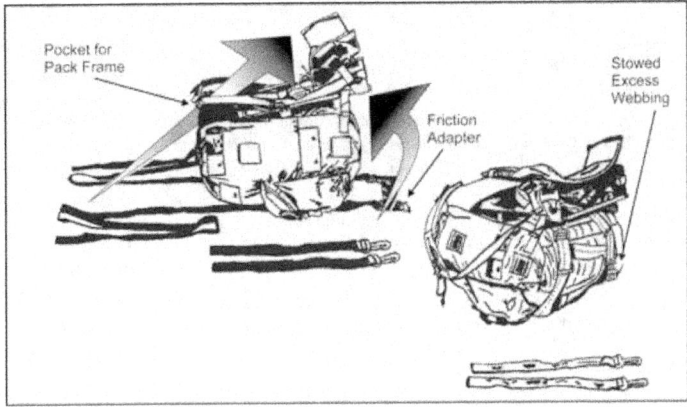

Figure 5-4. Combat pack and frame rigged with the modified H-harness

Chapter 5

IMPROVED EQUIPMENT ATTACHMENT SLING

5-17. The improved equipment attachment sling (Figure 5-5) was a component of the MC-3 MFF system. The parachutist modifies this sling by removing the leg straps with HPT closures or folds and tapes the leg straps so that he cannot use them. This sling is used to rig combat packs to the parachute harness. This system is also known as the spider harness.

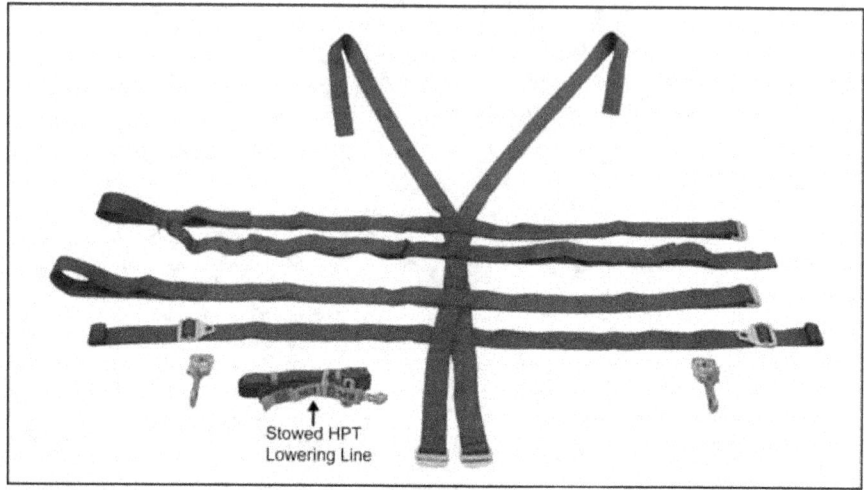

Figure 5-5. Improved equipment attachment sling and lowering line (spider harness)

Rigging the Large Combat Pack With the Improved Equipment Attachment Sling (Spider Harness) and Lowering Line

5-18. The parachutist—

- Tightens and secures all straps on the pack and positions the pack with the frame up (Figure 5-6A, page 5-9).
- Positions the harness on the frame with the friction adapters on the diagonal locking straps at the bottom of the frame and the running ends at the top of the frame.
- Routes the diagonal locking strap friction adapters under the pack frame's base.
- Routes the anchor straps (parachute harness attaching straps with adjustable quick-release lugs) and lateral locking straps under the shoulder straps and over the pack frame.
- Turns the pack over and routes the running ends of the diagonal locking straps around the long axis of the pack, across the straps at the center of the back.
- Secures the diagonal locking straps to the respective friction adapters that protrude beneath the bottom of the pack frame (Figure 5-6B).
- Tightens the lateral locking straps and secures them around the pack and to their respective friction adapters (Figure 5-6C).

Note: If the pack is small, the parachutist crosses and tightens the lateral locking straps and secures them around the pack and to their opposite friction adapters.

Equipment and Weapon Rigging Procedures

- Folds and secures the running ends of all straps to themselves with tape or ties them with 1/4-inch cotton webbing.
- Places the combat pack in an upright position.
- Attaches a quick-release snap hook to each adjustable lug so that the latch handles face away from his body when he attaches the combat pack to the equipment rings (Figure 5-6D).

> **WARNING**
>
> The parachutist tapes all combat pack shoulder strap quick-ejector releases to preclude inadvertent release in free fall, causing instability.

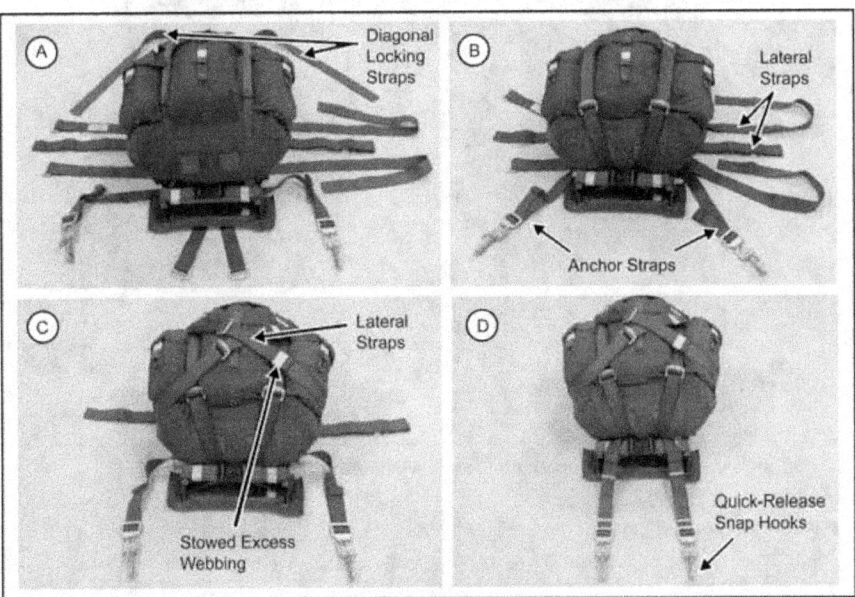

Figure 5-6. Combat pack and frame rigged with the improved equipment attachment sling

Attaching the Lowering Line

5-19. The parachutist—
- Routes the loop end of the lowering line under the crossed diagonal straps between the diagonal straps and the loop on the backside of the diagonal straps.
- Passes the running end of the lowering line through its own loop and tightens it (Figure 5-7, page 5-10).
- Makes S-folds with the remainder of the lowering line and places the S-folds into the retainer pocket.

Chapter 5

- Secures the retainer pocket to the appropriate side of the pack frame (right side for front mount, left side for rear mount) with retainer bands. He uses three retainer bands: two on the frame and one double-wrapped around the center of the lowering line.
- Removes the yellow release lanyard or, if it remains attached to the HPT lowering line, tapes it to the lowering line with one single wrap of masking tape the length of the lanyard, leaving 1 to 2 inches exposed at the top of the lanyard.
- Attaches the lowering line quick-ejector snap to the right-side lowering line attachment V-ring (Figure 5-7).

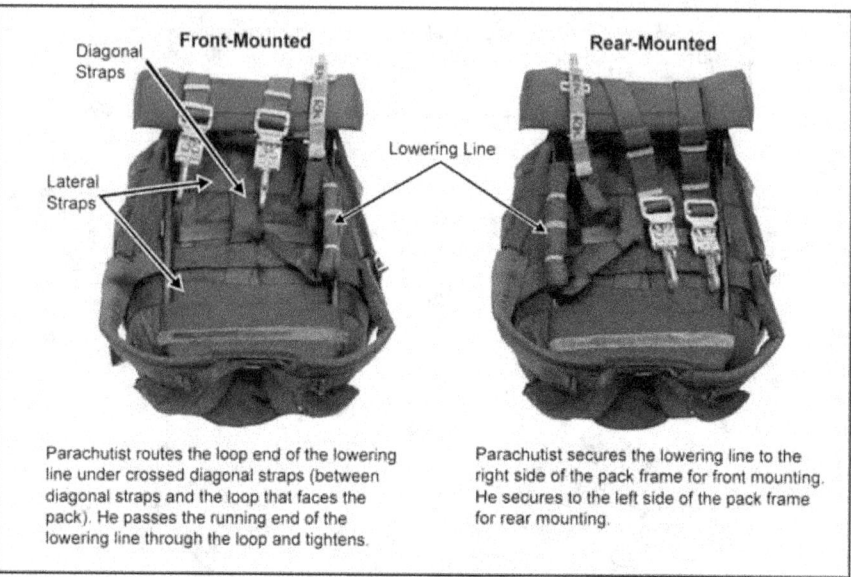

Figure 5-7. Attaching the lowering line to the combat pack

Attaching the Combat Pack

5-20. The parachutist attaches the combat pack with frame to himself in the same manner as the combat pack without frame.

Attaching the Rear-Mounted Combat Pack

5-21. The parachutist—
- Loosens the shoulder straps and steps through the shoulder straps, one leg through each strap (Figure 5-8A, page 5-11).
- Attaches the quick-release snap hooks to the large equipment attachment rings on the main lift webs (Figure 5-8B).
- No. 2 lifts up on the pack and jumper pulls the slack out (Figure 5-8C). In this last step, the parachutist could pull out the slack by himself by squatting and sitting on the pack.
- Attaches the lowering line to the right-side lowering line attachment V-ring on the parachute harness (Figure 5-8D and Figure 5-9, page 5-12).
- Tightens shoulder straps around legs (Figure 5-8E).

Equipment and Weapon Rigging Procedures

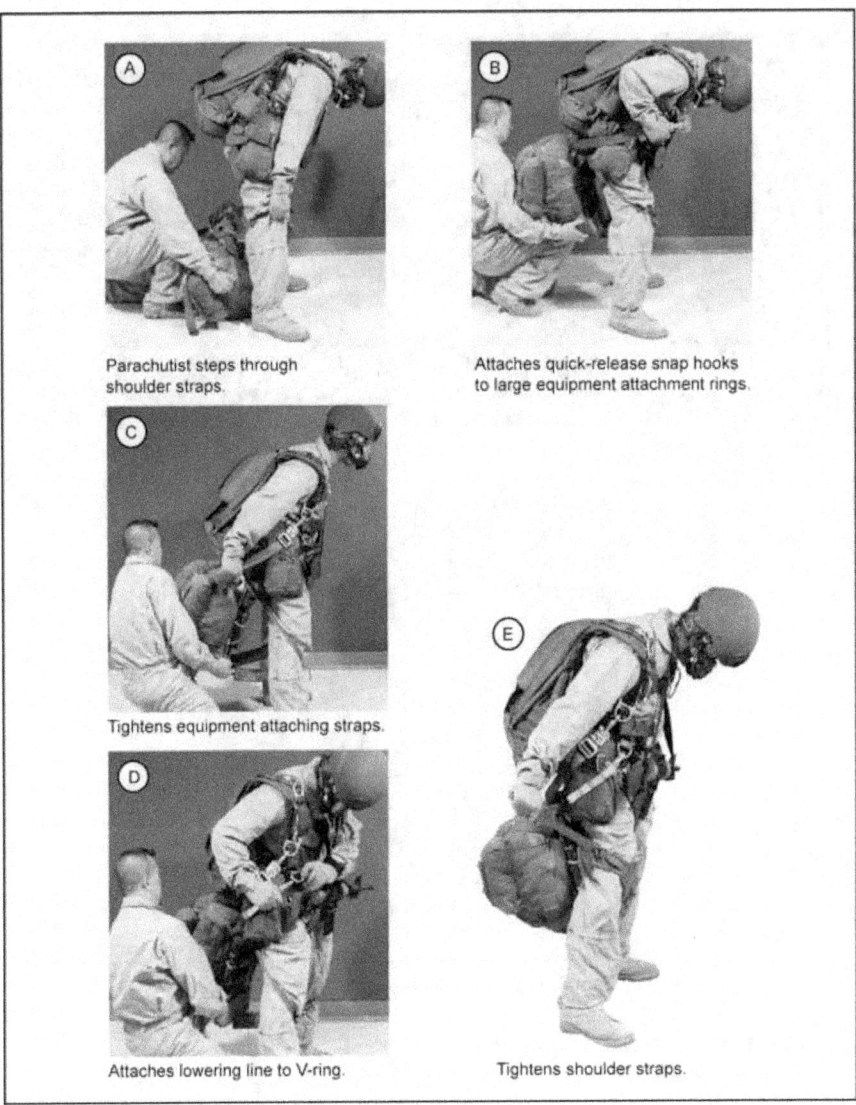

Figure 5-8. Attaching the rear-mounted combat pack

Chapter 5

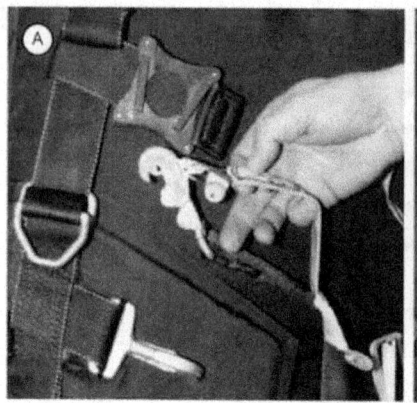
Parachutist levers the ejector snap hook to the open position.

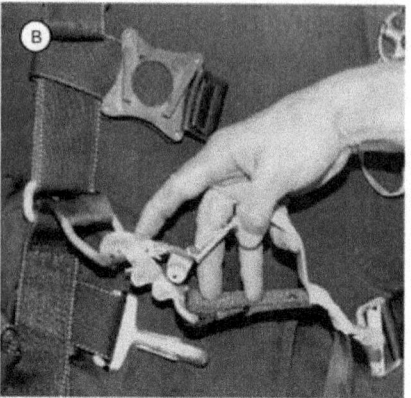
With the ejector snap hook open, he seats the snap hook into the V-ring.

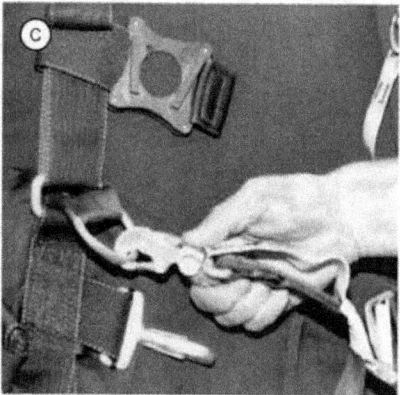
He levers the ejector snap hook closed and ensures the gate is closed.

Figure 5-9. Lowering line attached to the lowering line attachment V-ring

Attaching the Front-Mounted Combat Pack

5-22. The parachutist—
- Loosens the shoulder straps.
- Faces the combat pack and steps through the shoulder straps, one leg through each strap (Figure 5-10A, page 5-13).
- Attaches the quick-release snap hooks to the equipment attachment rings on the main lift webs (Figure 5-10B).
- Tightens equipment attachment straps (Figure 5-10C).

Equipment and Weapon Rigging Procedures

- Attaches the lowering line to the right-side lowering line attachment V-ring on the parachute harness (Figure 5-10D).
- Tightens shoulder straps (Figure 5-10E).

Figure 5-10F shows the parachutist rigged with the front-mounted combat pack.

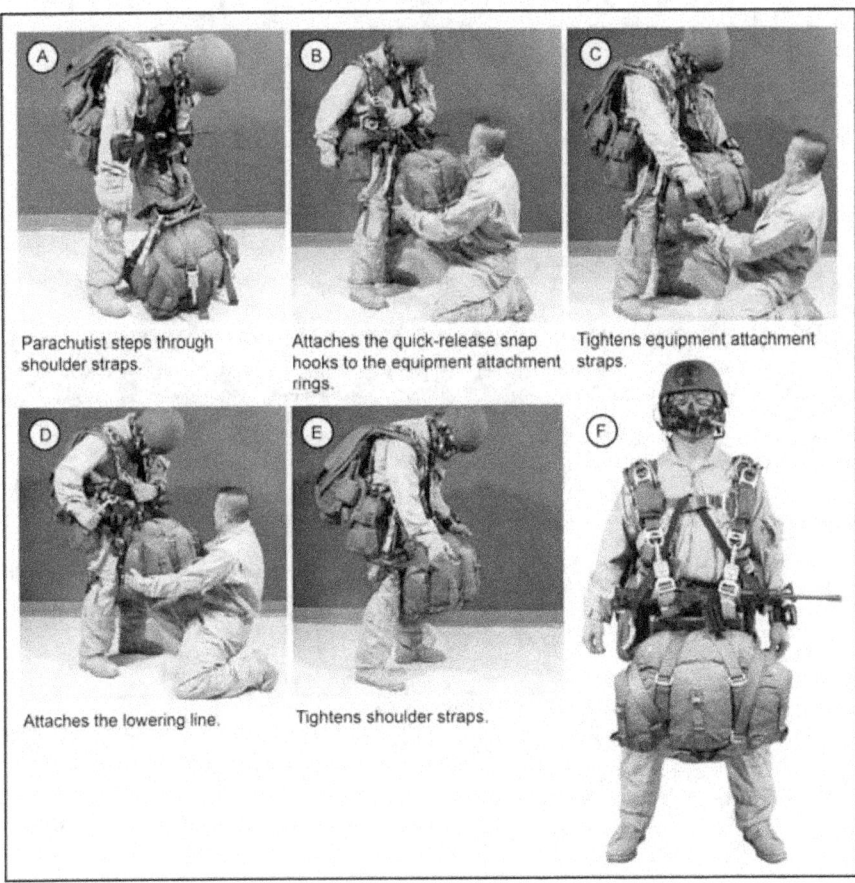

Figure 5-10. Attaching the front-mounted combat pack

Releasing the Combat Pack

5-23. After his canopy deploys and when he is clear of other parachutists and has canopy control, the parachutist will verify at 1,000 foot AGL that the HPT lowering line is still attached, cross fit (grip switch) and pull both brakes to the half position, loosen the combat pack's shoulder straps, and pull them clear of the kit bag. He then detaches the combat pack's left-side quick-release snap hook so that the pack falls cleanly when released. When on his final approach and 500 feet above the ground, he ensures that the ejector snap and HPT is still connected, then releases the second quick-release snap hook, catches his

Chapter 5

combat pack with his feet, and flies to the desired landing point. Between 500 to 200 feet after turning on final approach, the parachutist looks down and around to verify that airspace is clear and then ensures the combat pack is fully lowered and off the feet before he lands. To jettison the combat pack, he releases the lowering line's quick-ejector snap, allowing the pack to fall free.

> **WARNING**
>
> The parachutist lowers all rear-mounted combat packs with frames to avoid injury upon landing.

Note: If the combat pack is lowered before 200 feet AGL or the final approach and the parachutist has to make turns, the combat pack will start to swing and the parachutist may impact the ground the same time as the pack. No radical maneuvers should be made at the time or point before landing.

USMC SINGLE-ACTION RELEASE PERSONAL EQUIPMENT LOWERING SYSTEM

5-24. The Single-Action Release Personal Equipment Lowering System (SARPELS) is a complete lowering system authorized for use by the USMC for both static-line and MFF parachute operations. The SARPELS was designed to provide a single-point release capability for personal equipment carried by military parachutists. USMC TM 10121A-12&P, *Single-Action Release Personal Equipment Lowering System (SARPELS)*, provides further information.

5-25. The system consists of the SARPELS cargo carrier, two D-ring attaching straps, two leg strap cable retainers with buckle and grommet, the single-point release handle, a 15-foot static-line lowering line, and an 8-foot MFF lowering line. The complete system weighs 9.5 pounds empty. The SARPELS cargo carrier weighs 6 pounds empty and measures 22 inches long by 18 inches wide by 24 inches high. The 8-foot MFF lowering line is made of 1-inch tubular nylon with a maximum capacity of 1,000 pounds.

Loading the SARPELS Cargo Carrier

5-26. The parachutist—

- Loads personal supplies and equipment in the SARPELS cargo container in such a manner as to maintain the general shape of the cargo container (Figure 5-11, page 5-15). He does not load more than 110 pounds into the cargo container.
- Standing from the top of the SARPELS, folds over the left side flap and then the right side flap (Figure 5-12, page 5-15).
- Stows the unused webbing straps into the small side pockets on either side of the SARPELS cargo carrier (Figure 5-13, page 5-16).
- Folds the outer side flaps in half and over the side pockets and straps (Figure 5-14, page 5-16).
- Still standing at the top, folds the left top flap and then the right flap (Figure 5-15, page 5-17).
- Folds over the large top flap (Figure 5-15).
- Inserts the horizontal straps through the webbing and buckles (Figure 5-16, page 5-17). On the SARPELS model with three horizontal straps, the top strap is optional, depending on the size of the cargo.
- Roll-folds the excess webbing of the horizontal straps and secures them with the elastic retainer band.
- Inserts the vertical straps through the webbing and buckles (Figure 5-17, page 5-18).
- Roll-folds the excess webbing of the vertical straps and secures them with the elastic retainer band.

Equipment and Weapon Rigging Procedures

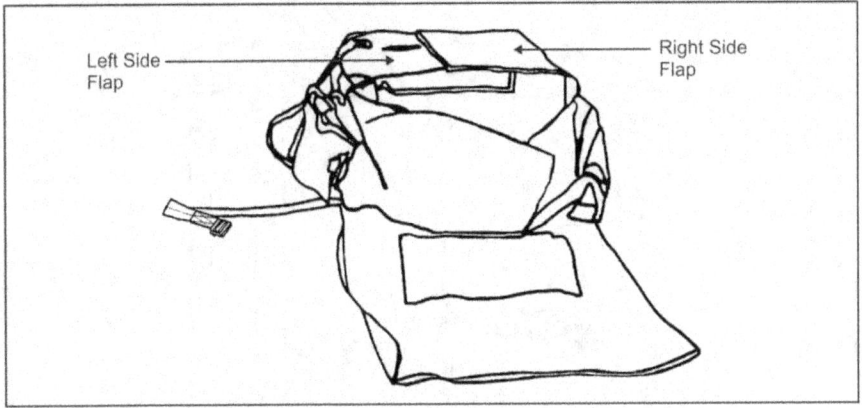

Figure 5-11. Opened SARPELS cargo carrier

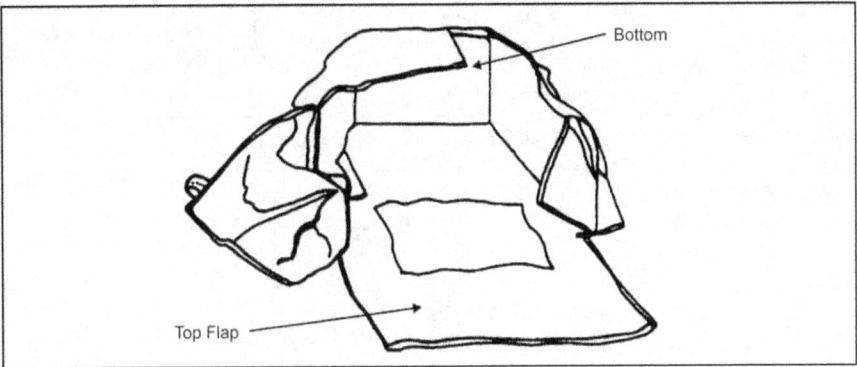

Figure 5-12. SARPELS with folded side flaps

WARNING

Parachutists should not overload the SARPELS cargo carrier. Personal injury may occur. The maximum safe load is 110 pounds.

Chapter 5

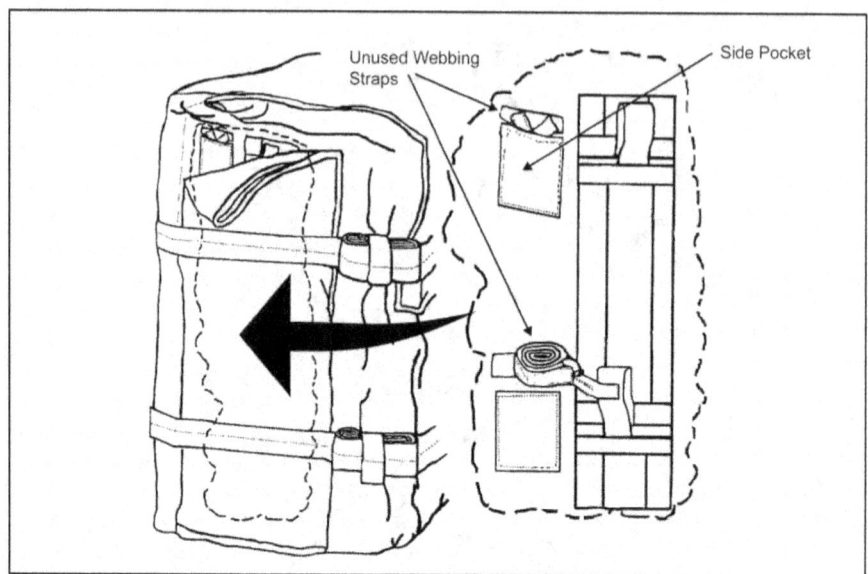

Figure 5-13. Stowage pockets

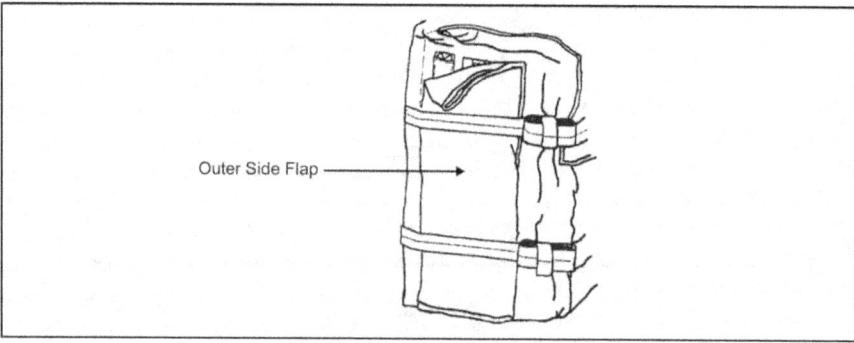

Figure 5-14. Side flap

Equipment and Weapon Rigging Procedures

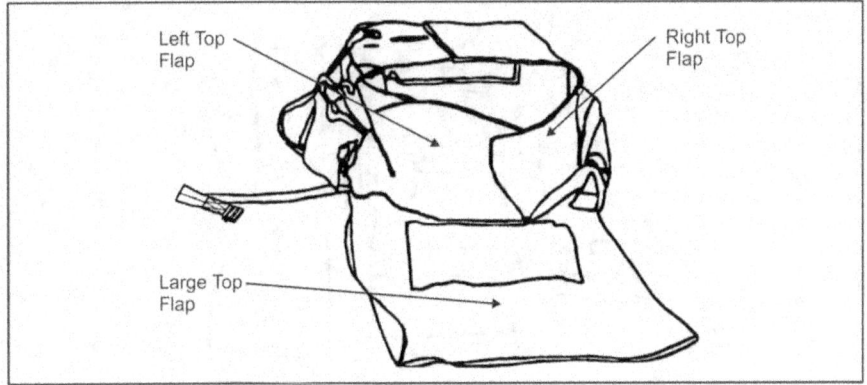

Figure 5-15. Top flaps

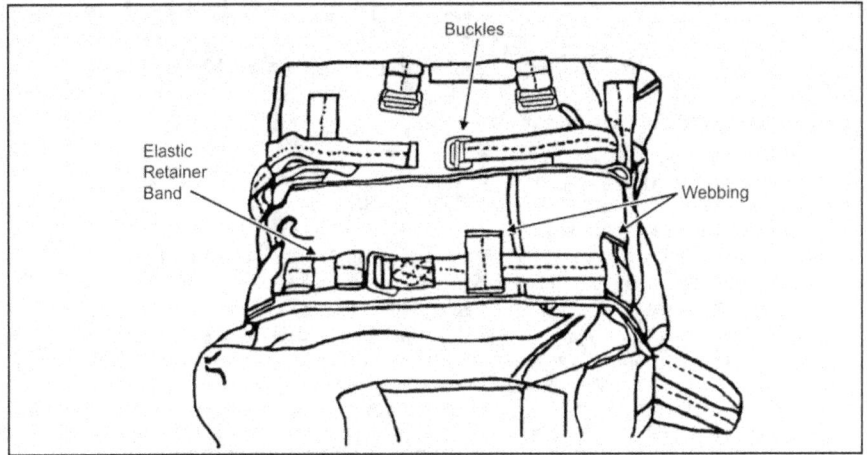

Figure 5-16. SARPELS with secured horizontal straps

Chapter 5

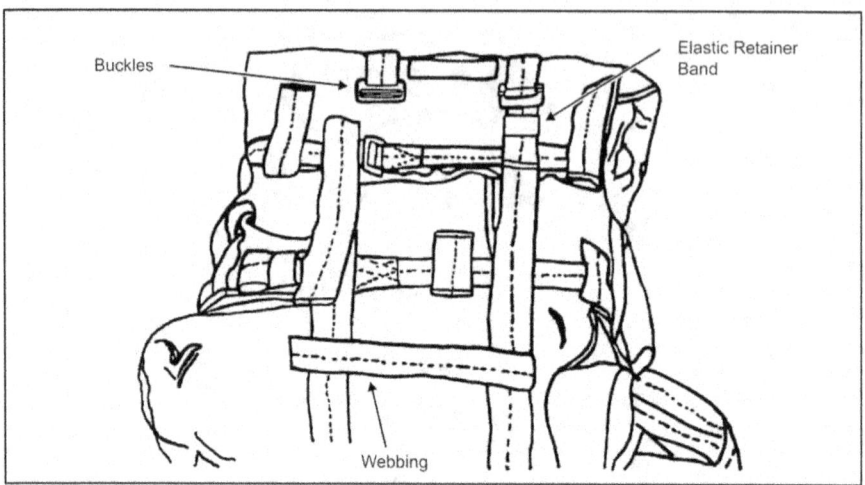

Figure 5-17. SARPELS with secured vertical straps

Rigging the SARPELS Cargo Carrier

5-27. The parachutist—

- Feeds the white webbing of the SARPELS cargo carrier through the parachute harness link of the D-ring attaching straps (Figure 5-18, page 5-19). He ensures that the opening to the snap hooks of the D-ring attaching straps are facing down toward him.
- Pulls the green 550 cord through the white webbing (Figure 5-19, page 5-19).
- Pulls the red 550 cord through the green 550 cord (Figure 5-20, page 5-20).
- Pulls the red 550 cord through the grommet of the leg strap (Figure 5-21, page 5-20).
- Runs the wire rope of the single-action release handle between the webbing handles (Figure 5-22, page 5-20). Then he runs the wire rope through the red 550 cord and into the retaining pouch of the leg strap cable retainers with buckle and grommet.

Note: The dotted line represents the wire rope hidden within the retainer pouch.

- Secures the single-point release strap with the Velcro within webbing handles (Figure 5-23, page 5-21).

Equipment and Weapon Rigging Procedures

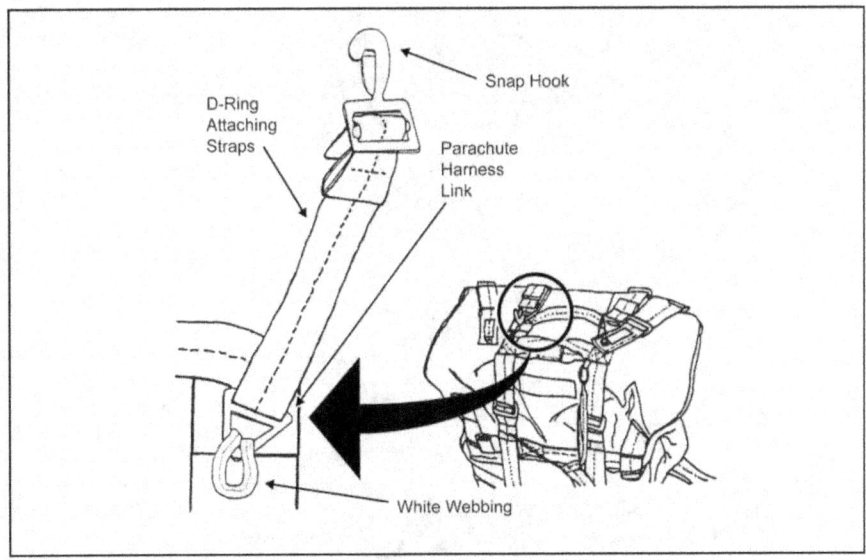

Figure 5-18. Inserting the white webbing through the parachute harness link

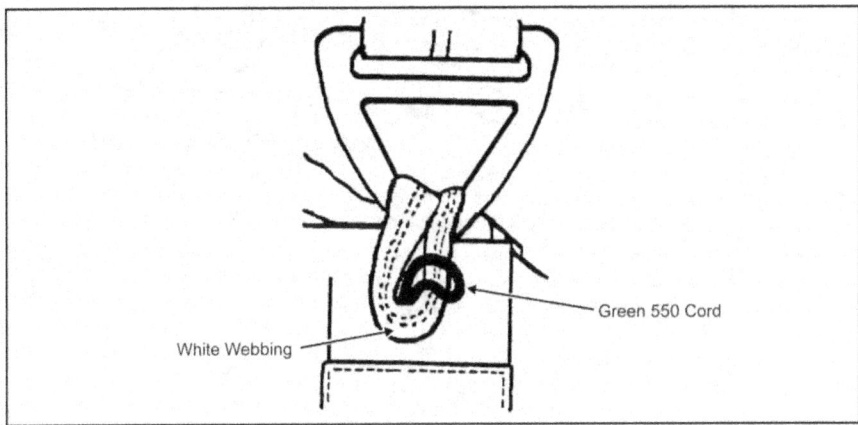

Figure 5-19. Inserting the green 550 cord through the white webbing

Chapter 5

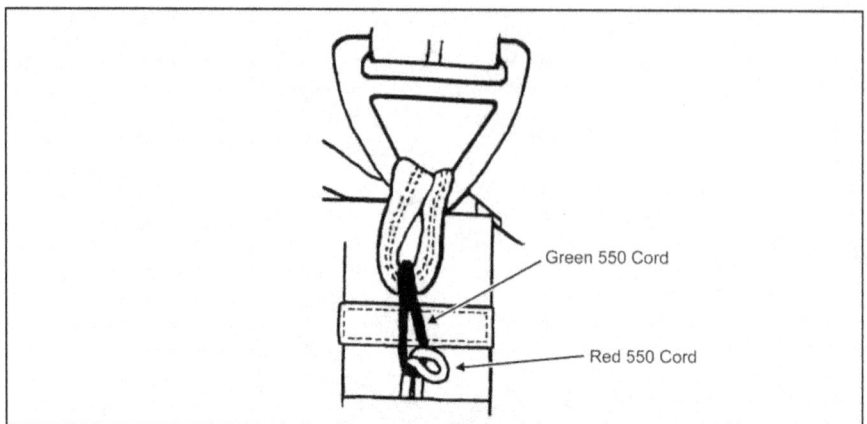

Figure 5-20. Inserting the red 550 cord through the green 550 cord

Figure 5-21. Inserting the red 550 cord through the grommet

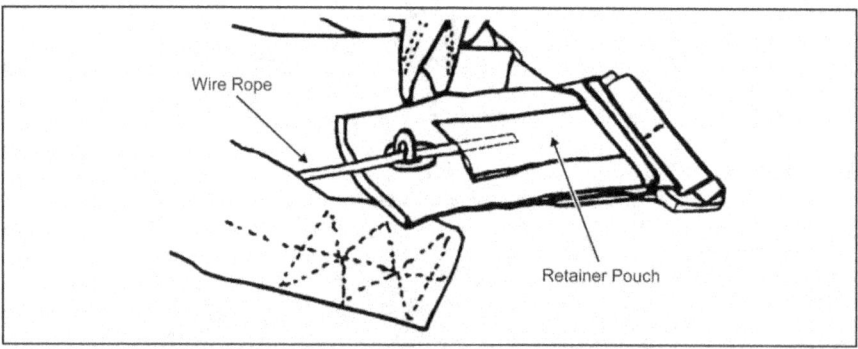

Figure 5-22. Leg strap cable retainer with buckle and grommet

Equipment and Weapon Rigging Procedures

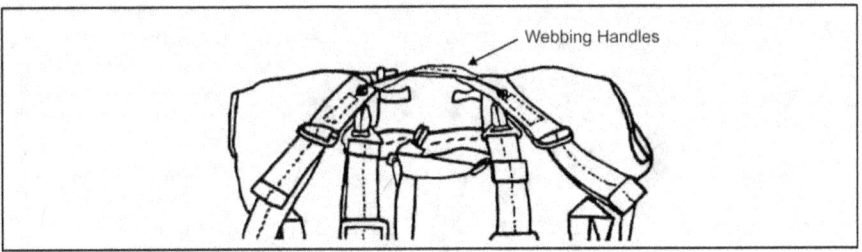

Figure 5-23. SARPELS release assembly

Installing the 8-Foot Military Free-Fall Lowering Line

5-28. The parachutist—
- Uses the 8-foot-long lowering line for MFF operations using a RAPPS MC-4.
- Attaches the lowering line through the sewn webbing loop on the backside of the SARPELS carrier at the top of the stowage pocket (Figure 5-24).
- Feeds the sewn loop of the lowering line through the sewn loop on the cargo carrier.
- Inserts the quick-ejector snap through the sewn loop of the lowering line (Figure 5-25, page 5-22).
- Pulls the entire lowering line through the loop and cinches it down.
- S-folds the excess lowering line, secures it with an elastic retainer band, and places it in the stowage pocket.
- Closes the stowage pocket with the Velcro.

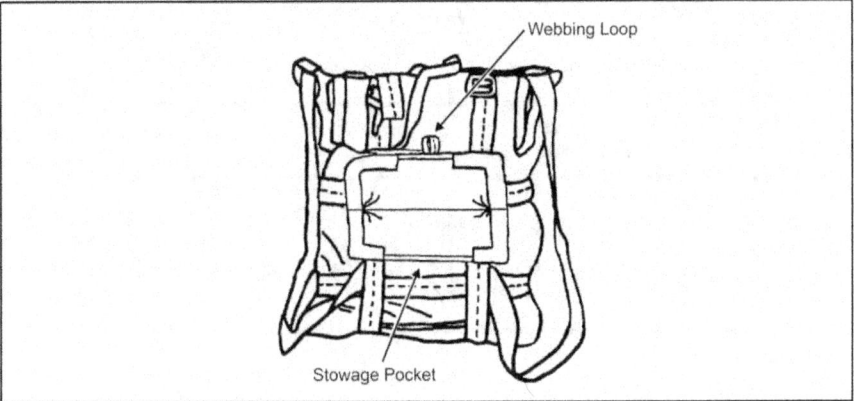

Figure 5-24. Stowage pocket with 8-foot lowering line

Chapter 5

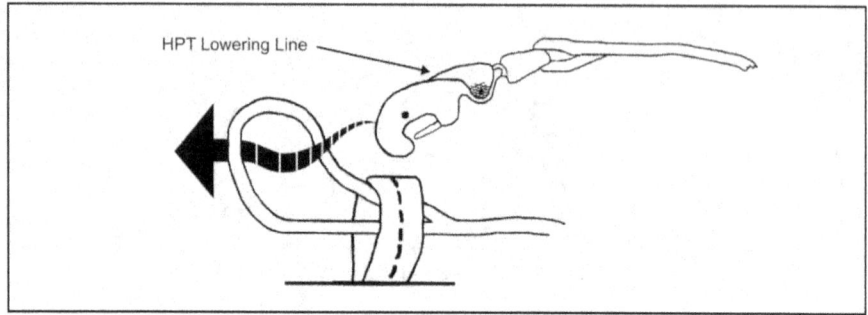

Figure 5-25. Securing the 8-foot lowering line to the cargo carrier

Mounting the SARPELS Cargo Container

5-29. The parachutist—
- Mounts the SARPELS cargo container to the large equipment attachment rings on the parachute harness using the snap hook of the D-ring attaching straps (Figure 5-26, page 5-23).
- Attaches the quick-ejector snap of the lowering line to the large equipment attachment rings on the parachute harness.
- Ensures the SARPELS cargo carrier is securely in place and is attached by the appropriate lowering line.

> **WARNING**
>
> Personal injury may occur if the wrong lowering line is used.

- Once under canopy, and IAW unit SOP, pulls the white webbing of the single-point release handle to lower the equipment load (Figure 5-27, page 5-23).

> **WARNING**
>
> Parachutists must release the SARPELS cargo carrier before landing to avoid personal injury.

Equipment and Weapon Rigging Procedures

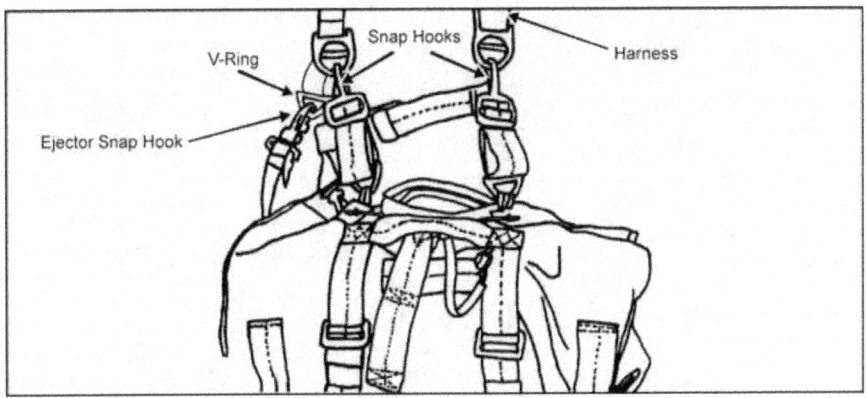

Figure 5-26. Mounted SARPELS

Figure 5-27. Single-point release handle

HARNESS, SINGLE-POINT RELEASE

5-30. The HSPR (Figure 5-28, page 5-24) is an H-type design for the parachutists. It is made of nylon webbing, has friction adapters to secure it around the load, and has two adjustable D-ring attaching straps. To stabilize the pack to the parachutist during movement in the aircraft, exit, free fall, and parachute deployment, two adjustable leg straps secure the pack to the parachutist's right and left legs. The leg straps are equipped with the male portion of the leg strap release assembly. The harness has a single-point release assembly that simultaneously releases the load and leg straps from the parachutist and parachute harness.

Chapter 5

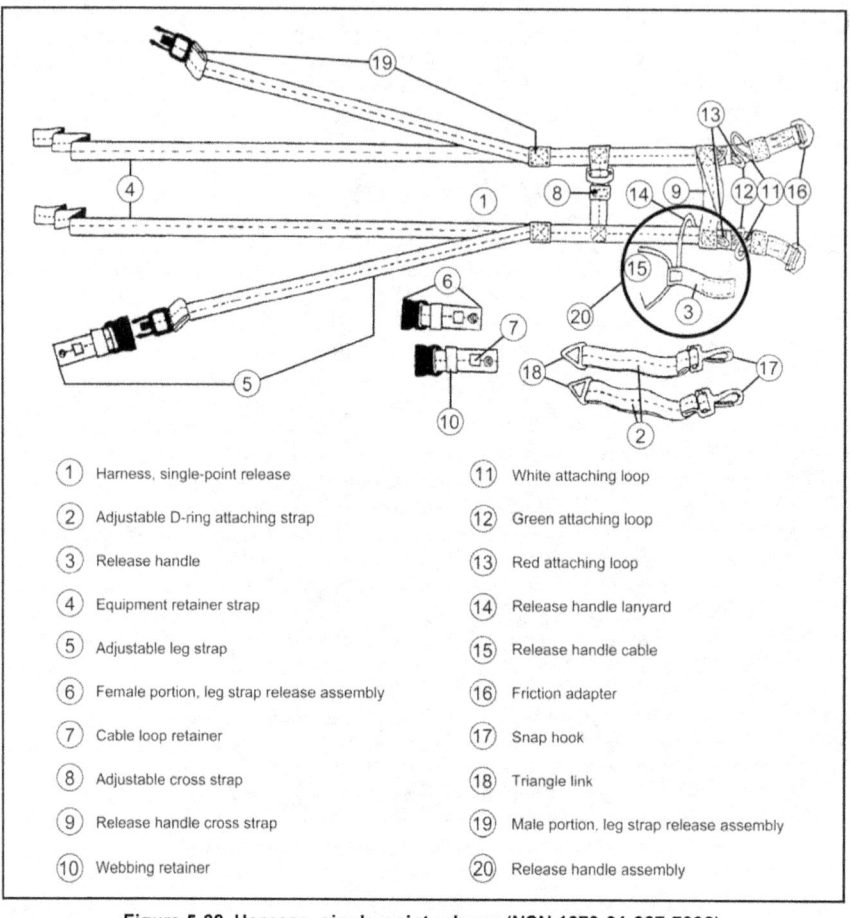

Figure 5-28. Harness, single-point release (NSN 1670-01-227-7992)

RIGGING THE ALL-PURPOSE, LIGHTWEIGHT, INDIVIDUAL, CARRYING EQUIPMENT PACK WITH THE HARNESS, SINGLE-POINT RELEASE

5-31. Before attaching the HSPR to the all-purpose, lightweight, individual, carrying equipment (ALICE) pack and Service-authorized combat pack, the release handle and adjustable D-ring attaching straps are attached to the HSPR (Figure 5-29, page 5-25).

Packing Procedures

5-32. When packing the combat container, the parachutist conforms to the following procedures:
- Pad and, if required, waterproof any fragile or sensitive gear, such as communications equipment. Place these items toward the rear of the container, locating the gear closest to the lowering line attachment point.

Equipment and Weapon Rigging Procedures

- Continue waterproofing and packing the personnel load IAW team or platoon SOPs.
- Place equipment in combat container and padding between the load and the portion of the container that will make contact with the ground first.
- Fill outside pockets with nonfragile items and tape snaps to prevent opening during free fall. If using the HSPR, full pockets help to keep harness in position.
- Close combat container by engaging drawstrings and tie-down straps.
- Roll and secure any excess webbing or drawstrings.
- Route the running ends of the waist straps behind the frame and secure them by tying or taping.
- If conducting a water jump, dip test equipment to assure positive buoyancy. If using the PDB, dip test bag again after it is completely rigged.

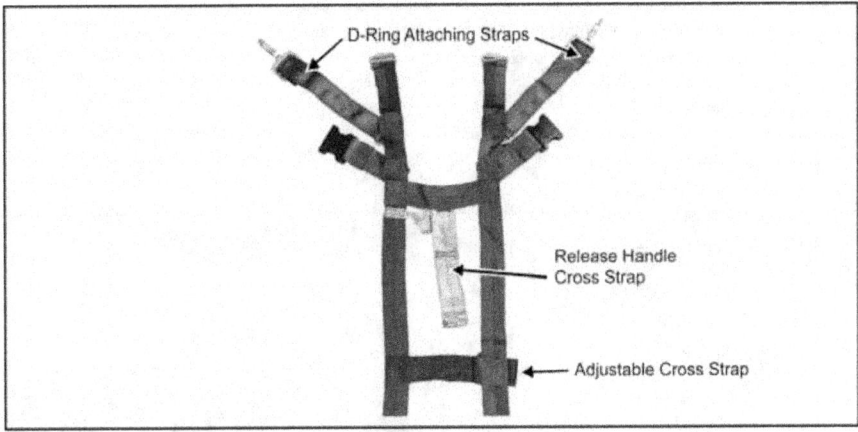

Figure 5-29. Release handle and D-ring attaching straps

Harness, Single-Point Release

5-33. The HSPR is an equipment attachment sling designed to be used as an alternative to the PDB. It is attached directly to the equipment being jumped. The parachutist pulls the single-release handle to lower equipment to the end of an 8- or 15-foot lowering line.

5-34. The HSPR (Figure 5-28, page 5-24) is an H-type design for MFF parachutists. The harness is made of nylon webbing, has friction adapters to secure it around the load, and has two adjustable D-ring attaching straps.

5-35. Two adjustable leg straps secure the equipment to the parachutist's right and left legs. This stabilizes the equipment to the parachutist during movement in the aircraft, exit, free fall, and parachute deployment. The leg straps are equipped with the male portion of the leg strap release assembly.

5-36. The harness has a single-point release assembly that simultaneously releases the load and leg straps from the parachutist and parachute harness.

Chapter 5

Rigging Procedures

5-37. The parachutist performs the steps discussed below to rig the HSPR to the ALICE pack. The parachutist—

- Routes the two release handle cables between the two plies of the release handle cross strap.
- Attaches the pile tape of the release handle to the hook tape attaching tab located between the plies of the release handle cross strap. He ensures that the release handle lanyard is not misrouted.
- Places the triangle links of the adjustable D-ring attaching straps on top of the white attaching loops.
- Routes the white attaching loop up through the triangle link (Figure 5-30A).
- Routes the green attaching loop up through the white attaching loop (Figure 5-30A).
- Routes the red attaching loop up through the green attaching loop (Figure 5-30A).
- Routes the red attaching loop through the grommet on the female portion of the leg strap release assembly. He ensures that the cable loop retainer on the female portion of the leg strap release assembly is facing up (Figure 5-30A).
- Routes the release handle cable through the red attaching loop and then through the cable loop retainer. He repeats the process for the other strap (Figure 5-30B).
- Turns the harness over so that the adjustable D-ring attaching straps are on the bottom.

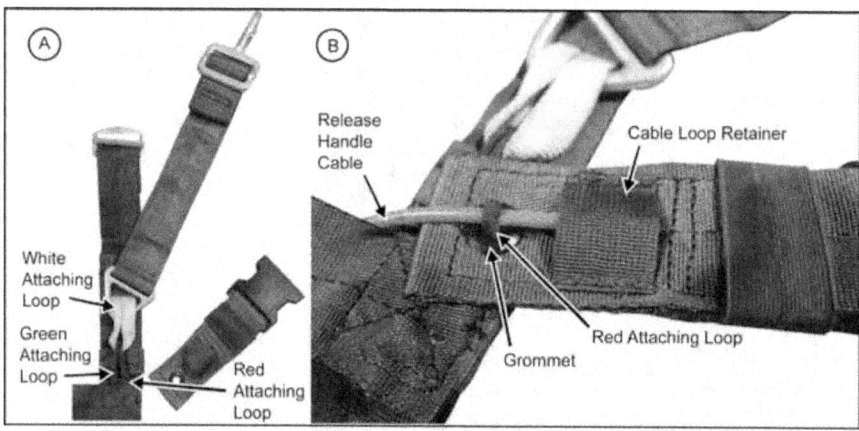

Figure 5-30. Attaching snap hooks and leg strap release assembly

- Places the ALICE pack on top of the harness so that the middle outer cargo pocket is placed between the release handle cross strap and the adjustable cross strap (Figure 5-31A, page 5-27).
- Ensures the top of the pack is facing the equipment retainer straps (Figure 5-31A).
- Routes the equipment retainer straps underneath the top of the frame, crosses them on the back of the pack to form an X, and then routes them underneath the frame and the backrest of the pack (Figure 5-31B).

Equipment and Weapon Rigging Procedures

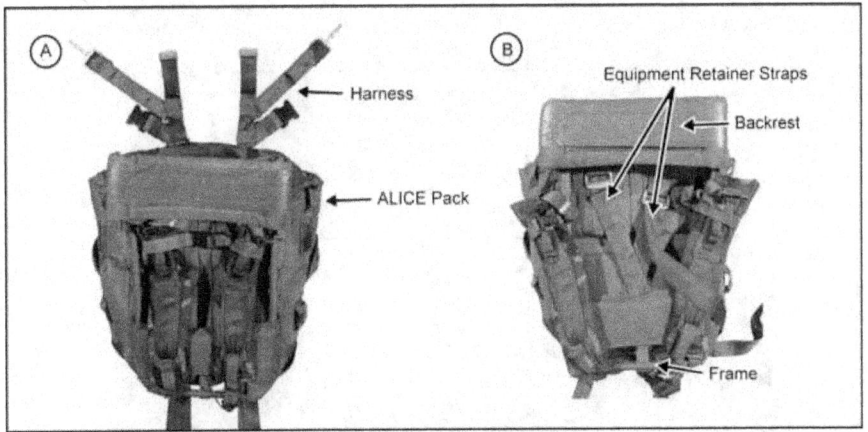

Figure 5-31. Rigging the HSPR

- Routes the equipment retainer straps through their appropriate friction adapters (a two- or three-finger quick release is optional; if used, the quick-release loop is secured to the harness with tape or a retainer band) (Figure 5-32A).
- S-rolls the excess webbing and secures it with retainer bands or tape (separates from the quick-release loop, if used) (Figure 5-32A).
- Tightens the shoulder straps.
- Routes the adjustable leg straps around the pack and attaches the male portion of the leg strap release assembly to the female portion of the leg strap release assembly, leaving it connected until it is time to attach the combat pack to the parachutist (Figure 5-32B). The HSPR leg strap release (male portion) may be routed through the pack, between the frame and pack, on shorter parachutists to allow tighter attachment of the rucksack.

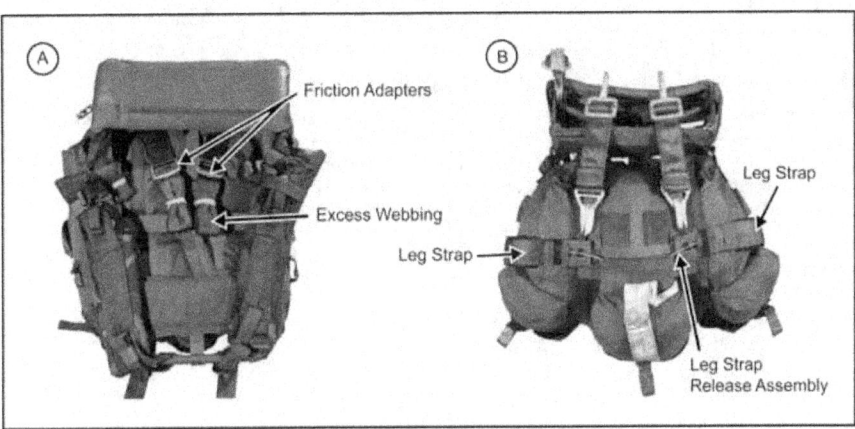

Figure 5-32. Completing rigging the HSPR

Chapter 5

Note: Oscillation under canopy is dramatically increased when using the 15-foot HPT lowering line.

Hook-Pile Tape Lowering Line Assembly

5-38. The parachutist attaches the HPT lowering line in the same way as with the modified H-harness for a front-mounted combat pack (Figure 5-33). The 8-foot HPT lowering line is normally used for MFF operations. Terrain considerations may require use of a 15-foot HPT and is authorized.

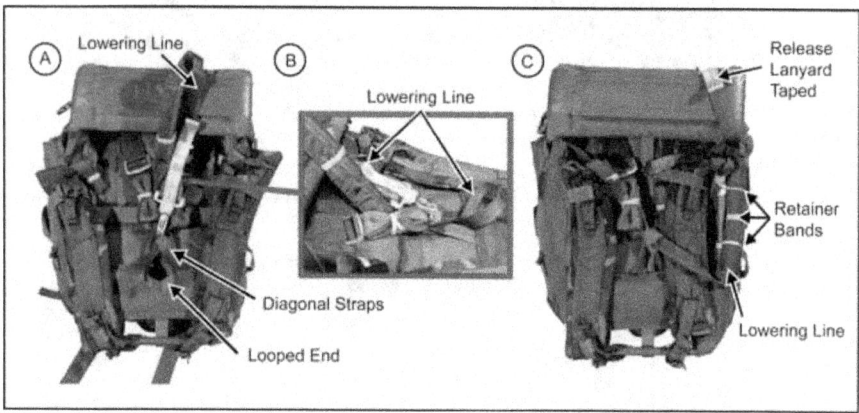

Figure 5-33. Attaching the hook-pile tape lowering line assembly

5-39. The current HPT lowering line assembly consists of an 8- or 15-foot lowering line made of 1-inch wide tubular nylon. The 8-foot lowering line is recommended for most equipment. A 9- by 7-inch nylon duck retainer is sewn to the upper end. The closing flaps have HPT sewn to the edges with a metal ejector snap on a yellow safety release.

ATTACHING THE HSPR-RIGGED COMBAT EQUIPMENT TO THE PARACHUTIST

5-40. The buddy system, or the seat of the aircraft, should be used to attach the HSPR to the parachutist.

Note: Parachutists can use either the shoulder straps or the leg straps to secure the combat equipment to the legs when using the HSPR. The same sequence is followed for attaching the front-mounted combat pack using the HPT.

The parachutist performs the steps discussed below to attach the HSPR-rigged combat equipment (Figure 5-34, page 5-29):

- The parachutist loosens shoulder straps and steps through them, if used in this configuration and not the HSPR leg straps. If not, continue to next step.
- Parachutist #1 grasps the harness by the two adjustable D-ring attaching straps and secures the snap hooks to the large equipment attachment rings directly below the three-ring release assemblies.
- Parachutist #1 attaches the quick-ejector snap on the HPT lowering line to the right-side lowering line attachment V-ring on the parachute harness.

Equipment and Weapon Rigging Procedures

- Parachutist #2 (if using the adjustable leg straps) routes the adjustable leg straps around the parachutist #1's legs and attaches the male portion to the female portion of the leg strap release assembly. If using the shoulder straps, this step is skipped.
- Parachutist #1 pulls on the free-running ends of the adjustable D-ring attaching straps and tightens the pack up to the large equipment attachment rings.
- Parachutist #1 folds the excess webbing and secures it in the webbing retainer.

Figure 5-35 shows the completed HSPR-rigging of the combat pack.

Figure 5-34. Attaching the HSPR-rigged combat equipment

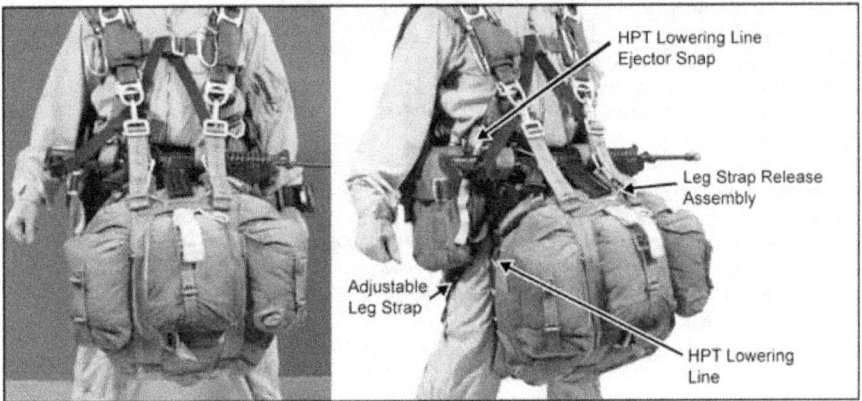

Figure 5-35. Parachutist with HSPR-rigged combat pack

PARACHUTIST DROP BAG

5-41. The PDB is a fast, easy, and secure way of carrying the parachutist's rucksack and load-bearing equipment (LBE) in free-fall or static-line operations. The bag opens and closes quickly so that the equipment can be secured efficiently on the DZ. There are exterior pockets for water and maps so that the parachutist does not have to get into his rucksack on the aircraft. There is an integral 8-foot lowering line attached to the bag. The bag is reversible with shoulder straps on both sides. The side with the hardware for

Chapter 5

dropping is camouflage in color, allowing the parachutist to put his parachute into it on the DZ for a hasty cache. The other side is dark gray, which presents a visually lower profile so that equipment can be carried through an airport. The standard size of the PDB is medium (Figure 5-36); this size will allow most parachutists to put a mission combat pack and LBE in the bag. The smallest bag possible should be used so the straps can compress the load to prevent the contents of the PDB from shifting.

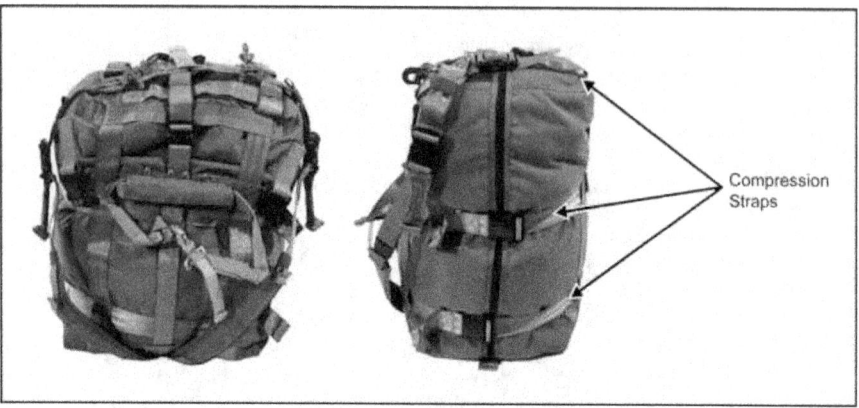

Figure 5-36. Compression straps connected and tightened

LOADING THE DROP BAG

5-42. The parachutist opens the bag completely, forming an "open clamshell." He places the rucksack and LBE on the open bag (Figure 5-37, page 5-31). The hip pad of the rucksack should be against the top of the side facing the parachutist (as the bag hangs on the harness). The parachutist then zips the bag shut and connects and tightens the compression straps (Figure 5-38, page 5-31).

5-43. After loading bag and securing any excess webbing, the parachutist girth hitches the lowering line to the attaching point and stows it in the pouch on the outside of the bag. After stowing is completed, he sets up the quick-release assembly using the same procedures as on the HSPR. The parachutist then does the following:

- Stows and mates the Velcro on the release handle with the cables facing toward the white loops.
- Ensures the release handle lanyard is not misrouted.
- Threads white attaching loop through the triangle link.
- Threads green loop through white loop and red loop through green loop.
- Threads red loop through the grommet on the female portion of the leg strap release assembly with the cable loop retainer facing up.
- Threads release cable through red loop and into the cable loop retainer.
- Repeats the same process on the other side.

Equipment and Weapon Rigging Procedures

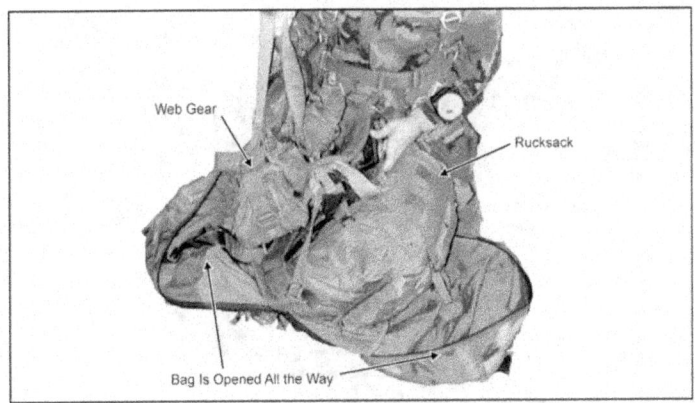

Figure 5-37. Loading the drop bag

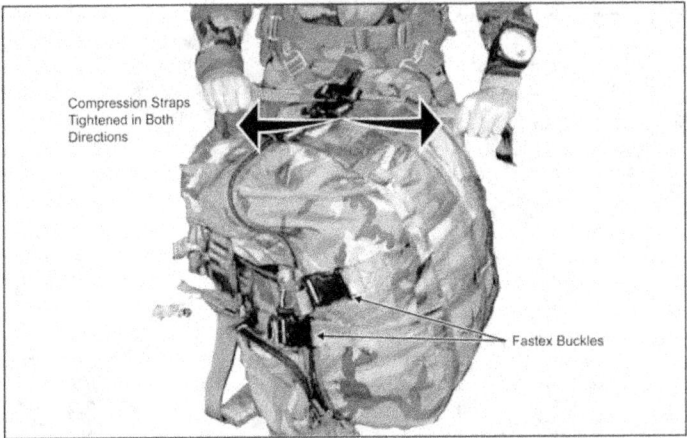

Figure 5-38. Drop bag zipped shut with compression straps connected and tightened

ATTACHING THE DROP BAG

5-44. The PDB can be attached to the front or rear of the parachutist. The PDB attaches to the parachutist by standard quick-release connectors (Figure 5-39, page 5-32) found in any rigger facility. The equipment attachment straps are long enough for the parachutist to connect the bag to his parachute harness's upper large equipment rings while the bag rests on the floor. The bag should be up against the bottom of the container, allowing approximately four inches between the bottom of the container and the top of the PDB when jumping the bag in the rear (Figure 5-40, page 5-32), and as close as possible to the equipment attachment rings when jumping the bag in front (Figure 5-41, page 5-33). The excess webbing on the attachment straps should be stowed in the elastic bands on the strap itself prior to jumping. The integral lowering line is identical to that already used by parachutists. It attaches in the same manner to the lower attachment point V-ring (Figure 5-41A). The integral lowering line may also be girth hitched to a Stubi-85 (locking carabiner) and the equipment lowering line V-ring to allow for quick derigging on the ground.

Chapter 5

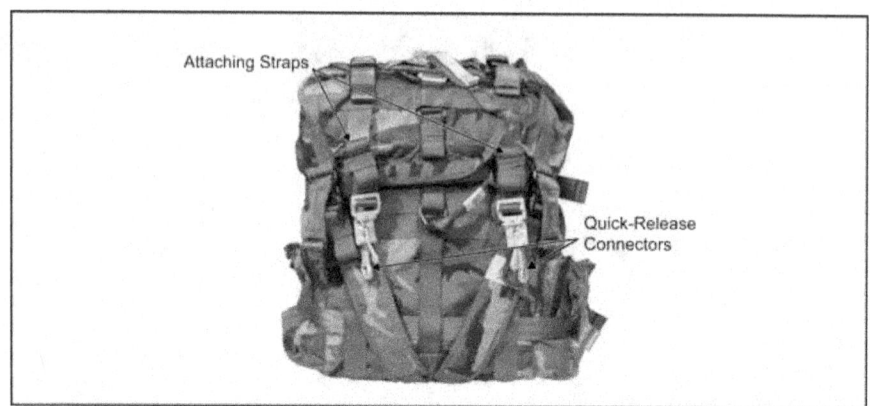

Figure 5-39. Drop bag attaches to the parachutist by standard quick-release connectors

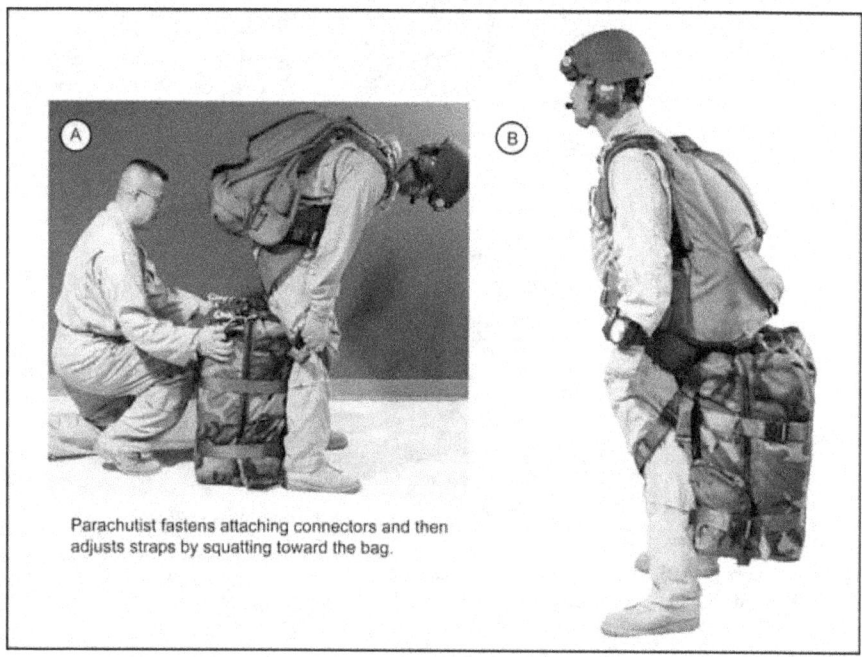

Figure 5-40. PDB rigged for rear-mounted jump

Equipment and Weapon Rigging Procedures

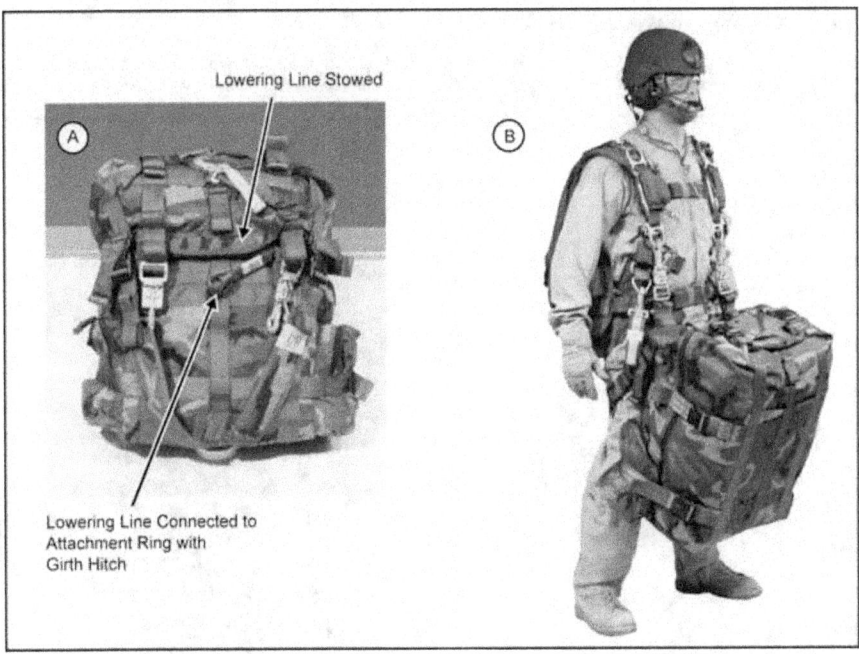

Figure 5-41. PDB rigged for front-mounted jump

JUMPING THE DROP BAG

5-45. The bag is jumped in an identical manner as the standard rucksack. The shoulder straps (used as leg straps while jumping) should be tightened around the thighs, but not so tight as to restrict movement. Once under canopy, the pull-tabs on the shoulder straps (leg straps) can be pulled to loosen the straps from around the legs.

DERIGGING THE DROP BAG

5-46. Once on the ground, the parachutist detaches the lowering line from the parachute. He then unbuckles all of the Fastex buckles securing the compression straps around the bag. The zipper closing the bag can be ripped open by pulling apart both sides of the bag, exposing the load. The parachutist removes the load and puts the parachute into the bag for storage or a hasty cache.

WEAPON-RIGGING PROCEDURES

5-47. An MFF parachutist can jump with his individual weapon exposed or inside a weapons container or another approved container. Weapons may be attached to the left or right side of the parachutist. When jumping with oxygen, the weapon should be rigged on the left side. If jumping with multiple weapons, the larger weapon should be attached to the left side or rigged horizontally on top of the combat pack with smaller weapon attached to the right side of the parachutist (Figure 5-42, page 5-34). The parachutist can jump with a pistol in a shoulder holster or in an equipment container. The parachutist should wear a shoulder holster under the jumpsuit or other protective clothing. The parachutist should secure the pistol in the holster by taping the holster closed using an airborne strap or by using a lanyard that will not interfere with the jumper or parachute system.

Chapter 5

CENTER-MOUNTED WEAPONS HARNESS

5-48. The center-mounted weapons harness (CMWH) (Figure 5-42) is in response to USASOC's need for a weapons harness that allows the parachutist ease of donning and doffing during MFF operations.

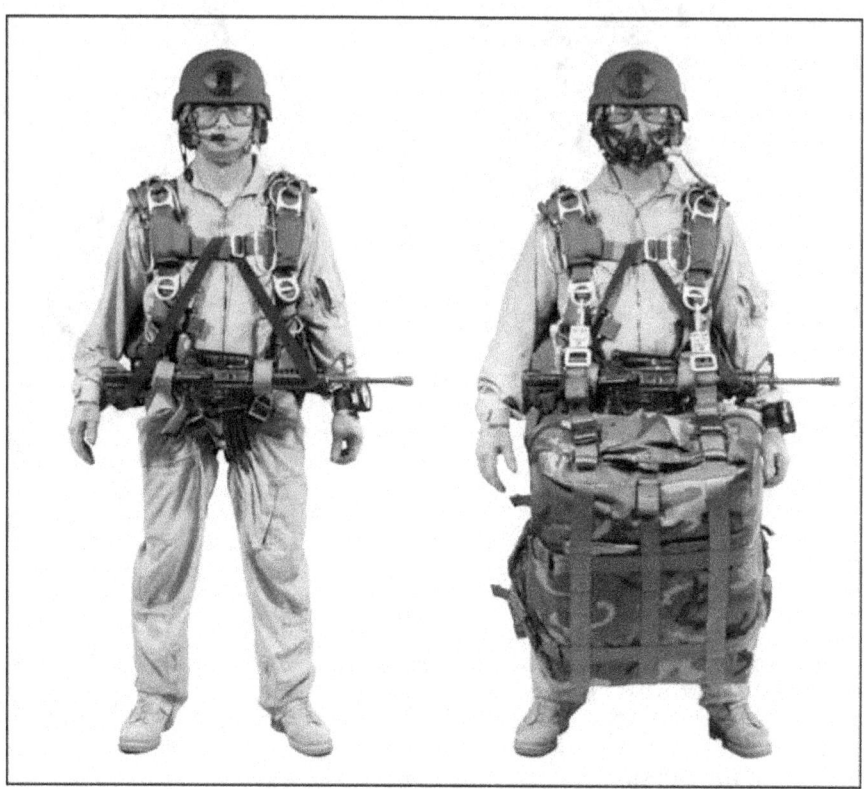

Figure 5-42. Center-mounted weapons harness

5-49. Over 150 MFF operations have been conducted using the CMWH at the MFF School and during user assessments as part of the Arc Angel MFF Training. The CMWH is manufactured using the following components (Figure 5-43, page 5-35):

- *Weapons harness* uses 1-3/4-inch nylon tape backing with pile sewed the length and two, non-nylon-based male end adapters. The additional straps on the weapons harness use 1-3/4-inch nylon tape with either pile or hook sewn to the center of the weapons harness. This forms a triple fold of hook and pile to secure weapons when worn in the vertical configuration. Length of the weapons harness is 24 inches.
- *Main lift web attaching points* use three sections of hook and pile sewn in a triple-fold configuration with 1-3/4-inch nylon tape as a backing with two female non-nylon-based adapters. Length of main lift web adapters, including female adapters, is 10 inches.

Equipment and Weapon Rigging Procedures

- *Horizontal attaching straps* use three sections of hook and pile sewn in a triple-fold configuration with 1-3/4-inch nylon tape as a backing. Length of horizontal attaching straps is approximately 20 inches.

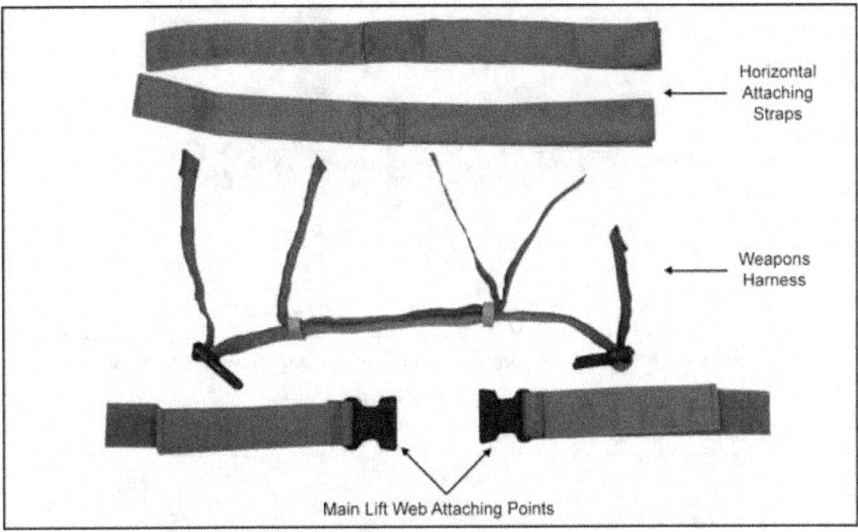

Figure 5-43. Center-mounted weapons harness components

5-50. To use the CMWH, the parachutist—
- Attaches main lift web attaching points to the main lift web of the MC-4 harness below the lowering line attachment point with the attachment buckle outboard (Figure 5-44).

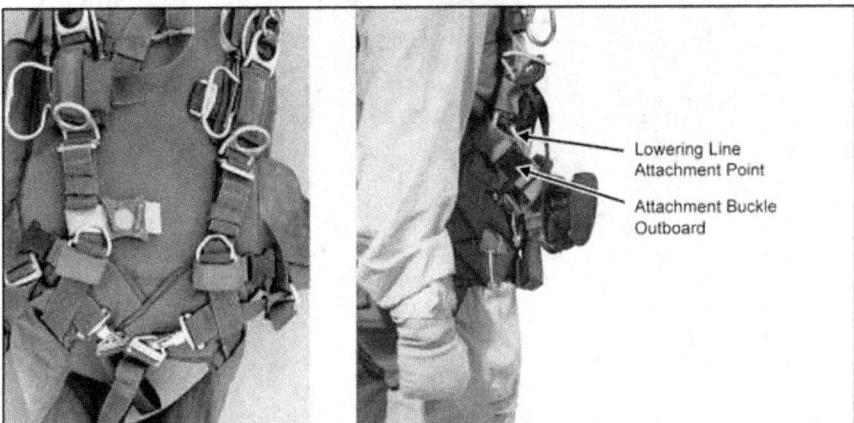

Figure 5-44. Main lift web attaching points

Chapter 5

- For horizontal configuration, attaches horizontal straps to the pile portion of the weapons harness as close as possible to the male portion of the Fastex buckles (Figure 5-45).

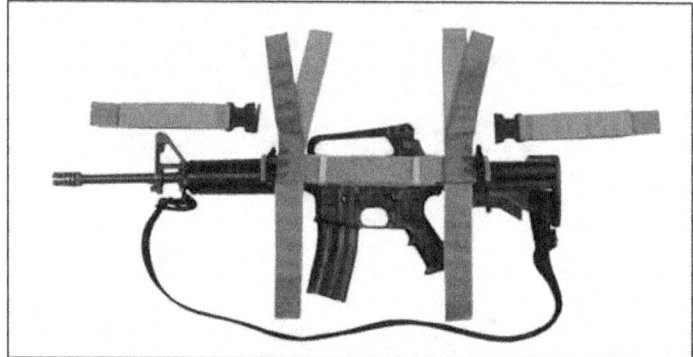

Figure 5-45. Attaching horizontal straps to the pile portion weapons harness

- Uses triple-fold hook and pile to secure weapon harness to weapon (Figure 5-46 and Figure 5-47, page 5-37).

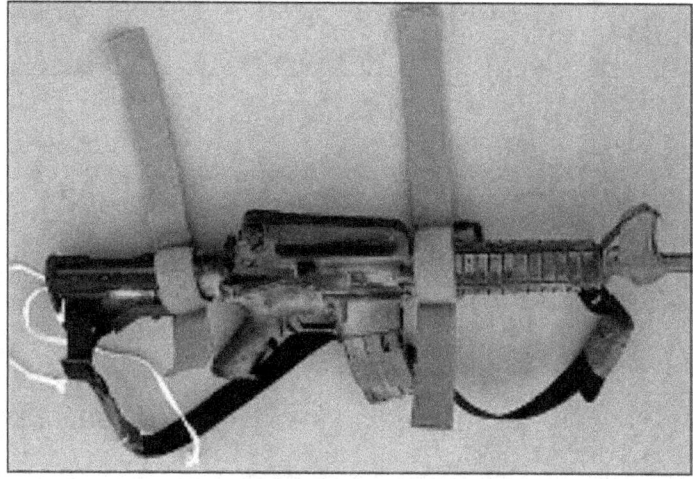

Figure 5-46. Securing triple-fold hook and pile

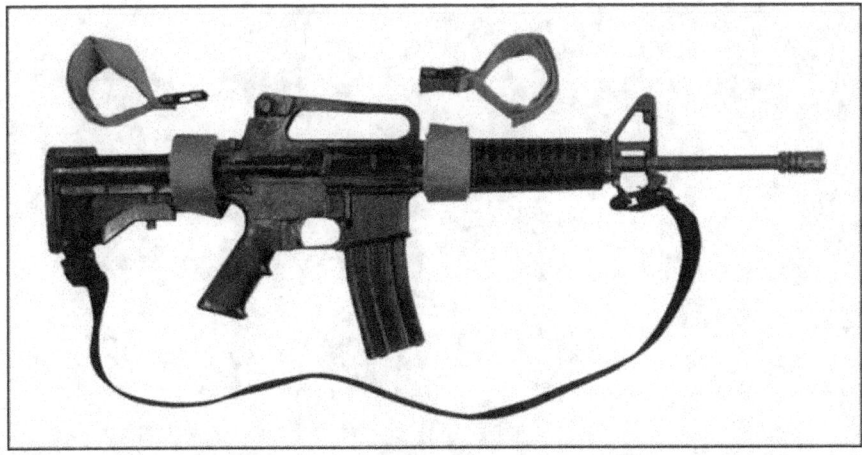

Figure 5-47. Securing weapon harness to weapon

- Attaches weapon to main lift web attaching points via the Fastex buckle (Figure 5-48) and routes sling (Figure 5-49, page 5-38) over chest strap to secure any excess sling.

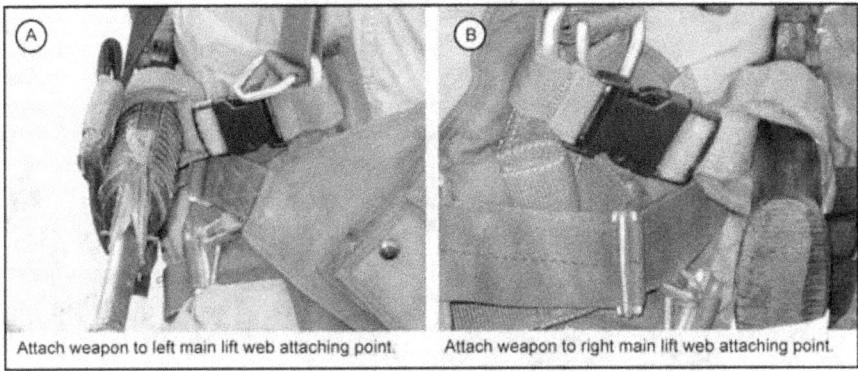

Figure 5-48. Attaching weapon to main lift web attaching points

Chapter 5

Figure 5-49. Sling over chest strap

5-51. When the parachutist jumps the PDB or ALICE pack, the attaching straps (Figure 5-50) are routed over the weapon for additional security.

Figure 5-50. Attaching straps routed over weapon

Equipment and Weapon Rigging Procedures

EXPOSED WEAPONS CONSIDERATIONS

5-52. If the commander decides that parachutists are to jump with weapons exposed, he must consider the increased risk of injury to the parachutists. To minimize the risks of jumping with exposed weapons, the commander should—

- Consider the proficiency and experience level of the parachutists.
- Conduct a thorough risk assessment that addresses the following risks associated with jumping exposed weapons:
    - Interference with the oxygen system or automatic opening device.
    - Interference with the parachutist's exit from the aircraft.
    - Stability of the parachutist while in free fall.
    - Ability of the parachutist to perform pull procedures.
    - Ability of the parachutist to perform emergency procedures.
    - Deployment of the parachute.
    - Entanglement of the weapon with another parachutist's parachute should a midair entanglement occur.
    - Ability of the parachutist to perform a parachute landing fall (PLF).
    - Injury to the parachutist during landing.
    - Damage to the weapon upon landing or when dragged on the ground.

M4 CARBINE-SERIES RIFLE

5-53. To prepare the M4 carbine-series rifle for jumping (Figure 5-51, page 5-40), the parachutist should—

- Adjust the sling to fit just over the shoulder and tape the sling keeper in place.
- Pad and tape the side-mounted bolt assist and the operating handle.
- Pad and tape the muzzle and the sights to avoid possible entanglement with the parachute suspension lines or dirt clogging the weapon upon landing.
- Insert the magazine and tape it to the receiver, including the ejector port cover, to prevent loss of the magazine and to keep debris from entering the bolt area.
- Tape the hand guards to prevent their loss during free fall or upon landing.
- Tape any accessories on the weapon, such as aim points, to ensure they do not come off during movement.

Note: A padded sling should not be used as it may interfere with emergency procedures when rigged on the parachutist.

Note: The M16-series rifle is rigged in the same manner as the M4 carbine-series rifle.

CAUTION

Any tape should have quick-release tabs and should not interfere with the weapon's operation in case weapon is needed immediately once on the DZ.

Chapter 5

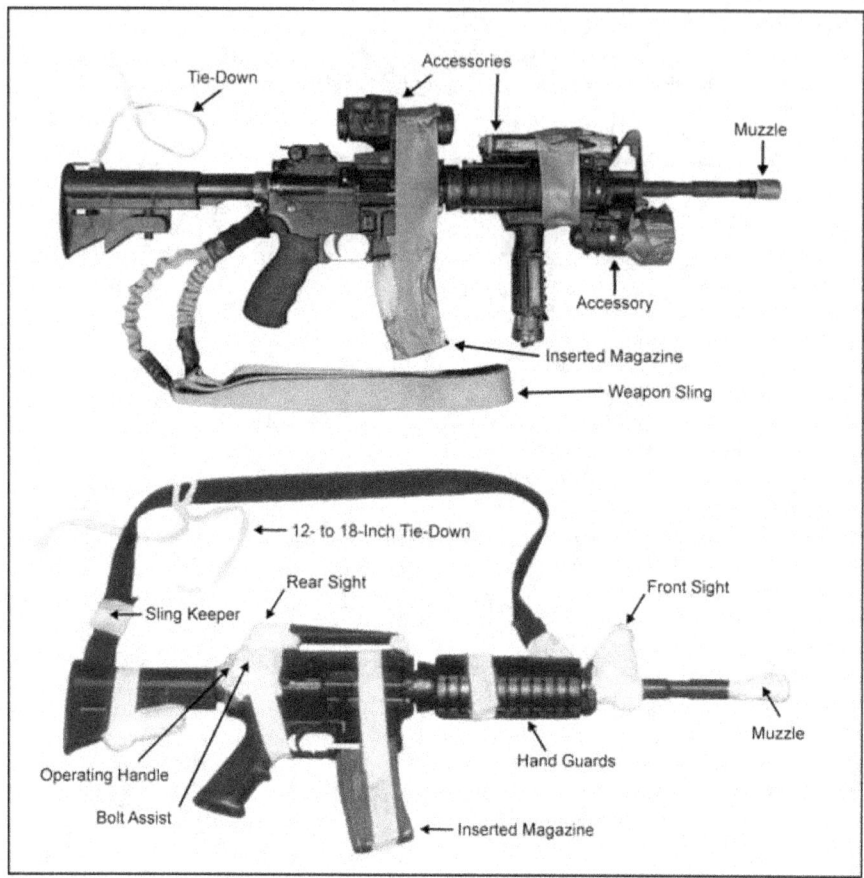

Figure 5-51. M4 carbine-series rifles rigged for jumping

Tie-Downs

5-54. The parachutist should use a 12- to 18-inch tie-down of 1/4-inch cotton webbing to secure the weapon. He should attach the tie-down to the weapon sling or to a hard point on the weapon with a girth hitch knot.

Positioning

5-55. With the help of a buddy, the parachutist should sling his weapon over his shoulder with the muzzle down, and rotate the pistol grip to his rear (Figure 5-52, page 5-41). The parachutist and his buddy should then—

- Place the sling from the lower keeper (butt stock) on the outside of the stock and over the parachutist's shoulder.

Equipment and Weapon Rigging Procedures

- Run the sling under the main lift web and route the chest strap through the sling. The buddy ties off the running ends of the 1/4-inch cotton webbing to a weapon tie-down loop on the harness with a soft knot (bowknot).
- Place the weapon between the wing flap and the parachutist with the waistband routed over or through the weapon-carrying handle.

Note: If optics are mounted on the weapon, they must be free and clear of the waistband.

- Tighten the waistband securely so that the weapon lies snugly against the parachutist's side.

The parachutist then assumes the basic free-fall position to test the fit of the weapon.

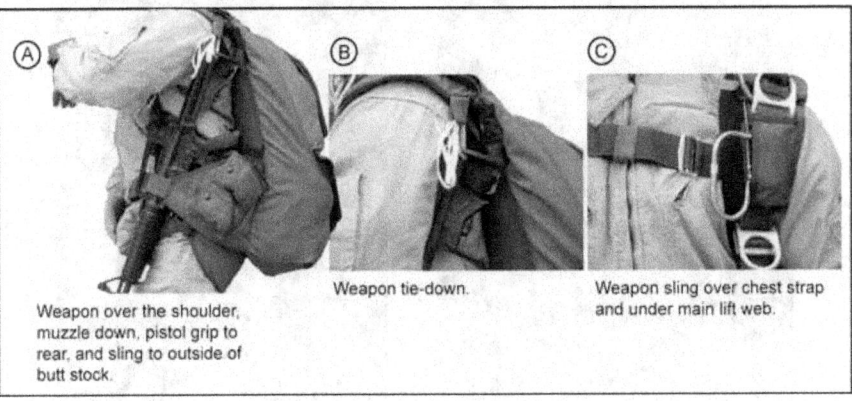

Figure 5-52. Positioning the weapon on the parachutist

SPECIAL OPERATIONS FORCES COMBAT ASSAULT RIFLE

5-56. The Special Operations Forces Combat Assault Rifle (SCAR), SCAR–L, or SCAR–H, can be rigged in the side-mounted configuration as well as in the horizontal position in the CMWH. To prepare the SCAR-L/H rifles for jumping, the parachutist should—

- Fully extend the weapon sling all the way and tape the sling keeper in place.
- Pad and tape the rear sight post and windage adjustment knob prior to conducting MFF operations to reduce the probability of inadvertent movement of the rear sight aperture. Continuous lengths of tape should cover the entire rear sight and wrap around folding-stock, forward of its hinge. The material used to pad and tape the SCAR for MFF operations will vary depending on what is locally available.
- Insert the magazine and tape it to the lower receiver, including the ejector port cover, to prevent loss of the magazine and to keep debris from entering the bolt area.
- Place the SCAR buttstock in the folded position (Figure 5-53A, page 5-42) when rigged in the side-mounted configuration to reduce contact with the risers during main parachute deployment. However, the SCAR can be jumped in the open position (Figure 5-53B).
- Tape any accessories on the weapon, such as aim points, to ensure they do not come off during movement.

Chapter 5

> **CAUTION**
>
> Any tape should have quick-release tabs and should not interfere with the weapon's operation in case weapon is needed immediately once on the DZ.

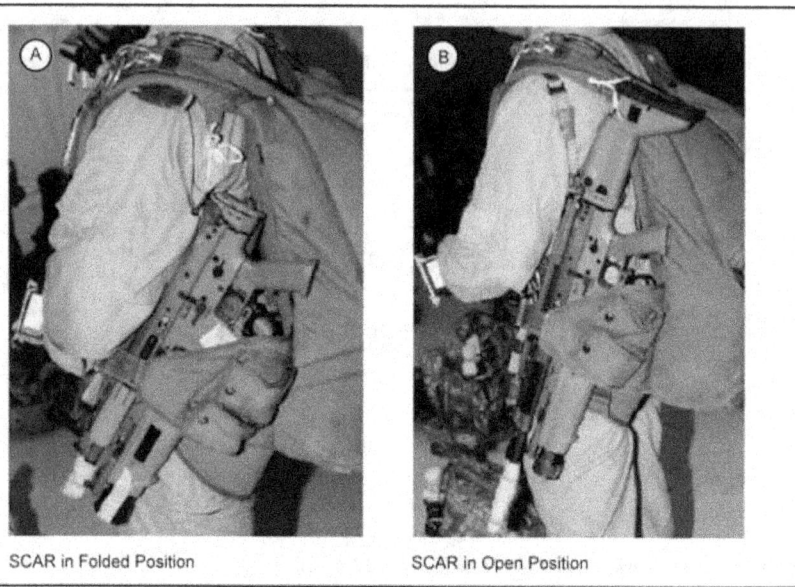

SCAR in Folded Position SCAR in Open Position

Figure 5-53. SCAR in folded and open positions

Tie-Downs

5-57. The parachutist should use a 12- to 18-inch tie-down of 1/4-inch cotton webbing to secure the SCAR. He should attach the tie-down to the weapon sling or to a hard point on the weapon with a girth hitch knot.

Positioning

5-58. With the help of a buddy, the parachutist should sling his weapon over his shoulder with the muzzle down, and rotate the pistol grip to his rear. The parachutist and his buddy should then—

- Place the sling from the lower keeper (butt stock) on the outside of the stock and over the parachutist's shoulder.
- Run the sling under the main lift web and route the chest strap through the sling. The buddy ties off the running ends of the 1/4-inch cotton webbing to a weapon tie-down loop on the harness with a soft knot (bowknot).
- Place the weapon between the wing flap and the parachutist with the waistband routed over the top of the weapon.
- Tighten the waistband securely so that the weapon lies snugly against the parachutist's side.

Equipment and Weapon Rigging Procedures

Note: If optics are mounted on the weapon, they must be free and clear of the waistband.

The parachutist then assumes the basic free-fall position to test the fit of the weapon.

Dual Rigging

5-59. When rigging weapons on both sides of the harness, the buddy should—
- Place the larger weapon on the parachutist's left side and the smaller one on his right side.

Note: This setup minimizes interference with the oxygen system, if used.

- Use standard weapon-rigging techniques to secure the weapon to the parachutist.

Note: When an oxygen system is used (Figure 5-54), the buddy should place the weapon behind the oxygen bottles and against the parachutist's body. The buddy should carefully route the medium-pressure delivery hose over or behind the weapon in a manner that does not restrict the flow of oxygen to the parachutist.

Figure 5-54. Right-side weapon rigging

M203 GRENADE LAUNCHER OR ENHANCED GRENADE LAUNCHER MODULE

5-60. The parachutist should prepare the M203 grenade launcher (Figure 5-55, page 5-44) in the same manner as he prepares the M16-series, M4, or SCAR carbine-series rifles. Additionally, he should—
- Tape the hand guards and the grenade launcher barrel together with the barrel latch covered.
- Remove the quadrant sight.
- Tape down the leaf sight.

Chapter 5

Tie-Downs

5-61. The parachutist should follow the same procedures used for the M16 series, M4, or SCAR carbine-series rifles.

Positioning

5-62. The parachutist and his buddy should follow the same procedures used for the M16 series, M4, or SCAR carbine-series rifles.

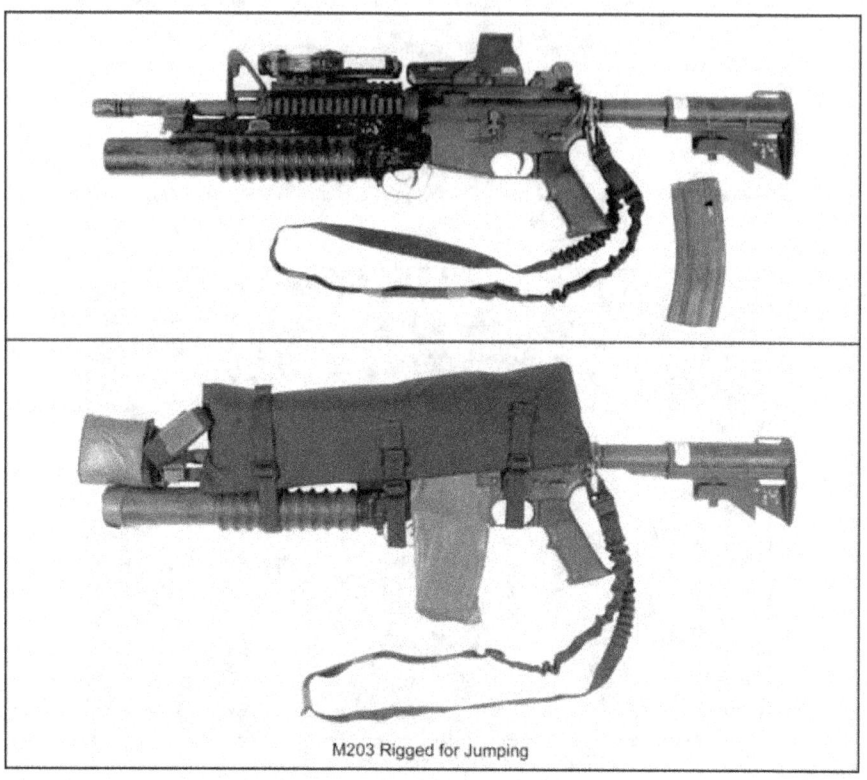

Figure 5-55. M203 rigged for jumping

M14, G3, AND FN FAL RIFLES

5-63. The parachutist should prepare the M14 (Figure 5-56, page 5-45), the G3, and the FN FAL rifles as follows:
- Remove the sling from the weapon and the sling keeper from the sling.
- Form a loop by running the sling through the sling hook.
- Replace the sling by placing the loop around the small of the stock.
- Replace the sling keeper and secure the sling to the barrel, just below the front sight, with a half hitch.

Equipment and Weapon Rigging Procedures

- Tape the butt plate closed.
- Pad and tape the flash suppressor, front sight, and bayonet lug.

Note: When jumping with larger weapons on the left side, the parachutist should position the HPT lowering line to the left side to facilitate a right-side PLF.

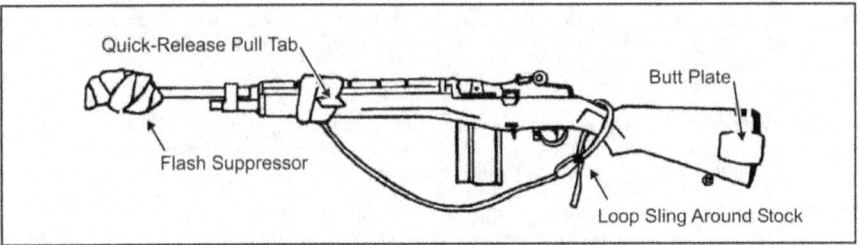

Figure 5-56. M14 rigged for jumping

Tie-Downs

5-64. The parachutist should use a 12- to 18-inch tie-down of 1/4-inch cotton webbing to secure the weapon. He should attach the tie-down to the weapon sling or to a hard point on the weapon with a girth hitch.

Positioning

5-65. With the help of a buddy, the parachutist should sling the weapon over his left shoulder, with the muzzle down, and rotate the operating handle away from his body. They then should secure the weapon to the parachutist in the same manner as for the M16-series and the M4 carbine-series rifles.

M110 SEMI-AUTOMATIC SNIPER SYSTEM AND M24 SNIPER RIFLE

5-66. The parachutist should prepare the sniper system as follows:
- Make an improvised sling. (*Note*: Parachutist must not use a standard marksmanship sling.)
- Make sure the rifle has a secure portion where the sling can be attached.
- Tape the muzzle to protect it from debris upon landing.
- Tape and secure the bolt (as needed).
- Insert the magazine and tape it to the receiver, including the ejector port cover, to prevent loss of the magazine and to keep debris from entering the bolt area.
- Tape and pad the scope to protect it as necessary.

Figure 5-57 shows the M110 Semi-Automatic Sniper System.

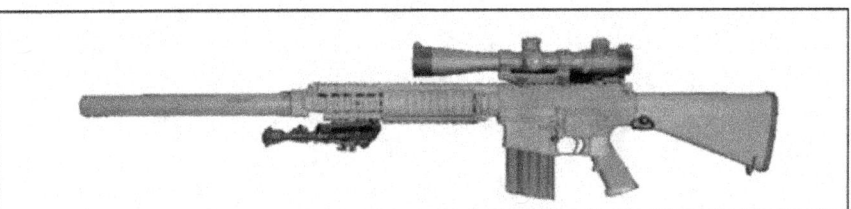

Figure 5-57. M110 Semi-Automatic Sniper System

Chapter 5

Note: If using a semiautomatic sniper system, the parachutist should follow the same procedures he used to prepare the M16-series, the M4-carbine series, and the M14 rifles.

Tie-Downs

5-67. The parachutist should use a 12- to 18-inch tie-down of 1/4-inch cotton webbing to secure the weapon. He should attach the tie-down to the weapon sling or to a hard point on the weapon with a girth hitch.

Positioning

5-68. With the help of a buddy, the parachutist should sling the weapon over his left shoulder, with the muzzle down, and the scope to the front or rear. They then should secure the weapon to the parachutist in the same manner as for the M16-series and the M4 carbine-series rifles.

Note: When jumping with larger weapons on the left side, the parachutist should position the HPT lowering line to the left side to facilitate a right-side PLF.

Note: If optics are mounted on the weapon, they must be positioned free and clear of the waistband.

CAUTION

All optics should be pressure tested to see if they will hold up under high altitude and pressure at these levels during flight and under canopy.

MP5, MP5A3, AND MP5K SUBMACHINE GUNS

5-69. The parachutist should prepare the MP5 (Figure 5-58, page 5-47), MP5A3, and MP5K submachine guns as follows:

- Remove the sling from the upper swivel.
- Fold the end of the sling and run the fold through the upper sling swivel.
- Pass the tip of the sling through the fold and fasten the snap.
- Close the ejector port cover and remove the magazine.
- Collapse the stock.
- Tape one magazine to the left of the receiver or carry it elsewhere.
- Cover and tape the muzzle.

Equipment and Weapon Rigging Procedures

NOTE: Magazine may be taped to either side of the weapon.

Figure 5-58. MP5 rigged for jumping

Tie-Downs

5-70. The parachutist should use a 12- to 18-inch tie-down of 1/4-inch cotton webbing to secure the weapon. He should attach the tie-down to the weapon sling or a hard point on the weapon with a girth hitch.

Positioning

5-71. With the help of a buddy, the parachutist should sling the weapon over his shoulder, with the muzzle down, and rotate the pistol grip to his rear. They then should secure the weapon to the parachutist in the same manner as for the M16-series and the M4 carbine-series rifles.

M249 AND PARA M249 SQUAD AUTOMATIC WEAPONS

5-72. The parachutist can jump with the M249 squad automatic weapon exposed or in an equipment container (Figure 5-59, page 5-48). To prepare the weapon, he should—

- Pad the optics as necessary. (*Note*: Parachutists must not insert the magazine and must not chamber rounds.)
- Tape the muzzle to avoid debris entering the weapon upon landing.
- Wrap one piece of the tape around the fore grip of the weapon, securing the carrying handle, hand guard, and bipod.

Note: The Para M249 squad automatic weapon requires an additional piece of tape forward of the vertical grip on the hand guard. The parachutist should consider padding the charging handle if the possibility of discomfort or injury exists. When jumping with larger weapons on the left side, the parachutist should position the HPT lowering line to the left side to facilitate a right-side PLF.

Chapter 5

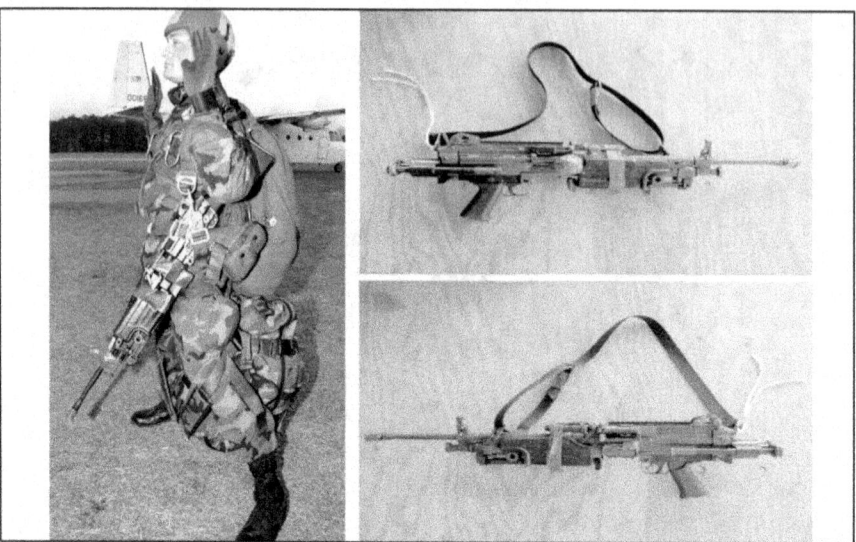

Figure 5-59. M249 and Para M249 squad automatic weapons rigged for jumping

Tie-Downs

5-73. The parachutist should use a 12- to 18-inch tie-down of 1/4-inch cotton webbing to secure the M249 squad automatic weapon. The parachutist should attach the tie-down to the weapon sling or to a hard point on the weapon with a girth hitch. On the Para M249 squad automatic weapon, he should attach the tie-down to a hard point on the rear of the weapon.

Positioning

5-74. With the help of a buddy, the parachutist should sling the weapon over his left shoulder, with the muzzle down, and rotate the pistol grip to his rear.

Note: The Para M249 may be rigged with the weapon pistol grip to the parachutist's front or rear.

They then run the weapon sling under the main lift web and route the chest strap through the weapon sling. Parachutist #2 ties off the running ends of the 1/4-inch cotton webbing to a weapon tie-down loop on the harness with a soft knot (bowknot). He places the weapon between the waistband and the parachutist with the waistband routed over or through the weapon-carrying handle. He tightens the waistband securely so that the weapon lies snugly against the parachutist's side. The parachutist then assumes the basic free-fall position to test the fit of the weapon.

Note: If optics are mounted on the weapon, they must be positioned free and clear of the waistband.

M60 AND M240 MACHINE GUNS, OTHER LIGHT MACHINE GUNS, AND .50-CALIBER SNIPER SYSTEMS

5-75. The parachutist must not jump these fully assembled and exposed weapons while they are attached to the parachute harness during MFF operations. The parachutist may break the weapons down and pack them

Equipment and Weapon Rigging Procedures

inside the combat pack, PDB, or a horizontally mounted kit bag with an H-harness. The parachutist pads optics and secures weapon in the PDB or jump pack. Figure 5-60 is an example of padding and preparing the M240G.

Figure 5-60. M240G disassembled and packed for jumping

PISTOLS

5-76. The parachutist can jump with a pistol in a shoulder holster or in an equipment container. The parachutist should wear a shoulder holster under his jumpsuit or other protective clothing. He should secure the pistol in the holster by taping the holster closed or by using a lanyard.

AT-4, 84-MM CARL GUSTAF RECOILLESS RIFLE, AND OTHER LIGHT ANTIARMOR WEAPONS

5-77. The recommended procedure for rigging the AT-4 and 84-millimeter (mm) Carl Gustaf recoilless rifle is to front-mount the weapon horizontally on top of a combat pack or a PDB, or in an H-harness. This arrangement limits the parachutist to a ramp-only exit because of the rigged width of the parachutist. To rig these weapons for a front-mounted jump on top of a combat pack or a PDB, the parachutist—

- Tapes and pads the end of the launch tube to prevent debris from entering the tube upon landing. If the launch tube has removable end caps, he tapes and secures them to the launch tube (Figure 5-61, page 5-50).
- Tapes the sling securely to the launch tube (Figure 5-61).
- Pads and tapes the weapon sights and the trigger mechanism (Figure 5-61).
- Mounts the weapon system on top of the combat pack or PDB. He routes the H-harness, spider harness, or PDB vertical compression straps over the weapon and through the carrying handle of the Carl Gustaf or through the sling of the AT-4, and tightens securely (Figure 5-62, page 5-50).
- Uses 1/4-inch cotton webbing to secure the weapon to the combat pack or PDB. This method prevents lateral shifting of the weapon system while in free fall (Figure 5-63, page 5-51).
- Has a buddy attach the combat pack or PDB to the parachutist (Figure 5-64, page 5-51).

Chapter 5

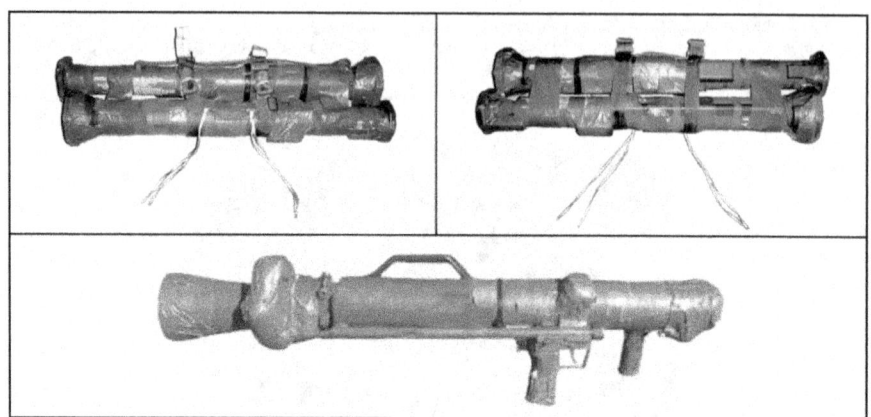

Figure 5-61. AT-4 and 84-mm Carl Gustaf rigged for jumping

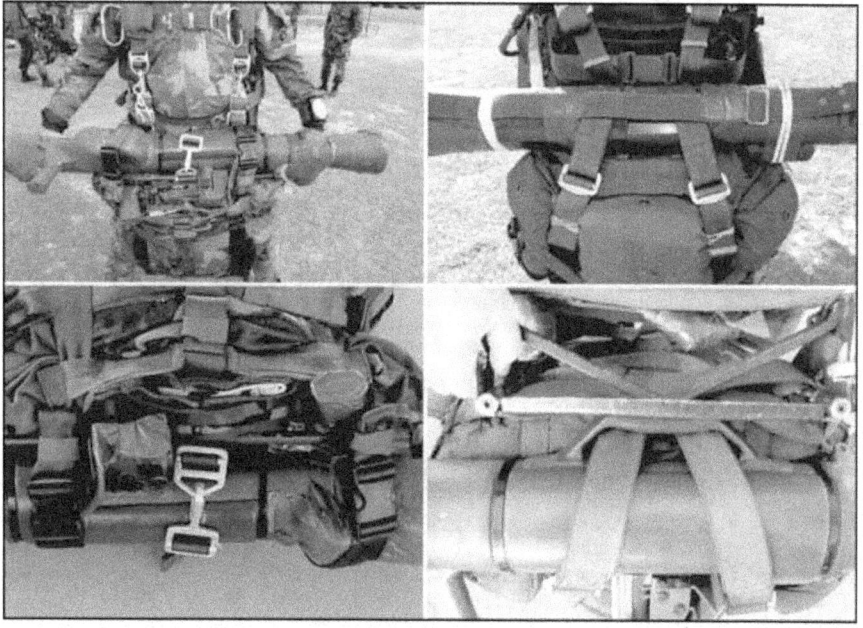

Figure 5-62. Routing of vertical compression straps

Equipment and Weapon Rigging Procedures

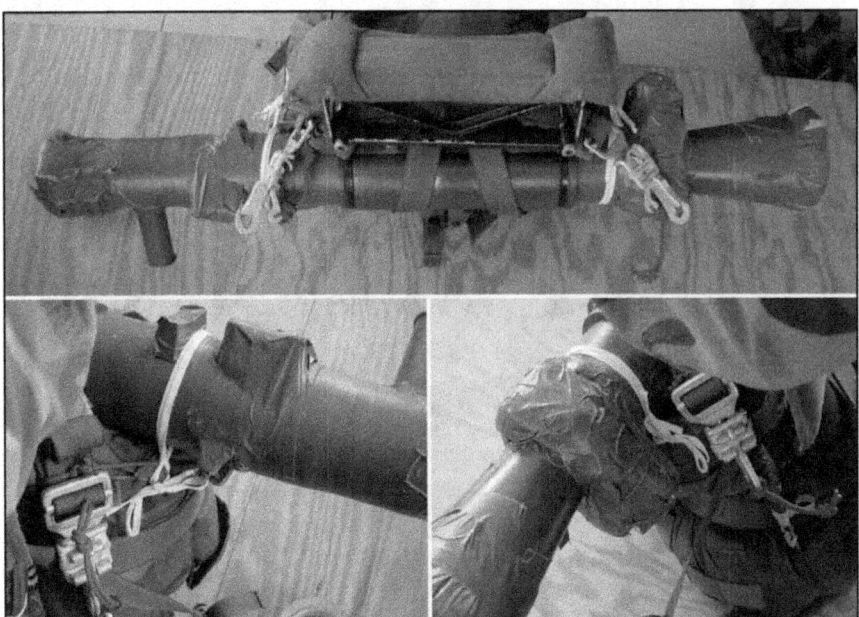

Figure 5-63. Antiarmor weapon tie-down locations

Figure 5-64. Parachutist rigged for jumping with an AT weapon mounted on top of combat pack

5-78. To rig the AT-4 or the 84-mm Carl Gustaf recoilless rifle inside an H-harness for a front-mounted jump (Figure 5-65, page 5-52), the parachutist—
- Tapes and pads the end of the launch tube to prevent debris from entering the tube upon landing. If the launch tube has removable end caps, he tapes and secures them to the launch tube.
- Tapes the sling securely to the launch tube.
- Pads and tapes the weapon sights and the trigger mechanism.

Chapter 5

- Places the weapon system on top of the H-harness, then routes the vertical compression straps around the launch tube and through the sling or carrying handle, and tightens securely.
- Has a buddy attach the H-harness or weapon system to the parachutist as follows:
  - If jumping a rear-mounted combat pack or PDB, uses a 12- to 18-inch piece of 1/4-inch cotton webbing to secure the H-harness or weapon to the leg straps. This method prevents the weapon from flying up into the parachutist's face during free fall (Figure 5-65).
  - If jumping a front-mounted combat pack or PDB, attaches the H-harness or weapon to the equipment attachment rings first, and then attaches the combat pack or PDB. The attachment straps of the combat pack or container hold the H-harness or weapon system securely in place (Figure 5-66).

Figure 5-65. Front-mounted weapon with rear-mounted rucksack

Figure 5-66. Front-mounted weapon with front-mounted rucksack

Equipment and Weapon Rigging Procedures

M224 60-MM MORTAR

5-79. The recommended procedure for rigging the M224 60-mm mortar is to mount it horizontally on a combat pack or PDB in an H-harness. This arrangement limits the parachutist to a ramp-only exit from the aircraft. The parachutist should rig the M224 60-mm mortar for a front-mounted jump (Figure 5-67) as follows:

- Tape and pad the end of the barrel to prevent debris from entering the barrel upon landing.
- Tape the entire trigger guard and carrying handle, making a smooth surface to prevent any entanglement of the parachute during deployment.
- Mount the weapon on top of the combat pack or container and route the H-harness, spider harness, or container vertical compression straps over the weapon and through the improvised sling.

Figure 5-67. M224 60-mm mortar rigged for front mount

5-80. The parachutist should rig the M224 60-mm mortar (Figure 5-68, page 5-54) for a side-mounted jump as follows:

- Make an improvised sling with 1-inch tubular webbing.
- Run the webbing around the end of the mortar and through the trigger guard.

Note: A secured clove hitch on the mortar barrel between the lower sections for the bipod mount holds the webbing in place.

- Tape the webbing in place to make sure it does not slide on the barrel.
- Tape the entire trigger guard and carrying handle, making a smooth surface to prevent any entanglement of the parachute during deployment.
- Tape and pad the end of the barrel to prevent debris from entering the barrel upon landing. When jumping with larger weapons on the left side, the parachutist should position the HPT lowering line to the left side to facilitate a right-side PLF.

Tie-Downs

5-81. The parachutist should use a 12- to 18-inch tie-down of 1/4-inch cotton webbing to secure the weapon. He should attach the tie-down to the weapon sling or a hard point on the weapon with a girth hitch.

Chapter 5

Positioning

5-82. With the help of a buddy, the parachutist should sling the mortar over his left shoulder, with the muzzle down, and rotate the trigger to his rear. He should run the sling around the mortar to make sure tension pulls the trigger assembly toward the container. The parachutist and his buddy should secure the mortar in the same manner as for the M16-series and the M4 carbine-series rifles.

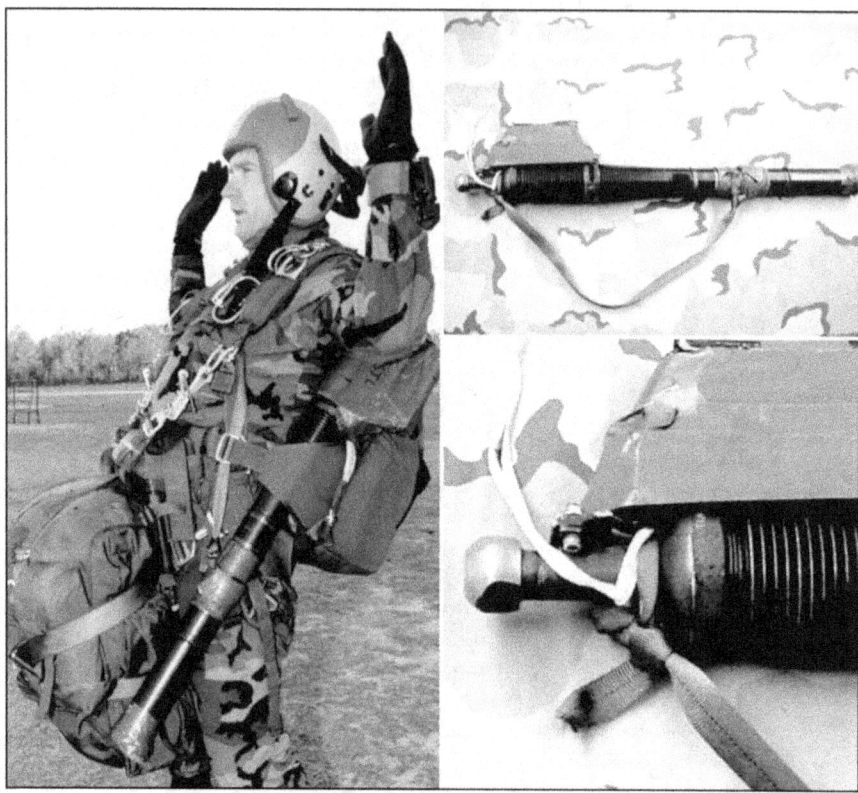

Figure 5-68. Left-side mount for M224 60-mm mortar

OTHER WEAPONS

5-83. The parachutist can rig other weapons using the methods previously described. User unit SOPs should specify ways to pack or rig similar types of weapons consistent with safety requirements. Units requiring technical help should contact B Company, 2d Battalion, 1st Special Warfare Training Group, USAJFKSWCS, Yuma, Arizona; Defense Switched Network (DSN) 899-3626/3639.

FLOTATION DEVICES/LIFE PRESERVERS

5-84. Parachutists must wear military-approved flotation devices (Tactical Flotation Support System [TFSS]-5326, B-7, Life Preserver Unit [LPU]-10/P, or underwater demolition team [UDT] life vest

Equipment and Weapon Rigging Procedures

[Figure 5-69]) whenever the planned flight path is over open bodies of water large enough to be unavoidable with a maneuverable chute for one third or more of the distance under canopy. They also wear them when an open body of water is within 1,000 meters of the planned impact point.

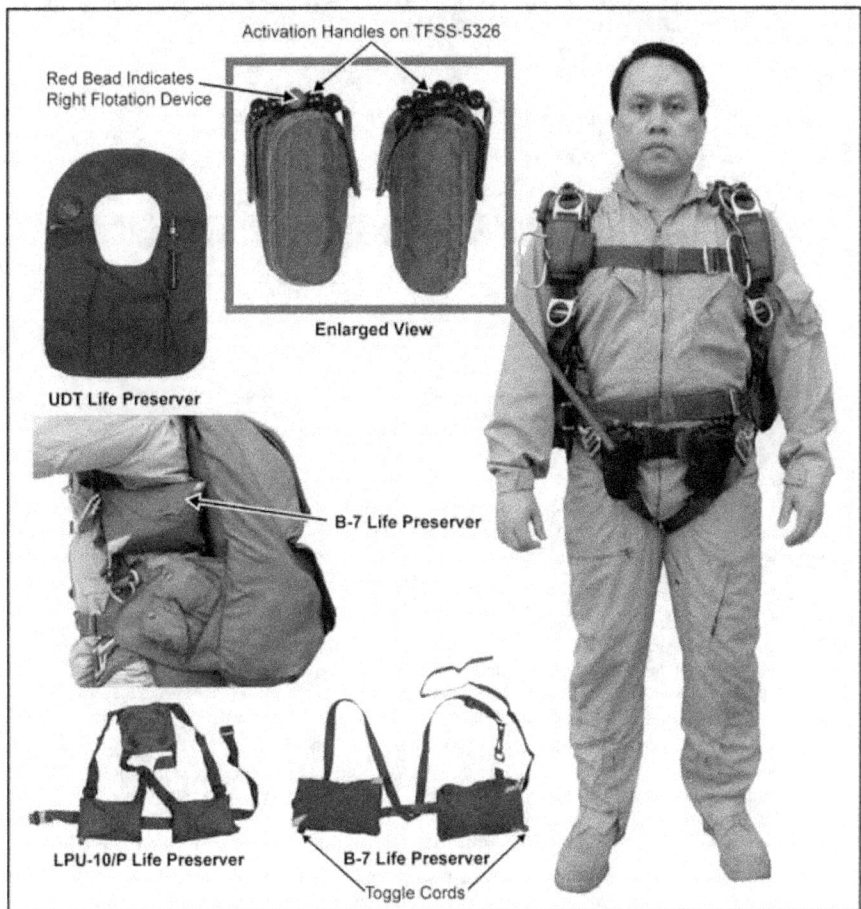

Figure 5-69. Underwater demolition team life preservers

TACTICAL FLOTATION SUPPORT SYSTEM

5-85. The TFSS-5326 is an inflatable aid to flotation device specifically designed for SOF warfighters, combat swimmers, and/or maritime airborne operations personnel. Each system consists of one each independent left- and right-hand units, which can be mounted on a belt. Each unit includes a welded flotation bladder, an inflation system, a pouch closure system, a pouch, and a firing handle. The bladder is a reusable welded fabric enclosure that deploys from a belt on the waist and can be placed under the arms of the parachutist while floating in the water. It is readily collapsed and stowed for future use. The inflation system uses a manually actuated carbon dioxide (CO_2) cartridge (two Leland 38-gram CO_2 cartridges) for primary inflation and an oral inflation tube for secondary inflation. The pouch contains and protects the bladder, inflation system, and closure system. It includes a waist belt loop and clip loops to secure the pouch to the webbing belt. The firing handle attaches to the outside of the pouch and uses color-coded beads to help distinguish left- and right-hand units. The handle serves to release the closure system and actuate the CO2 inflation system.

Buoyancy

5-86. The TFSS-5326 is designed to provide a minimum of 45 pounds of positive buoyancy in seawater at 33 feet and 80 pounds at the surface (Table 5-5). The overt system comes with reflective tape on the yellow-colored bladder to aid in recovery in sea operations.

Table 5-5. Lift capabilities

Depth (Feet)	Lift (Pounds)
50	35
33	45
15	57
3	80
NOTE: All test data is in seawater.	

Wearing

5-87. The TFSS-5326 is a one-size-fits-all system that has been designed to accommodate the personal preference of the user for ease of wearing and comfort. When wearing the TFSS-5326 for MFF operations, the units should be worn to the front of the parachutist's body (Figure 5-69, page 5-55) to avoid interference with the parachute or any emergency procedures that could arise that the parachutist might have to correct. The TFSS-5326 is designed for simplicity and ease of wear. However, an improperly mounted flotation device could interfere with airborne operations, causing injury to the parachutist or damage to equipment. The parachutist must ensure he does not wear the TFSS-5326 flotation packets between the parachute harness and his body. Serious injury may result if inflated when worn incorrectly. To mitigate this risk, MFF parachutists should properly mount the TFSS-5326 ensuring the system is not placed under the parachute harness at any location.

5-88. The CO2 cartridge cover does not extend the entire length of the CO2 cartridge cylinder, which poses a potential for a cold burn to unprotected skin (frostbite burn) when the cartridge is activated. To mitigate this risk, parachutists must ensure the TFSS-5326 mounting belt is worn tightly around their waist to limit CO2 cartridge movement, and ensure there is a layer of clothing between the CO2 cartridge and the parachutist's skin. All parachutists should participate in the mounting procedures for the TFSS-5326.

5-89. To manually actuate a TFSS-5326 unit, the user pulls upward on the firing handle. This motion initiates two sequential actions. First, the pouch closure pins are released, allowing the pouches to open freely. Second, the manual inflator lever is actuated, causing the firing pin to puncture the seal on the CO2 cartridges to release the gas and completely fill the bladder. Should the CO2 inflation system fail to operate, the bladder is filled through an oral inflation tube. This is accomplished by depressing the Oralock valve (Figure 5-70A, page 5-57), and then breathing into the tube (Figure 5-70B). Gas is released from the TFSS-5326 bladder by pressing downward on the Oralock valve and forcing the air out of the oral inflation tube. Once all of the gas is evacuated from the bladder, the CO2 cartridges are replaced, maintenance is performed, and the units are repacked for future use.

Equipment and Weapon Rigging Procedures

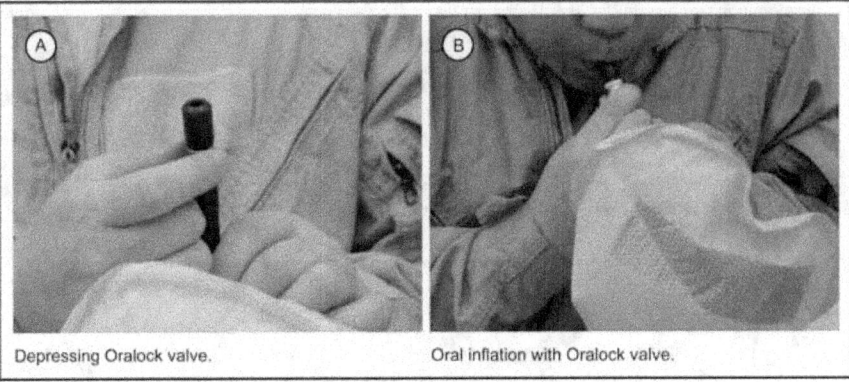

Depressing Oralock valve.   Oral inflation with Oralock valve.

Figure 5-70. Oralock valve

Note: Test results indicate wear of the TFSS-5326 poses no greater risk to Soldiers when compared to currently issued flotation devices worn during airborne operations (for example, the B-7 and LPU-10 types of life preservers), provided guidance contained in the referenced documentation and the safety releases are adhered to.

Periodic Maintenance

5-90. After each use, the parachutist—
- Rinses entire unit in fresh water.
- Allows unit to completely air dry.
- Applies a small amount of a silicon lubricant to all valves.
- Visually inspects all bladders for any damage.
- Replaces unit if damaged.
- Weighs and ensures the minimum gram weight for replacement of 38-gram CO_2 cartridges is 147 grams.
- Repacks unit IAW procedures.

Inspections

5-91. The parachutist conducts an annual inspection consisting of the following:
- Orally inflate unit and allow to sit for 24 hours.
- Thoroughly inspect unit for leaks. Replace if damaged.
- Inspect all valves to ensure they are in good working order.
- Reweigh and ensure the minimum gram weight for each 38-gram CO_2 cartridge is 147 grams.

Upon initial issue, the user is required to conduct an annual inspection and report any damage to the cognizant authority.

B-7 LIFE PRESERVER

5-92. The parachutist wears the B-7 over his uniform or jumpsuit and under his parachute harness (Figure 5-69, page 5-55). He fits the B-7 by placing one flotation packet under each arm, making sure the packet flaps are to the outside and the toggle cords are down and to the front. He routes the shoulder strap from front to rear over his left shoulder, under the back strap, then from rear to front over his right shoulder and attaches it to the ring on the right flotation packet.

5-93. The parachutist adjusts the shoulder strap so that the flotation packets fit snugly against his armpits. Before donning the parachute, he attaches the chest strap to the attachment ring on the left flotation packet, forming a quick release. If there is a water emergency, the parachutist inflates the life preserver by pulling the toggle cords located on each flotation packet. He can also manually inflate it by blowing into the rubber hose located on each flotation packet.

Note: The parachutist uses manual inflation only if the $CO2$ inflation system fails to operate.

> **WARNING**
>
> **The parachutist makes sure he does not wear the B-7 life preserver with the flotation packets between the parachute harness and his body. Serious injury may result if inflated when worn incorrectly.**

LPU-10/P LIFE PRESERVER

5-94. The LPU-10/P is a standard USAF $CO2$ cartridge-activated life preserver assembly worn during flights over water or during airdrops when water obstacles are near or on the intended DZ. It has an adjustable harness and underarm inflation bladders. The LPU-10/P is designed to keep the wearer's head above water at weights up to 250 pounds for up to 10 minutes. It is compatible with the USAF C-9, T-10, and MC-4 parachute harness assemblies. It must be maintained IAW TM 1-4220-252, *Maintenance Instructions With Illustrated Parts Breakdown USAF Flotation Equipment LRU-1/P, F-2B, 20-Man VPLR, and 25-Man Life Rafts LPU-3/P, LPU-6/P, LPU-10/P, A-A-50652, and MB-1 Life Preservers*, and TO 14S-1-102-21, *USAF Flotation Equipment LRU-1/P, F-2B, 20-Man VPLR, and 25-Man Life Rafts LPU-3/P, LPU-6/P, LPU-10/P, A-A-50652, and MB-1 Life Preservers*.

5-95. The LPU-10/P is worn under the parachute harness. The harness is worn so that the inflatable packets are under the parachutist's arms. The manual inflating valves should be completely closed when donning the life vest. The shoulder and waist straps are then adjusted to ensure the inflation packet is one hand width beneath the armpit and not constrained by the parachute harness.

> **WARNING**
>
> **If the inflation packets are too snug under the armpit, or if they are between the harness and the parachutist's body, the parachutist may experience severe pain or crushed ribs during inflation.**

5-96. The parachutist inflates the flotation bladders by pulling two toggle cords (at the bottom of the vest), which activate $CO2$ cartridges that fill the flotation bladders with gas. An alternate way to inflate the vest is by blowing into the manual inflation valve rubber hoses located on the bottom side of the wings. Manual inflation should only be used if the $CO2$ inflation valves fail to operate.

Equipment and Weapon Rigging Procedures

UNDERWATER DEMOLITION TEAM LIFE VEST

5-97. The UDT life vest is put on over the uniform before donning the parachute (Figure 5-71). The UDT vest is worn around the neck, with the straps passing under the arms and fastened to the vest. The parachutist fits the UDT life vest by performing the following steps:

- Adjust the straps until snug to prevent movement of the vest during free fall and interference with the red cutaway handle or the reserve rip cord handle.
- Pass the parachute chest strap between the UDT life vest and the body.
- Secure the UDT life vest with a lightweight retaining band around the middle to prevent interference with the green cutaway handle and red reserve rip cord handle.
- Route the oral inflation tube through its retainer loop.
- Screw the oral inflation tube knurled nut down in the open position to allow inflation.

The UDT life vest has a manually activated CO_2 actuator for immediate inflation or an oral inflation tube that can be used to inflate the vest or to manage a slow leak.

> **WARNING**
>
> The parachutist must not wear the UDT life vest with the flotation chamber worn between the parachute chest strap and his body. Serious injury may result if inflated when worn incorrectly. Parachutists must protect the activation lanyard of the UDT vest. Accidental inflation by the CO_2 cartridges may result in obstruction of the reserve rip cord and cutaway handles.

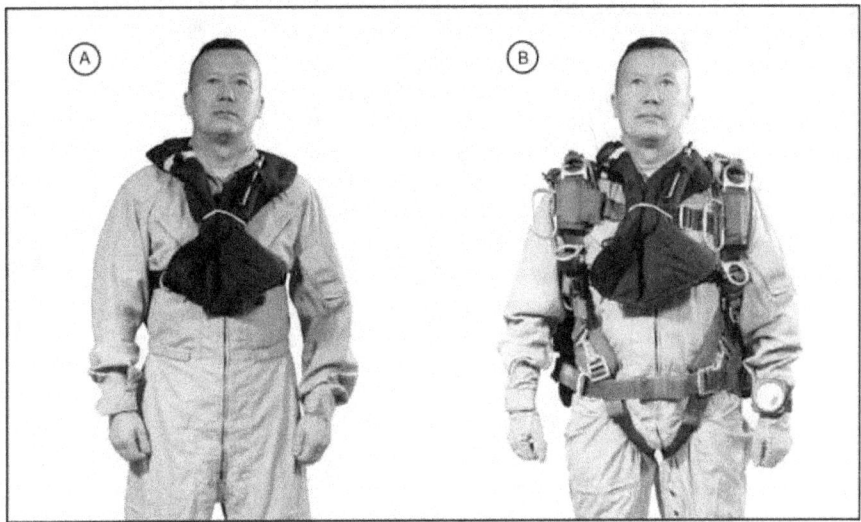

Figure 5-71. Parachutist with UDT life vest and MC-4 parachute harness

This page intentionally left blank.

Chapter 6

Aircraft Procedure Signals and Jump Commands

Aircraft noise, the MFF parachutist helmet, and the oxygen mask make verbal communication extremely difficult. Therefore, the parachutist receives aircraft procedure signals (Table 6-1, pages 6-1 and 6-2) and jump commands (Table 6-2, pages 6-2 and 6-3) by hand-and-arm signals. The MFF parachutist must be thoroughly familiar with all signals, and the commands and required actions for each one. Standardization of procedural signals and jump commands permits interoperability of all MFF-capable units. Safety significantly increases when the parachutist understands the jumpmaster's intent and the jumpmaster understands the parachutist's desired response.

AIRCRAFT PROCEDURE SIGNALS

6-1. Signals used between aircraft boarding and the jump command STAND UP are procedure signals. The aircraft procedure signals discussed in the following paragraphs begin before takeoff. The jumpmaster gives these signals.

Table 6-1. Aircraft procedure signals (oxygen and nonoxygen jumps)

Aircraft Procedure Signals	Jumpmaster Actions	Parachutist Actions
DON HELMETS	Gives command before takeoff or landing. *CAUTION If the helmet is removed after the JMPI, the jumpmaster ensures there is no twist in the oxygen delivery hose.	Dons helmet, fastens chin strap, and fastens seat belt.
UNFASTEN SEAT BELTS	Normally gives command on reaching an altitude of 1,000 feet AGL or when notified by the flight crew that it is safe to do so.	Disconnects seat belts and stows it to the left and right for easy retrieval.
*MASK	*Turns on own console and masks.	*Turns on console. *Secures mask to face and assures proper attachment and seal. Checks delivery of oxygen.
*CHECK OXYGEN	*Gives signal immediately following the command MASK and then periodically. *Gives signal after the 20- and 10-minute warnings.	*Checks own oxygen and returns the thumbs-up signal to the jumpmaster. In the event of an oxygen problem, extends arm straight forward, palm down.
*These signals, commands, and actions are used only during oxygen jumps with prebreather. *NOTE: Mask and oxygen checks will be determined by flight plan and mission profile when given. *NOTE: Oxygen safety technician checks gauges.		

Chapter 6

Table 6-1. Aircraft procedure signals (oxygen and nonoxygen jumps) (continued)

Aircraft Procedure Signals	Jumpmaster Actions	Parachutist Actions
TIME WARNING 20-Minute Warning		*All parachutists must be awake. First pass attaches combat equipment.
*CHECK OXYGEN		*Checks own oxygen and returns the thumbs-up signal to the jumpmaster. In the event of an oxygen problem, extends arm straight forward, palm down.
TIME WARNING 10-Minute Warning	Ensures red jump/caution light is on.	Second pass attaches combat equipment.
WIND SPEED	Normally gives signal immediately after the 10-minute warning, if known, and updates to remain current with the DZ party's information.	Receives a Military CYPRES 2 and pin check, and passes the thumbs-up signal from the last parachutist in the front of the aircraft to the rear, and then to the jumpmaster.

*These signals, commands, and actions are used only during oxygen jumps with prebreather.

Table 6-2. Aircraft jump commands (oxygen and nonoxygen jumps)

Jump Commands	Jumpmaster Actions	Parachutist Actions
STAND UP	Gives command about 2 minutes before time on target (TOT). (Oxygen or equipment jumps may require additional time for this command only; all other commands remain the same.)	Stands, faces the rear, and checks own equipment. Checks the pins *and oxygen pressure gauge of the man in front and taps him to indicate he is OK. The last two parachutists check each other. *NOTE: During an oxygen jump, the right hand should be on the ON/OFF switch of the oxygen bailout bottle and the left hand on the disconnect for the console hose.
MOVE TO THE REAR	Gives command about 1 minute before TOT.	Tightens shoulder straps of rucksack and puts goggles down. *Turns on oxygen bailout bottle and disconnects from the console. Moves to within 1 meter of the jump door or to the hinge of the ramp.
STAND BY	Gives command about 15 seconds before TOT.	Returns thumbs-up signal and moves to 1 foot of edge of ramp or door. Focuses attention on jumpmaster.
GO	Ensures green jump/caution light is on. Ensures aircraft is over release point. Gives command and proper hand-and-arm signals.	Exits the aircraft.
ABORT	Gives command any time an unsafe condition exists inside the aircraft or on the DZ. Gives command when the red jump/caution light is on. *Reconnects own console and turns off own oxygen bailout bottle.	Returns to seat. *Reconnects to console and turns off oxygen bailout bottle.

*These signals, commands, and actions are used only during oxygen jumps with prebreather.

Aircraft Procedure Signals and Jump Commands

Table 6-2. Aircraft jump commands (oxygen and nonoxygen jumps) (continued)

Jump Commands	Jumpmaster Actions	Parachutist Actions
*CHECK OXYGEN		*Checks own oxygen and returns the thumbs-up signal to the jumpmaster. In the event of an oxygen problem, extends arm straight forward, palm down.

*These signals, commands, and actions are used only during oxygen jumps with prebreather.
NOTE: There are no disarm procedures for the Military CYPRES 2. The jumpmaster must brief the pilots not to exceed 5,000 feet per minute as this descent rate is easy to remember and covers all three Military CYPRES 2 models.

DON HELMETS

6-2. The jumpmaster gives the signal DON HELMETS before takeoff (Figure 6-1). He may also give it during the flight or if the aircraft is landing with the jumpers (for example, winds out of limits on the DZ or mission aborted). Upon receiving this signal, the parachutist dons his helmet, fastens his chin strap, and fastens his seat belt.

Jumpmaster extends his arms to his sides at shoulder height and parallel to the floor, with palms up. Then he bends arms at the elbows so that his fingers tap the sides of his helmet.

Figure 6-1. DON HELMETS signal

UNFASTEN SEAT BELTS

6-3. The jumpmaster normally gives the signal UNFASTEN SEAT BELTS upon reaching an altitude of 1,000 feet AGL or when the flight crew chief indicates that it is safe to do so (Figure 6-2, page 6-4). If the aircraft descends back through 1,000 feet AGL later in the flight, the parachutist refastens his seat belt upon receiving the command DON HELMETS.

Chapter 6

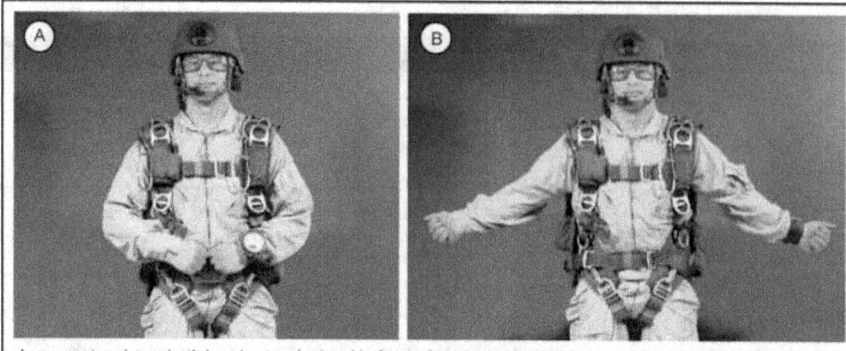

Jumpmaster places both hands at waist level in front of his body simulating grasping a seat belt. He extends both arms to his sides.

Figure 6-2. UNFASTEN SEAT BELTS signal

EMERGENCY BAILOUT

6-4. The jumpmaster gives the EMERGENCY BAILOUT signal for an emergency exit during flight (Figure 6-3, pages 6-4 and 6-5). Jump commands may be given if time permits. If there is no time for the full jump command sequence, he gives abbreviated signals immediately after the bailout signal:

- For exits from 1,000 to 3,000 feet AGL, the jumpmaster signals to IMMEDIATELY EXIT, CLEAR, AND PULL THE RESERVE RIP CORD HANDLE.
- For exits at 3,000 feet AGL and above, the jumpmaster signals to IMMEDIATELY EXIT, CLEAR, AND PULL THE MAIN RIP CORD HANDLE.

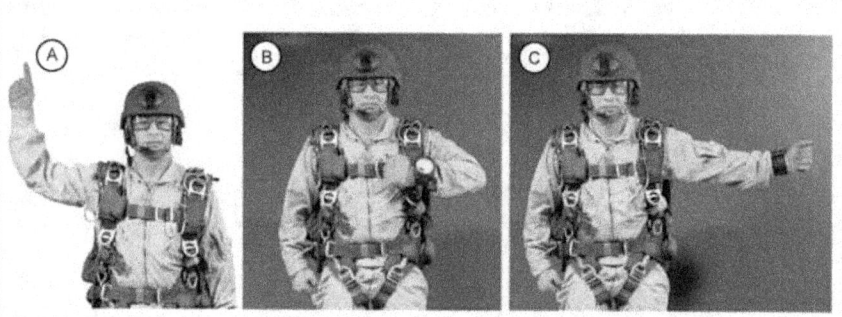

Jumpmaster extends one arm straight up, with the index finger extended, and moves it in a circular motion.

From 1,000 to 3,000 feet AGL: He places the clenched left fist by the reserve rip cord handle and thrusts it out to the side.

Figure 6-3. EMERGENCY BAILOUT signal

Aircraft Procedure Signals and Jump Commands

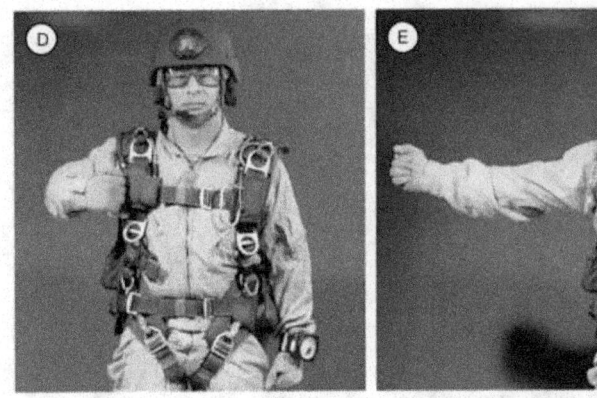

Above 3,000 feet AGL: He places the clenched right fist by the main rip cord handle and thrusts it out to the side.

Figure 6-3. EMERGENCY BAILOUT signal (continued)

MASK

6-5. The jumpmaster signals MASK when the parachutist must begin using supplemental oxygen (Figure 6-4). Upon receiving this signal, the parachutist masks and checks to make sure the oxygen system is functioning properly.

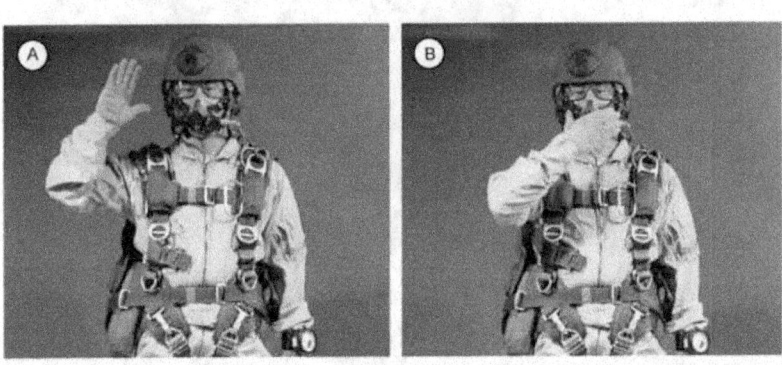

Jumpmaster places his right thumb on his right cheek and rotates the palm and fingers of his right hand across his nose and mouth.

Figure 6-4. MASK signal

Chapter 6

CHECK OXYGEN

6-6. The jumpmaster signals CHECK OXYGEN immediately after the signal to mask and periodically after that (Figure 6-5). At a minimum, he gives it following the 20- and 10-minute time warnings. Upon receiving this signal, the parachutist returns the signal if everything is functioning correctly. If there is a problem, the parachutist (Figure 6-6) extends an arm in front of his body with his hand open, palm down.

Jumpmaster extends his right arm in front of his body with closed fist and thumb extended upward.

Figure 6-5. CHECK OXYGEN signal

If there is a problem, he extends his arm in front of his body with his hand open, palm down.

Figure 6-6. OXYGEN PROBLEM signal

Aircraft Procedure Signals and Jump Commands

TIME WARNINGS

6-7. The jumpmaster receives time warnings from the flight crew. The jumpmaster signals the TIME WARNINGS to the parachutist to allow him adequate time to prepare for the jump (Figure 6-7). The parachutist normally receives the time warnings 20 minutes and 10 minutes before TOT.

Jumpmaster raises his left arm, elbow at the shoulder level and forearm perpendicular to the ground.

Next, he taps left wrist with his index and middle fingers of the right hand.

He brings both hands and arms back with hands closed preparing to indicate time warnings.

Finally, he indicates the time by extending his fingers indicating the appropriate time (20 minutes/10 minutes).

NOTE: If jumping with combat pack, the parachutist dons it after the 20-minute warning.

Figure 6-7. TIME WARNINGS signal

WIND SPEED

6-8. The jumpmaster signals WIND SPEED after the 10-minute time warning (Figure 6-8, page 6-8). In gusting wind conditions, the jumpmaster gives the wind speed signal first to indicate the lower wind speed. He follows with the GUSTING WINDS signal to indicate the higher wind speed (Figure 6-9, page 6-9).

Chapter 6

Next, he turns his upper body at the waist from right to left to right.

Jumpmaster extends his right arm in front of his body at waist level, palm up, forearm parallel to ground, and elbow locked at his side.

(D) He returns to center.

Then he extends one or both arms to his side at about head level, hands open, palms forward, with the correct number of fingers held up to indicate knots of wind speed.

Wind speed at 5 knots.

Figure 6-8. WIND SPEED signal

Aircraft Procedure Signals and Jump Commands

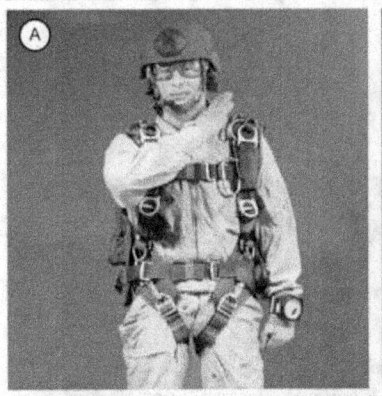 

Jumpmaster moves his right arm out and down in a diagonal slash across his body from left to right.

Then he displays the appropriate number of fingers to indicate the higher wind speed.

Figure 6-9. GUSTING WINDS signal

Note: The ARM ARR signal has been eliminated from the jump commands when jumping the Military CYPRES 2, but all jumpers must receive a Military CYPRES 2 and pin check after the WIND SPEED or GUSTING WINDS signal is given and then pass the THUMBS-UP signal.

JUMP COMMANDS

6-9. The jump commands discussed in the following paragraphs begin as early as 2 minutes before the actual jump is made. The jumpmaster gives these commands.

Note: The 2 MINUTE, 1 MINUTE, 15 SECOND, and GO commands can be given with either hand, depending upon which side of the aircraft the MFF jumpmaster is on.

Chapter 6

STAND UP

6-10. The jumpmaster commands STAND UP about 2 minutes before TOT (Figure 6-10). (Oxygen or equipment jumps may require additional time for this command only; all other commands remain the same.) Upon receiving this command, the parachutist stands up, receives pin check, faces the jumpmaster, and checks his equipment. If jumping oxygen, the parachutist also places his right hand on the ON/OFF valve of the bailout bottles and grasps with his left hand the console hose at the AIROX VIII.

Jumpmaster raises his arm in an arc, palm facing up, from his side to shoulder level (standing or kneeling).

Figure 6-10. STAND UP command

MOVE TO THE REAR

6-11. The jumpmaster commands MOVE TO THE REAR about 1 minute before TOT (Figure 6-11). Upon receiving this command, the parachutist tightens the combat pack's shoulder straps around his legs, adjusts his goggles, and moves to within 1 meter of the jump door or to the hinge of the cargo ramp. If jumping oxygen, the parachutist must activate the bailout oxygen system, check the flow indicator of the AIROX VIII, and disconnect from the oxygen console before moving to the rear of the aircraft.

Jumpmaster extends his arm straight out to the side at shoulder level, palm up. Then he bends his arm at the elbow until his hand touches his helmet (standing or kneeling).

Figure 6-11. MOVE TO THE REAR command

STAND BY

6-12. The jumpmaster commands STAND BY about 15 seconds before the exit (Figure 6-12, page 6-11). Upon receiving this signal, the parachutist signifies readiness by returning the jumpmaster's signal and then moves to the jump door or the cargo ramp.

Aircraft Procedure Signals and Jump Commands

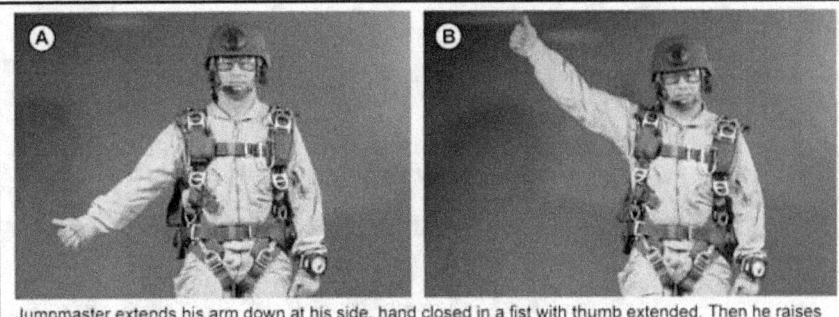

Figure 6-12. STAND BY command

Go

6-13. The jumpmaster commands GO when the aircraft is over the release point and the green jump light is on (Figure 6-13).

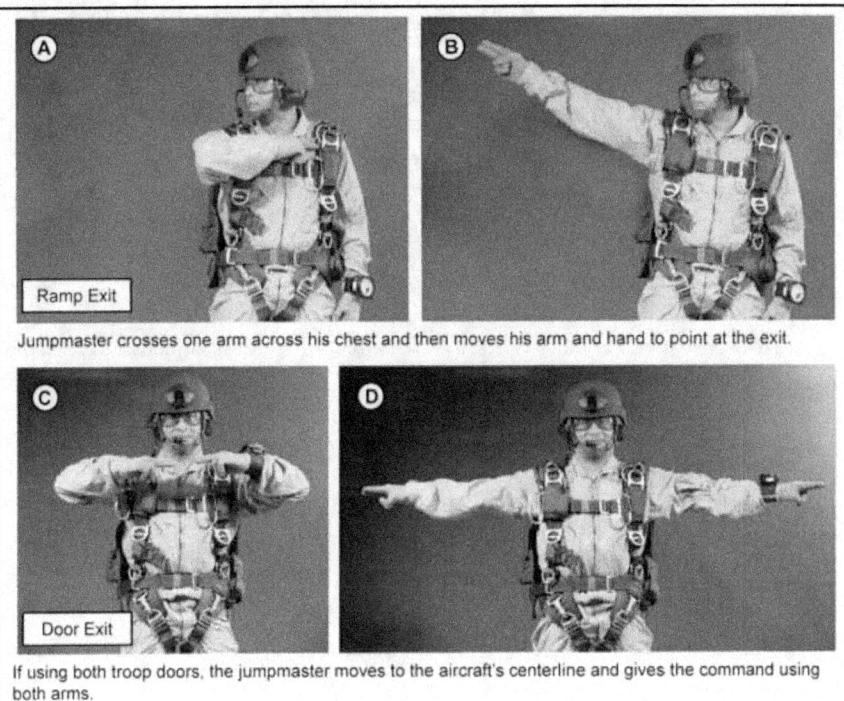

Figure 6-13. GO command

Chapter 6

ABORT

6-14. The jumpmaster commands ABORT anytime an unsafe condition exists inside or outside the aircraft (red jump light comes on) or on the DZ (Figure 6-14). Upon receiving this command, the parachutist returns to his seat and sits down. If jumping oxygen, the parachutist reconnects to the oxygen console, turns off the bailout system, and then sits down.

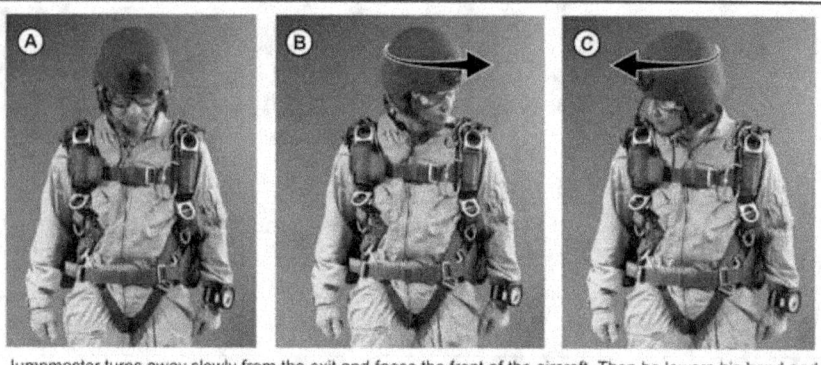

Jumpmaster turns away slowly from the exit and faces the front of the aircraft. Then he lowers his head and shakes it from side to side while walking toward the front of the aircraft.

Figure 6-14. ABORT command

CAUTION

If the jumpmaster has cocked his arm to give the command GO, he must NOT move it when he gives the ABORT command. The parachutists may exit if the jumpmaster moves his cocked arm.

Chapter 7

Body Stabilization

The MFF parachutist must be able to exit an aircraft with his combat equipment, fall on a designated heading, and manually deploy his main parachute without losing stability. Body stabilization skills allow the parachutist to group in free fall, cover small lateral distances with a rucksack, move off a lower parachutist's back in free fall, and turn to keep the DZ or group leader in sight. The MFF parachutist maintains these skills through regular MFF jumps and periodic refresher training. This chapter addresses the body stabilization skills needed to make a night tactical MFF jump with combat equipment from oxygen altitudes. Appendixes B and C provide recommendations for MFF proficiency training programs, and Appendix D covers suggested sustained airborne training.

TABLETOP BODY STABILIZATION TRAINING

7-1. Any stable tabletop or flat surface can be used for body stabilization training. The parachutist lies on his stomach on the tabletop. At the command GO, he lifts his arms and legs from the tabletop, assumes the poised, box man, or diving exit position, then moves to the stable free-fall position (Figures 7-1 through 7-4, pages 7-1 through 7-4). Controlled movement positions during free fall include turns, gliding, and altimeter check (Figures 7-5 through 7-7, pages 7-4 and 7-5).

Upon exiting the aircraft, parachutist faces in the direction of flight and extends his body with his back arched. He positions legs comfortably apart and knees slightly bent. He holds his head back and extends arms to the rear and away from his body at a 45-degree angle.

Figure 7-1. Poised exit position

Chapter 7

The "box man method" of poising out of an aircraft, by ramp or side door, is simply to exit in the normal box man free-fall position: chin up, hands and arms in the lazy "W," legs relaxed, and arch. This method eliminates the need for the jumper to transition from a "poise" position to a normal free-fall position.

Figure 7-2. Box man method

Body Stabilization

Figure 7-3. Diving exit position

Chapter 7

In the stable free-fall position, parachutist arches his back and holds his head up and back. He extends his arms horizontally, elbows bent at 90-degree angles. He holds his hands at eye level, palms down and slightly cupped, with fingers spread. He holds his legs about shoulder-width apart, with knees bent at a 45-degree angle. His knees should be slightly higher than his thighs.

Figure 7-4. Stable free-fall position

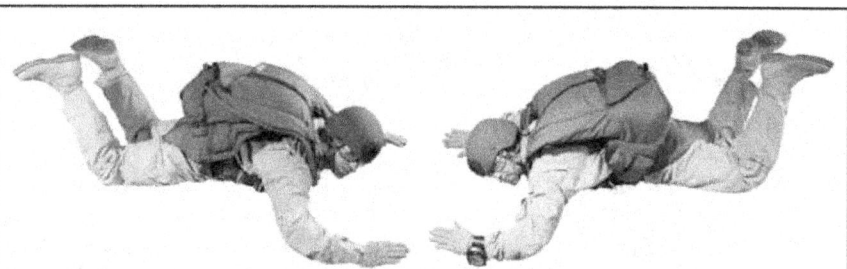

To execute right or left turn, parachutist arches, looks in direction of turn while rotating his upper body (shoulder) in direction of turn to start the turn. Before reaching desired heading, he counters and stops turn on desired heading. He resumes the stable free-fall position.

Figure 7-5. Body turn

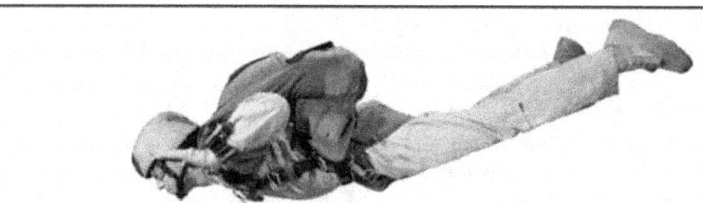

The glide is a controlled lateral movement. It allows the parachutist to maintain relative position with a designated group leader or bundle. To glide, the parachutist brings his arms back with elbows held near his sides and forearms at about a 90-degree angle to his body. He rotates his shoulders up and forward to cup his upper body. Then he straightens his legs from the knees. The straighter he holds his legs, the faster the glide will be. To stop the glide, he returns to the stable free-fall position.

Figure 7-6. Gliding

Body Stabilization

Parachutist should check his altimeter before and after performing any maneuver during free-fall descent to maintain altitude awareness. He normally wears the altimeter on his left wrist. He reads the altimeter by glancing toward his wrist without altering his stable free-fall body position.

Figure 7-7. Altimeter check

MAIN RIP CORD PULL

7-2. The parachutist executes the main rip cord pull at the predesignated altitude. He looks at the main rip cord handle on the right main lift web. He extends his left arm beyond his head with his hand held palm down. He traces the main rip cord cable housing. He grasps the main rip cord handle with his right hand, pulls the handle from the rip cord pocket, pulling the main rip cord cable to full-arm extension. Then he raises his right shoulder to disrupt the partial vacuum while looking straight down. After canopy deployment, he slips the main rip cord handle over his wrist.

7-3. When the parachutist uses night vision goggles (NVGs) to assist him during a night jump, the pull sequence during opening will remain the same as without NVGs (Figure 7-8, page 7-6). The parachutist will—

- Look below the NVGs to identify and read the altimeter, rip cord, cutaway pillow, and reserve rip cord handle when required during free-fall and canopy descent.
- "Arch-Look-Trace-Grab" as normal. The counter hand will extend slightly more forward to allow the head to be cradled under the parachutist's arm in a manner that protects the NVGs from inadvertent contact with the deploying pilot chute.
- Pull the rip cord handle to full-arm extension while continuing to look straight down with the counter arm in place to continue protecting the NVGs from a deploying pilot chute.
- **Not** conduct a visual check of the pilot chute; the parachutist will raise his right shoulder in an attempt to disrupt the partial vacuum while continuing to look straight down. He will attempt this twice and only twice; if this is unsuccessful, he will perform cutaway procedures for a total malfunction.

Note: Before pulling the main rip cord, the parachutist should ensure that he is not backsliding. Backsliding could cause the pilot chute to move forward of the parachutist's head and body.

Note: There are no special emergency procedures associated with the use of NVGs during MFF operations. If a horseshoe malfunction occurs, the parachutist should make no attempt to clear the malfunction and should immediately execute cutaway procedures.

Chapter 7

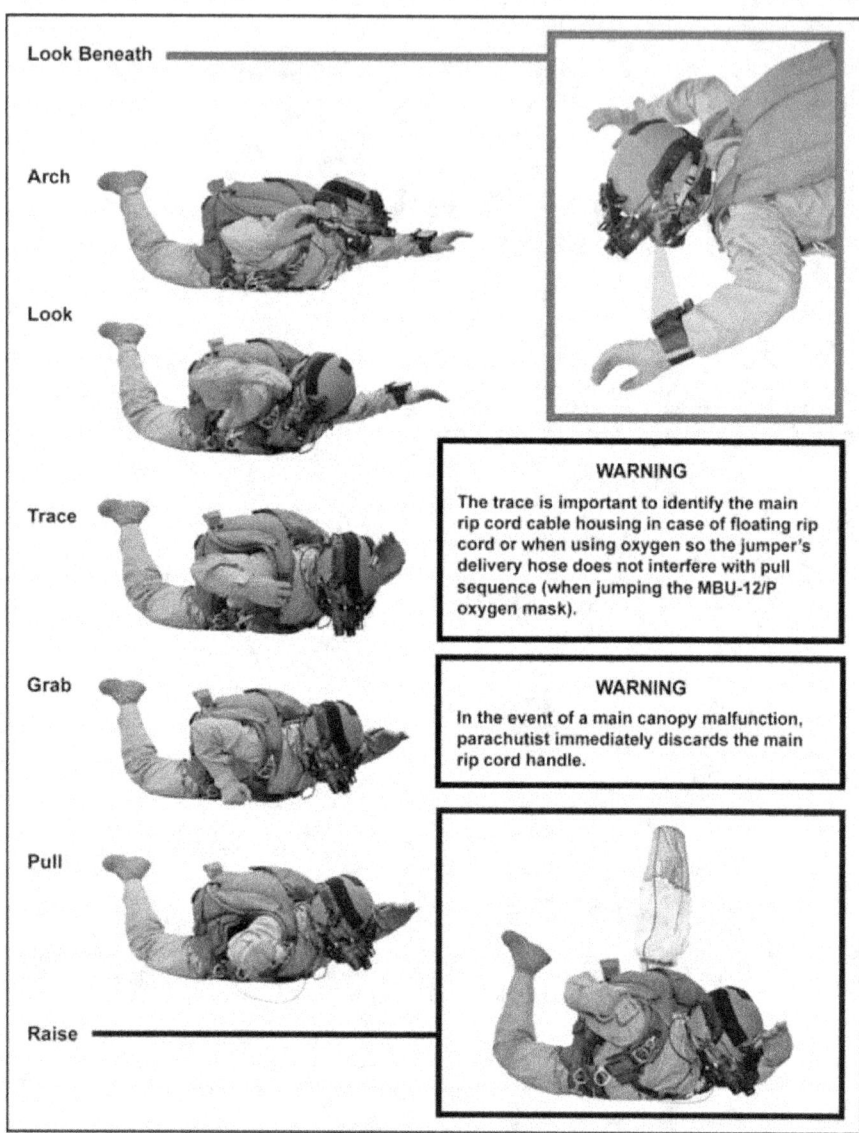

Figure 7-8. Main rip cord pull

TRACKING

7-4. Tracking is the technique of assuming a body position that allows the jumper to move horizontally while free falling. Although there are many variations of the basic body position, it essentially involves the jumper moving out of the traditional face-to-earth arched position, straightening the legs, bringing the arms to the sides at a 45-degree angle, rolling the shoulders forward, and cupping the air to provide maximum lift. There is, however, debate over what exactly constitutes the most efficient tracking position (providing the best glide ratio), especially concerning how far (if at all) the jumper's legs should be spread. An example of the tracking position (Figure 7-9) works well for some individuals and not so well for others. It is claimed that good trackers can cover nearly as much ground as the distance they fall (approaching a glide ratio of 1:1). It is known that the fall rate of a jumper in an efficient track is significantly lower than that of a jumper falling in a traditional face-to-earth position; the former reaching speeds as low as 90 mph, the latter averaging around the 120-mph mark. Inexperienced jumpers would expect such a position to increase the fall rate. If two jumpers begin with the same fall rate, one will appear to float at the same place relative to the other—neither appearing to drop below nor float above the other. If one jumper assumed a very good track position, the other jumper would see the tracking jumper not only accelerate quickly away but quickly upwards (relative to one another) as well.

7-5. Tracking is regarded as an essential lifesaving skill for all MFF jumpers engaging in grouping exercises, allowing the jumpers to gain horizontal separation before opening their parachutes or when the jumpers must cover great distance because of an incorrect spot determined by the jumpmaster. Accordingly, the greater the number of MFF jumpers on a jump, the better their tracking skills must be. In addition to having to track a longer distance after break off (tracking away for separation before opening), they also have to be more aware of other jumpers around them and have to be able to track in a straight line away from the center of the formation (Figure 7-10, page 7-8).

7-6. Because a good track body position can lead to significant horizontal speed and because the body's curved and slightly head-down position can cause less-experienced jumpers to be aware of a reduced area around them, novice jumpers should train themselves to be aware of what is going on around them in all directions for a greater distance while tracking.

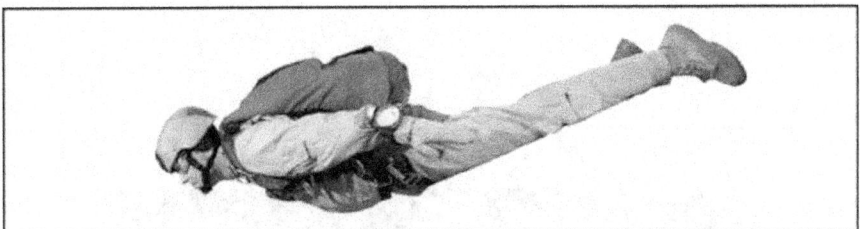

Figure 7-9. Tracking position

Chapter 7

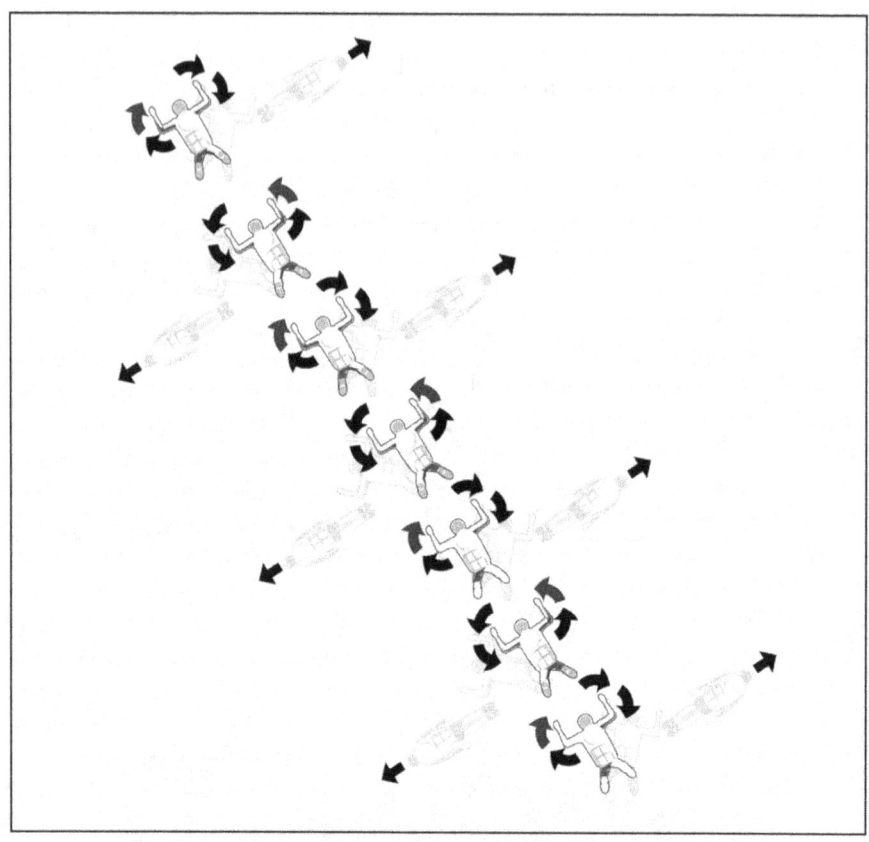

Figure 7-10. Example of tracking away for separation

> **DANGER**
>
> The dangers associated with tracking should not be underestimated. An efficient track can reach horizontal speeds of nearly 100 mph; collisions with other MFF jumpers could result in serious injury or DEATH. For this reason, the number of inexperienced jumpers should be limited per jump.

RECOVERY FROM INSTABILITY

7-7. Instability creates a hazard to the parachutist and to other parachutists in the air. Instability is the primary cause of MFF malfunctions. There are a variety of reasons for instability. In most cases, it is caused by a parachutist who does not present a symmetrical body position to the relative wind, either on exit or in free fall. A contributing factor to instability in free fall is the inadvertent shift or release of

Body Stabilization

combat equipment. A flat spinning or tumbling body motion characterizes instability. Instability is dangerous not only to the parachutist experiencing it, but often to other parachutists in free fall with him. Instability prevents tactical grouping.

Note: If a parachutist encounters any or all of these situations, he should maintain altitude awareness and pull at the prescribed pull altitude.

RECOVERY FROM A FLAT (HORIZONTAL) SPIN

7-8. If the parachutist is spinning or falling on his back, he must first return to a face-to-earth free-fall attitude by arching his body. Depending upon the speed of his spin, sometimes this movement alone is enough to slow or stop a flat spin. If he is still spinning after facing the earth, he must counter the direction of the spin. He does this movement by looking in the opposite direction of the spin (for example, if spinning clockwise, he looks counterclockwise) and making a hard body turn in that direction. He holds this body position until the spin slows and stops. Depending on the amount of momentum he developed before he started countering the spin, he may have to hold this body position for several revolutions. Once the spin has stopped, he checks his body position, makes an altimeter check, gets back on heading, and continues with the mission.

7-9. If a shift of the combat pack causes a flat spin, the parachutist may have to adjust his body position to obtain stability or maintain a heading. The severity of the shift (versus an inadvertent release) determines how much adjustment of the knees, the angle of the lower leg, hand and arm placement, or cocking of the hips he must make to counter the effect of a combat pack that is now not symmetrical or square to the relative wind.

RECOVERY FROM TUMBLING

7-10. A bump during a group exit or breaking the arched body position normally causes tumbling. If tumbling, the parachutist assumes the hard arch body position until facing the earth. Then, he relaxes the hard arch and assumes a stable free-fall body position. The time it takes to return to a face-to-earth position will vary with the severity of the tumble, the body area surface, and the parachutist's combat equipment configuration. Presenting a symmetrical body position to the relative wind on exit from the aircraft is the most significant factor in preventing tumbling.

ALTITUDE AWARENESS

7-11. A parachutist who is unstable must remain altitude-aware. The stress created by instability can cause a normal human phenomenon of temporal (time) distortion. The resultant effect varies from individual to individual. It can appear to be either time compression or a slowing down of perceived time passage. He must not get so caught up in his attempts to recover stability that he loses altitude awareness and forgets to manually activate his parachute. He must never sacrifice the pull altitude for stability or the continued attempts to obtain stability before the pull. An unstable parachutist must remember that as he is falling, an area of low pressure is created above him. Any altimeter reading while in this low-pressure area will not reflect the correct altitude AGL. An example is a parachutist falling back to earth who looks at his altimeter while holding it in front of his face. Due to the low-pressure zone in which the altimeter is located, the parachutist will read a higher altitude than where he actually is in feet AGL.

Note: Parachutists must remember that this pressure differential can cause the altimeter to be off as much as 1,000 feet.

Chapter 7

CORRECTIVE ACTIONS DURING FREE FALL

7-12. These actions are movements used to get off of a fellow parachutist's back. Primary movements include—

- Left or right turns into a safe direction.
- Forward glides (elbows into lazy "W," legs extended) to clear airspace.
- Track (arms back to side, legs straight).

Note: A modification to a forward glide is the high-lift track. Only experienced HALO-qualified parachutists should use this technique, and only qualified MFF instructors will train parachutists on this technique to gain separation before canopy deployment.

Chapter 8

Ram-Air Parachute Flight Characteristics and Canopy Control

This chapter describes the RAPPS canopy, its components, deployment sequence, theory of flight, flight characteristics, and canopy control procedures.

RAM-AIR PARACHUTE CHARACTERISTICS

8-1. The ram-air parachute canopy's design is similar to an aircraft's wings, with curved upper surfaces (top skin) and flat lower surfaces (bottom skin). Support ribs maintain the airfoil shape of the canopy (Figure 8-1).

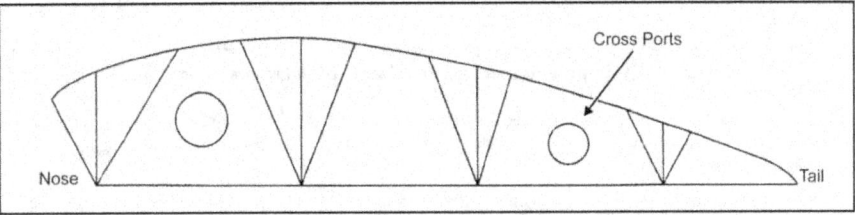

Figure 8-1. Shape of the ram-air parachute canopy

8-2. Reinforced, load-bearing support ribs serve as attaching points for the suspension lines, and non-load-bearing ribs separate a cell into two compartments. Cross-port vent holes in the support ribs equalize the internal air pressure in a canopy. Figure 8-2 shows the structure of the ram-air parachute canopy.

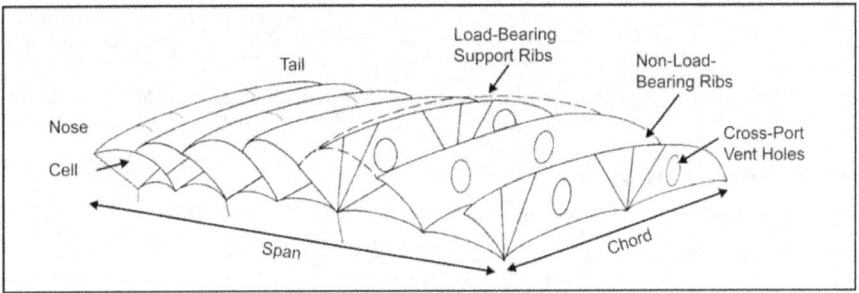

Figure 8-2. Structure of the ram-air parachute canopy

8-3. Nose, tail, chord, and span are terms of reference applied to ram-air parachutes. The open portion at the front is called the nose, with the rear being the tail. The distance from left to right is the span, and from nose to tail is the chord. Figure 8-3, page 8-2, shows the components of the ram-air parachute.

Chapter 8

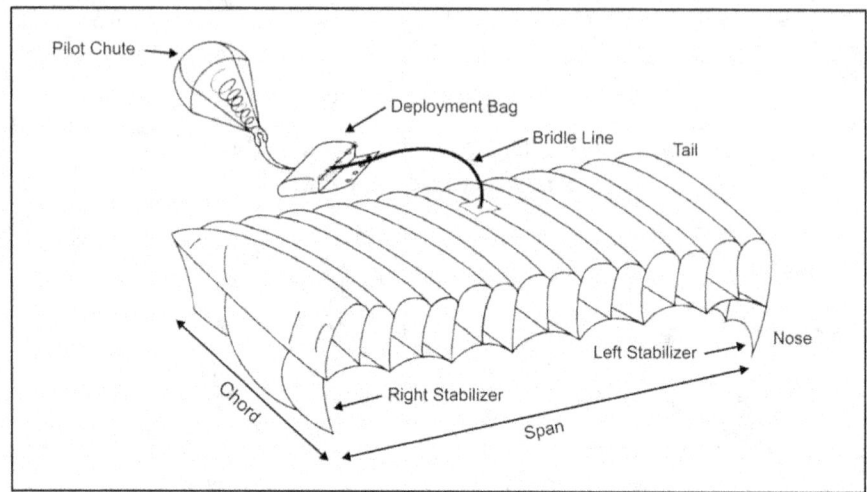

Figure 8-3. Components and nomenclature of the ram-air parachute

8-4. The stabilizers are single-layered extensions of the canopy on the left and right sides of the parachute. The stabilizers channel the airflow across the chord and help to maintain straight and stable flight.

8-5. The military ram-air canopy has four suspension line groups. They are identified from nose to tail as A, B, C, and D. A continuous line group is a line attached to the parachute's bottom skin that runs directly to the connector link without having another line attached to it. The suspension lines distribute a suspended load under the canopy without distorting the canopy's airfoil shape. Figure 8-4, page 8-3, shows the location of the ram-air parachute components.

8-6. Upper control lines converge from points of attachment on the left and right trailing edges of the tail, respectively, to common connection points with the lower control lines. The lower control lines are attached to the upper control lines and have a soft steering toggle secured to the lower end. Deployment brake loops sewn into the lower control lines set the canopy brakes for deployment. Figure 8-5, page 8-4, shows the components of the lower portion of the ram-air parachute.

8-7. The sail slider is a rectangular piece of reinforced fabric with a large grommet in each corner. The sail slider is a deployment device that retards the opening of a ram-air parachute.

8-8. Plastic disks called slider stops are sewn to the stabilizers at suspension line attachment points. These slider stops limit the upward travel of the sail slider.

8-9. The suspension lines are attached to a connector link on each riser (Figure 8-5). Trim tabs on the main parachute's front risers shorten the risers to create an artificial decrease in the canopy's angle of attack into the wind. Guide rings sewn to the rear risers function as anchor points for the deployment brakes and guides for the lower control lines (Figure 8-5).

Ram-Air Parachute Flight Characteristics
and Canopy Control

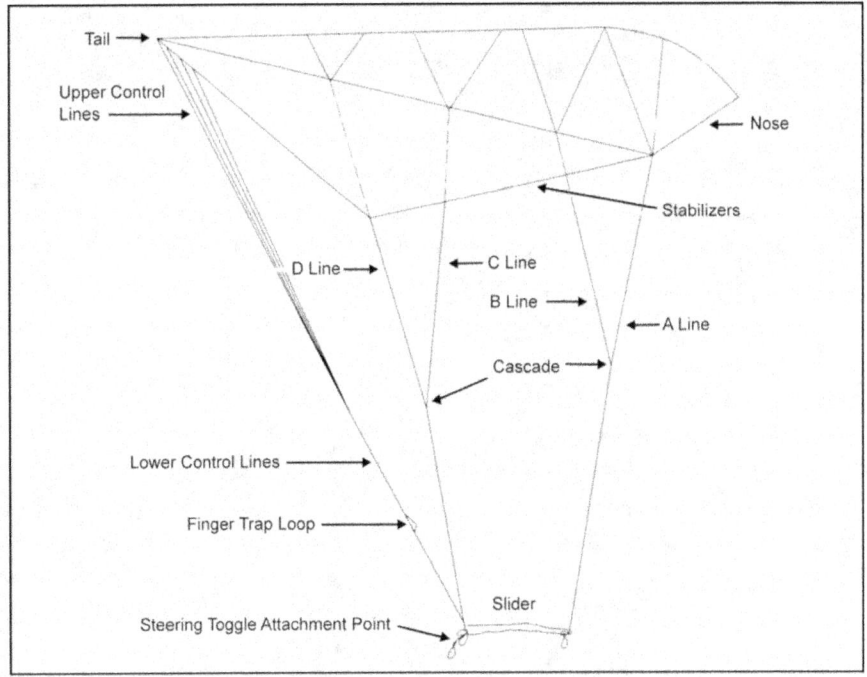

Figure 8-4. Location of components of the ram-air parachute

RAM-AIR PARACHUTE DEPLOYMENT SEQUENCE

8-10. At the prescribed parachute deployment altitude, the parachutist manually activates his parachute. He grabs and unseats the main rip cord handle in his right hand and fully extends his arm.

8-11. When the main rip cord pin clears the closing loop, the main pilot chute opens the closing flaps, launches from the main parachute container, and extends the pilot chute bridle. The bridle extracts the deployment bag from the main container, and the suspension lines unstow from their retainer bands. When the lines are fully extended, they pull the main parachute from the deployment bag, and the canopy begins to inflate (Figure 8-6, page 8-5). The sail slider retards the canopy's deployment. As the canopy inflates, it forces the sail slider down toward the risers as the suspension lines spread apart. After complete canopy deployment, the parachutist pulls the steering toggles from the deployment brake loops to release the control lines from the deployment brakes setting to the full flight setting.

Chapter 8

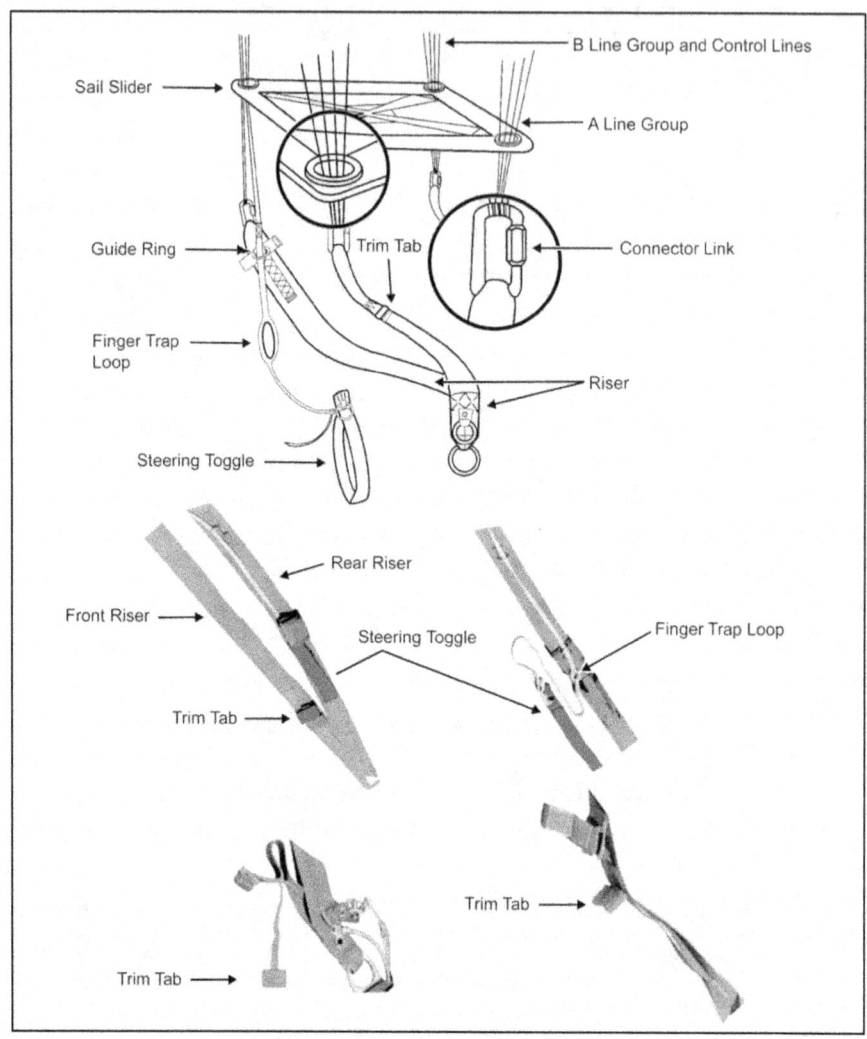

Figure 8-5. Detailed lower portion of the ram-air parachute

Ram-Air Parachute Flight Characteristics and Canopy Control

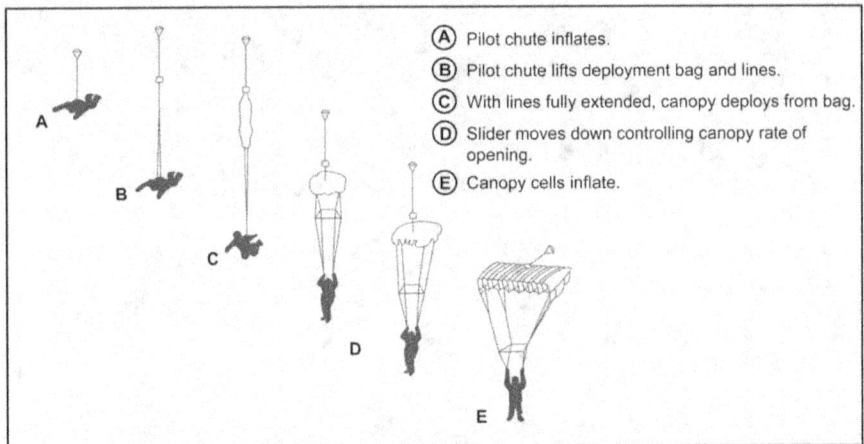

(A) Pilot chute inflates.
(B) Pilot chute lifts deployment bag and lines.
(C) With lines fully extended, canopy deploys from bag.
(D) Slider moves down controlling canopy rate of opening.
(E) Canopy cells inflate.

Figure 8-6. Deployment sequence

8-12. The parachutist follows the below procedures should he encounter an uncontrollable situation requiring the initiation of emergency procedures:
- Discards the main rip cord handle.
- Looks at and grabs the cutaway handle with his right hand.
- Looks at and grabs the reserve rip cord handle with his left hand.
- Arches vigorously.
- Pulls the cutaway handle to full-arm extension and releases it.
- Immediately pulls the reserve rip cord handle to full-arm extension and releases it.
- Performs postopening procedures.

8-13. This action allows the cutaway cables to clear the release loops threaded through the small rings of the canopy release assembly. The three-ring system activates the right side a moment before the left side to prevent an entanglement. As the left riser set is jettisoned, it pulls the reserve static line (RSL), usually deploying the reserve before manual activation of the reserve rip cord. Figure 8-7, page 8-6, identifies the cutaway sequence and deployment of the reserve parachute.

WARNING

The parachutist must first pull the cutaway handle AND THEN the reserve rip cord handle to full-arm extension and discard them to make sure complete emergency procedures are followed.

8-14. As the reserve rip cord pins clear the closing loops, the pilot chute opens the closing flaps. The pilot chute deploys from the reserve parachute container and, as it catches air, extends the 2-inch-wide high-drag bridle. Upon extraction of the reserve free bag from the container, the free-stowed suspension lines deploy from a pocket on the free bag and extract the reserve parachute from the free bag. The free bag then completely separates from the reserve parachute. As the canopy deploys, it forces the sail slider down the suspension lines. When the parachutist releases the toggles from the deployment brake loops, he releases the control lines from the deployment brake setting to the full flight setting.

Chapter 8

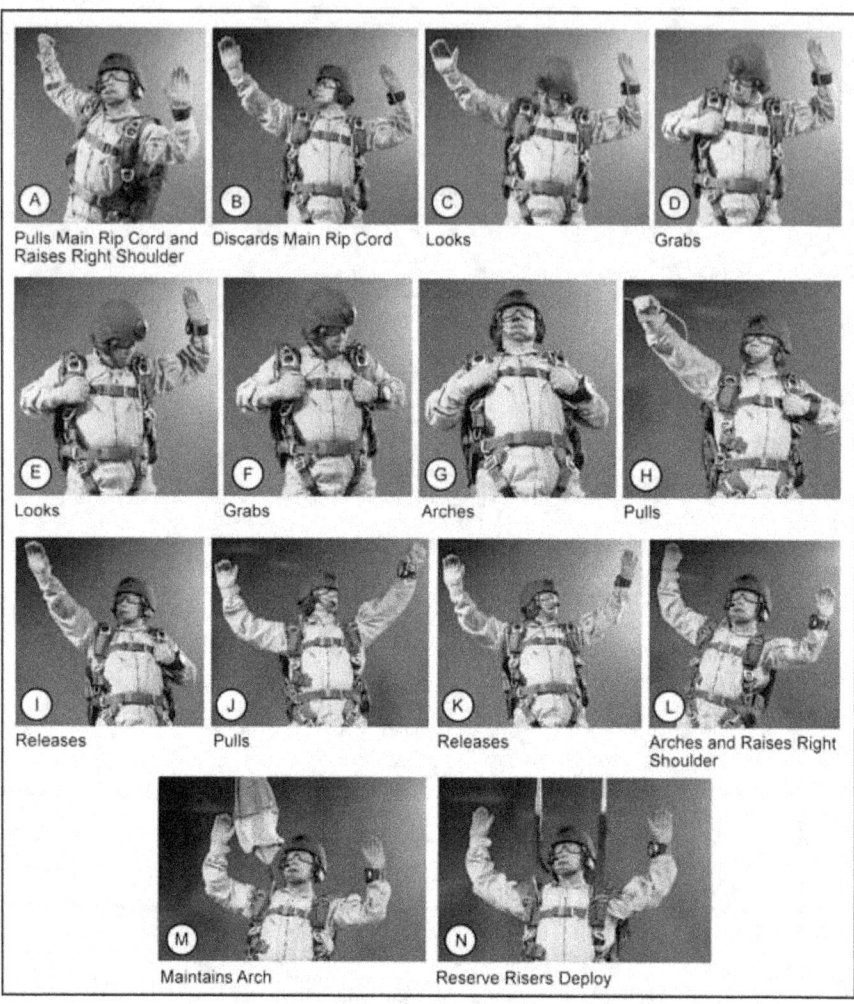

Figure 8-7. Cutaway sequence and deployment of the reserve parachute

RAM-AIR PARACHUTE THEORY OF FLIGHT

8-15. The ram-air parachute is an inflated and pressurized fabric airfoil that generates lift by moving forward through the air. The relative lengths of the suspension lines maintain the airfoil's trim angle. In flight, the parachutist keeps the wing's leading edge at a slightly lower angle than the trailing edge. Thus, this angle forces the canopy's airfoil-shaped surface to glide or plane through the air, very much like a glider in descending flight. The wing-shaped ram-air parachute generates lift caused by the reduced pressure of the airflow over the curved upper surface.

8-16. The ram-air parachute's leading edge is open or physically missing, forming intakes that allow the cells to be ram-air inflated. Internal air pressure pushes a small amount of stagnant air ahead of the airfoil, forming an artificial leading edge. The focal point of this stagnant air acts as a true leading edge, deflecting the relative air above and below. Drag is the only force that retards the wing's forward motion through the air. Drag is created by the friction of air passing over the canopy fabric, the suspension lines, and the parachutist and his equipment. Gravity, plus the resultant sum of these aerodynamic forces on the upper surface, acts to pull the ram-air parachute through the air and contributes to the flat glide angle of the canopy (Figure 8-8).

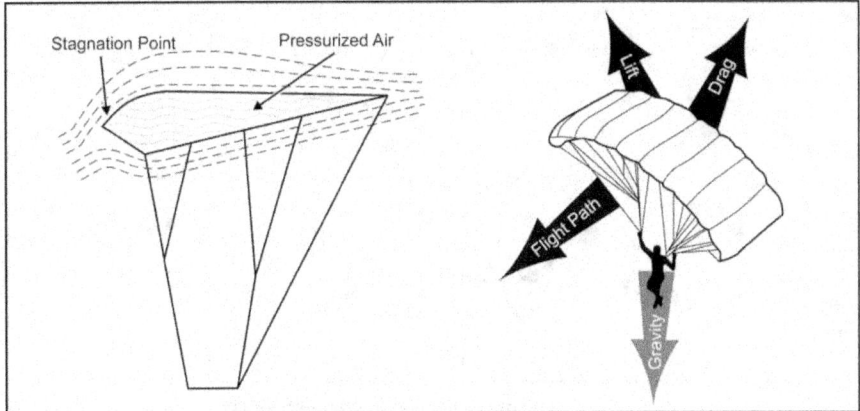

Figure 8-8. Ram-air parachute theory of flight

8-17. Applying brakes on the ram-air parachute causes the trailing edge to deflect downward, creating additional drag (Figure 8-9, page 8-8). This drag produces a proportionate loss of airspeed but generates a slower vertical descent. The glide angle increases with the application of toggles. As full brakes are reached, the wing ceases to generate dynamic lift, resulting in an increased rate of descent at an almost vertical descent angle. Depressing the toggles beyond full brakes causes the parachute to cease flying and enter a stall.

8-18. Differential application of brakes (one side only, or one side more than the other) produces an unbalanced drag force at the trailing edge. This drag results in a yaw-type turn toward the side with the highest drag.

8-19. Because the slow side generates less lift, it tends to drop slightly in a shallow banking motion, much like an airplane. This bank angle increases as differential toggle displacement increases.

CANOPY PERFORMANCE FACTORS

8-20. The performance of a ram-air canopy is primarily affected by the weight of the jumper, density altitude, and movement of the toggles.

WEIGHT OF JUMPER AND GEAR

8-21. The forward speed and descent rate of the ram-air is affected by the weight of the jumper and his equipment. A heavier jumper will have greater forward speed and a higher descent rate than a light jumper. A jumper has some control over how much gear is carried and thus his jump weight.

Chapter 8

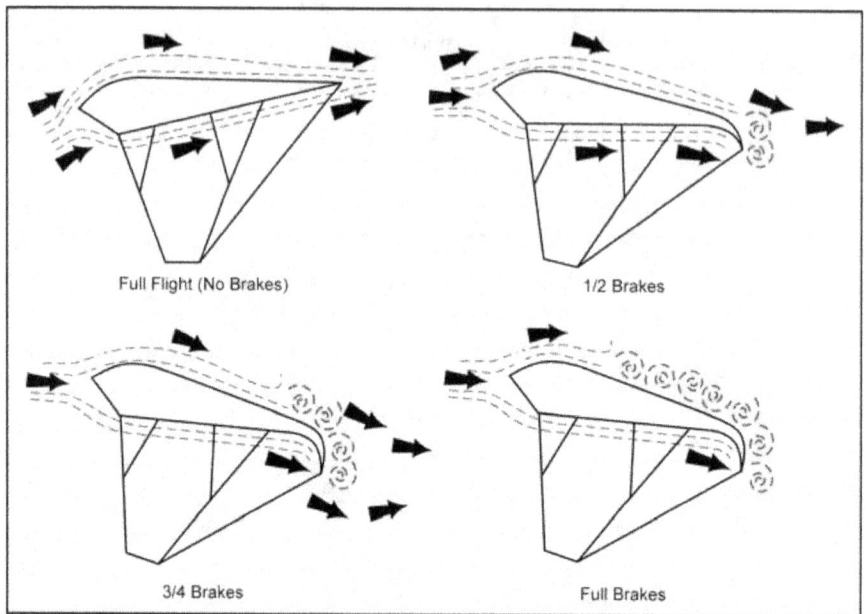

Figure 8-9. Applying brakes on the ram-air parachute

DENSITY ALTITUDE

8-22. Hot temperatures and/or high altitude will result in the lifting capacity being reduced, the glide slope decreases requiring an adjusted K-factor, and the overall rate of descent increases.

TOGGLE MOVEMENT

8-23. The ram-air parachute is very responsive to commands. It will do what the parachutist makes it do. When he decides to perform a flight maneuver, he makes the decision based on what he sees and feels. If he is in a steeply banked turn or is swinging under the canopy, it is difficult to make a good flight decision because canopy response lags behind his commands. Thus, if the parachutist flies from an unstable platform or if he over-controls, he will not get maximum performance. Slow, gentle toggle movements help a jumper avoid disorientation, and in the case of deep brake flight and stall recovery, gentle toggle movements are required to avoid hard landings. It is better to fly with the toggles kept fairly close to the body than with outstretched arms. This technique helps reduce arm fatigue, provides more accurate identification of the stall point, and helps prevent outstretched arms when performing PLFs.

PARACHUTE FLIGHT CHARACTERISTICS

8-24. The parachutist must remember that the ram-air parachute is a high-performance gliding system. Because of its high performance, the ram-air parachute is potentially dangerous in the hands of an inexperienced parachutist. The parachutist must possess a working knowledge of the flight capabilities and limitations of the ram-air parachute and must fully understand the canopy control techniques.

Ram-Air Parachute Flight Characteristics and Canopy Control

8-25. The ram-air parachute is not overly complicated. It is basically a fabric wing section. The parachutist must have a very basic knowledge of aerodynamics to better understand its flight and handling characteristics.

8-26. The ram-air parachute planes or glides through the air at about 20 to 30 mph. It always flies at this speed regardless of wind conditions, except when the parachutist applies brakes.

8-27. The flying speed is called airspeed and remains constant regardless of whether the parachute is headed upwind, downwind, or crosswind. The only variation in flying upwind or downwind is a change in ground speed that is often mistaken for a change in airspeed.

8-28. Wind affects ground speed only and has no effect on airspeed. Brakes applied with conventional control lines and toggles control the ram-air parachute's airspeed. Fifty percent of toggle travel on a ram-air parachute will cause a speed reduction of close to 12 mph.

8-29. There is almost no surge on deployment, and there is no wind noise at all until after releasing the brakes. A parachutist who has not been previously exposed to the ram-air parachute's flight characteristics can use the wind noise created by forward speed as a rough airspeed indicator. A reduction in the wind noise level can provide a stall warning.

8-30. After the parachutist becomes accustomed to the canopy, he may fail to notice the wind noise. By this time he should have learned to fly the canopy by feel, and he should notice the stall warning point and determine this point at altitude under his canopy controllability check. The parachutist will feel the canopy shudder as it loses lift and begins to stall. The parachutist should remember that angle of attack, cross wind, and wind turbulence can increase the stall point without warning.

8-31. The parachutist must remember that, in controlling the canopy's flight, how fast he moves the toggles from one position to another is as critical as the relative position of the toggles. As a rule, rapid and generous (more than 30 percent) application of both toggles will cause a rapid decrease in airspeed, decelerating into the stall range at about 0 to 3 mph. (Depending on the wind speed, the ground speed could still be very high.)

8-32. Due to the penetrating ability of the ram-air parachute, parachutists often find it difficult to determine wind direction without the aid of a wind sock, streamer, or smoke on the ground. All landings should be made facing into the wind.

8-33. The ram-air parachute has a constant airspeed of 20 to 30 mph. If the parachutist points the ram-air parachute downwind with a 10-mph wind, the ground speed will be 30 to 40 mph. If he turns the ram-air parachute into the wind and the winds are 10 mph, the airspeed remains the same but the ground speed reduces by 10 mph. If the ram-air parachute faces into 20-mph winds, the ground speed will be 0 mph (Figure 8-10, page 8-10).

CANOPY CONTROL

8-34. The overall objective of MFF parachuting is to land personnel and equipment intact to accomplish the assigned mission. The free-fall parachutist must know and employ the principles of canopy control as they relate to the use of the ram-air parachute.

8-35. Wind action, direction of canopy flight, and manipulation of the control toggles primarily control the movement of the ram-air parachute. Upon canopy deployment, the parachutist grabs the control toggles and performs a controllability check of the parachute. The purpose of this check is to determine if the parachutist's canopy is capable of landing him safely. Figure 8-11, page 8-10, contains a condensed guide to good canopy control.

8-36. The parachutist must first know wind direction and approximate speed since the direction of his canopy's flight, as determined by his toggle manipulation, is in relation to wind action. The canopy's shape, design, span, and chord generate the ram-air parachute's 20- to 30-mph glide. The flow of air over and under the canopy's wing shape provides the lift and forward flight of the parachute. By specific

Chapter 8

manipulation of the toggles, the parachutist may distort the trailing edge and cause the canopy to turn, to vary forward speed, and to increase the rate of descent.

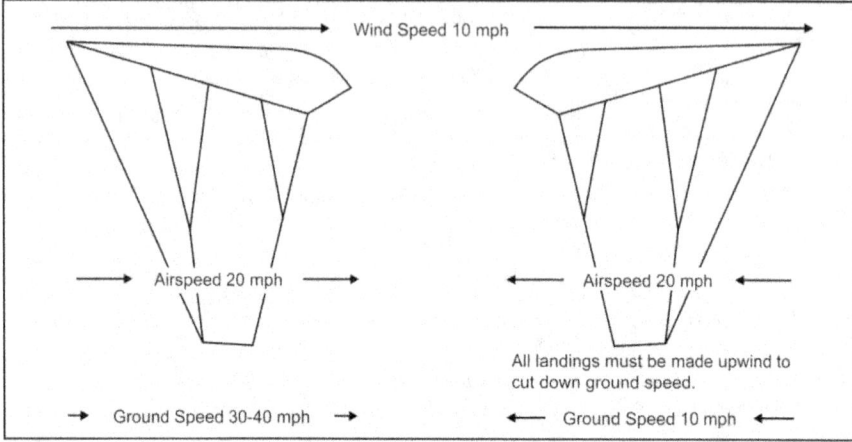

Figure 8-10. Controlling ground speed

8-37. Canopy control involves the coordination of wind direction and speed, canopy flight and penetration, and the parachutist's own selective manipulation and distortion of the canopy. Maneuvering the parachute requires more than simply turning the canopy. A properly executed parachute maneuver requires correct canopy manipulation to combine the wind's force and the canopy's flight to move the parachute in a given direction. The parachutist may have to hold into the wind, run with the wind, or crab to the left or right while holding or running.

- Checks canopy and ground position after opening.
- Keeps a sharp lookout for other parachutists.
- Checks his altitude and his first ground reference point.
- Picks out intermediate ground references between him and the target.
- Determines wind direction (on the ground and at altitude).
- Checks the holding pattern and penetration of his canopy.
- Uses the upwind toggle to turn his canopy.
- Locates the wind line and determines the direction in which he wants to move.
- Always maneuvers toward the wind line.
- Checks his progress at halfway and three-quarter-way points and makes necessary adjustments.
- Turns into the wind at a minimum altitude of 500 feet.
- Controls his canopy all the way to the ground.
- Always lands facing into the wind.

Figure 8-11. Parachutist guide to good canopy control

HOLDING MANEUVER

8-38. Pointing the canopy into the wind, or "holding," aims the canopy flight directly into the wind (Figure 8-12). This maneuver has the same effect as reduced wind speed and slows the canopy's forward movement. The parachutist manipulates the toggles to maintain the position. To crab to either direction while holding, he turns the canopy slightly in the direction in which he wants to move. Turning the canopy too far may cause it to become wind-cocked and move with the wind. As the parachutist's canopy begins to move in the desired direction, he manipulates the toggles to keep it in position until he completes the maneuver.

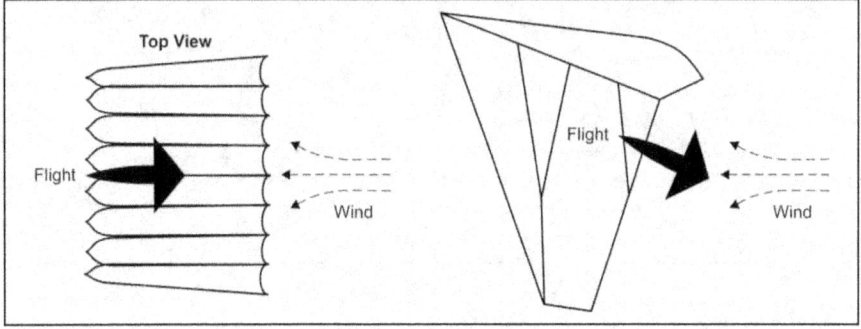

Figure 8-12. Holding maneuver

RUNNING MANEUVER

8-39. "Running" is when the parachutist points the canopy with the wind; the combined glide speed of the canopy and the wind speed produce an increased overall ground speed (Figure 8-13). He manipulates the toggles to maintain the canopy in position. To crab while running, the parachutist turns the canopy slightly in the desired direction and maintains the position until he completes the maneuver.

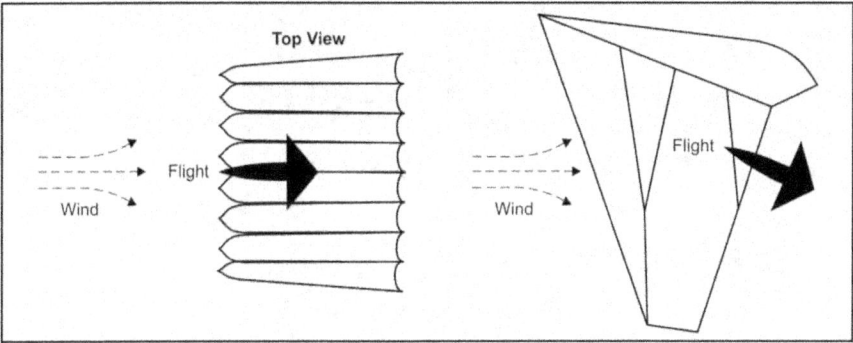

Figure 8-13. Running maneuver

CRABBING MANEUVER

8-40. The parachutist performs a "crabbing" movement by pointing the canopy at any given angle to the wind direction (Figure 8-14). The force of the wind from one direction and the flight of the canopy at an angle to it move the canopy at an angle to the direction of flight. The direction of flight varies with the wind speed and the angle at which the parachutist points the canopy. A canopy pointed at a downwind angle makes a sharper angle than one pointed upwind.

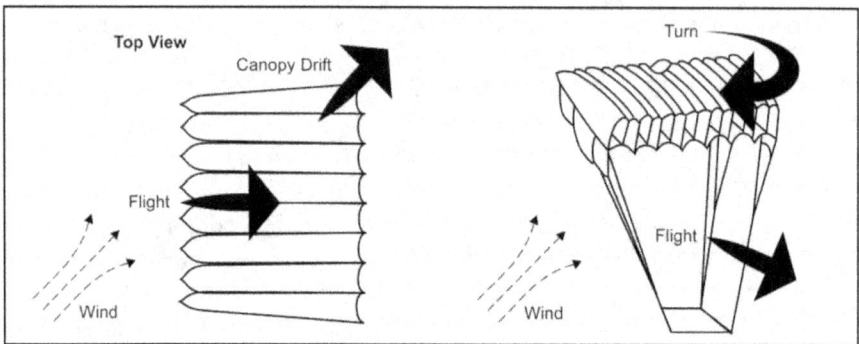

Figure 8-14. Crabbing maneuver

8-41. The effective canopy range and the wind line determine the course (direction of movement) the parachutist follows in maneuvering toward the target area. The effective canopy range is the maximum distance from which the parachutist can maneuver the canopy into the target area from a given altitude. It is greater at high altitudes and decreases proportionately at lower altitudes, forming a cone- or funnel-shaped area (Figure 8-15). Changes in wind direction and conditions may cause this range to shift in any direction.

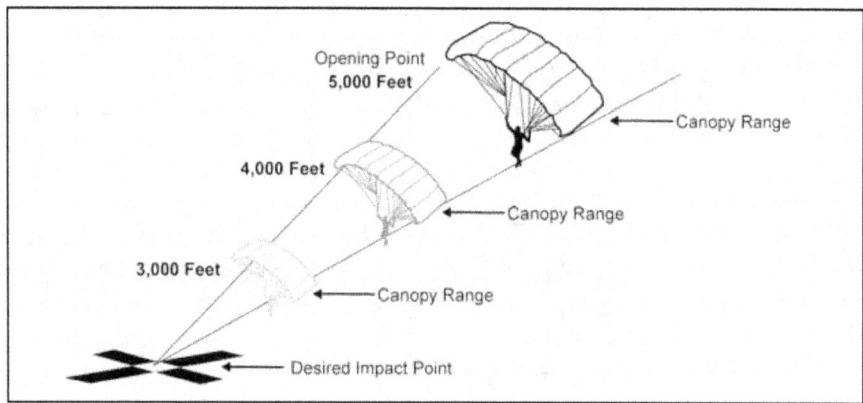

Figure 8-15. Effective canopy range

8-42. A wind line is an imaginary line extending upwind from the target area into the prevailing wind. A wind line can be marked by ground references. Accurate reference points are essential to effective parachute maneuver.

8-43. The parachutist checks his movement in relation to the ground. Winds at altitude may be from different directions than those at the desired impact point.

Ram-Air Parachute Flight Characteristics and Canopy Control

8-44. For the "half method," the parachutist picks a ground reference point on the wind line, halfway between the opening point and the target area. This point is the first checkpoint that he can reach in half the opening altitude with correct canopy manipulation. The second checkpoint is a reference point halfway between the first checkpoint and the target area that he should reach in half the remaining altitude.

8-45. The "horizon method" allows a parachutist to determine his flight progress by looking at his target and watching if it rises or descends in his line of sight. If it is rising, he will not make it to that point; if it is descending, he will probably have enough altitude to get back. It is also important to note that the parachutist always tries to maintain the "upwind advantage." This advantage is a margin in his canopy range where he will not be blown behind his target area and become unable to recover and land with his group.

8-46. The ram-air parachute is a highly maneuverable canopy capable of 360-degree turns in 3 to 5 seconds under normal conditions. Its maneuverability comes from the parachutist's use of its capabilities to vary forward speed, rate of descent, turn, and crosswind movement.

8-47. Under normal conditions, the parachutist varies his forward speed and rate of descent by using the canopy's toggles. Immediately upon canopy deployment, he clears the toggles from the deployment brakes setting and performs a controllability check. His toggle position at the stall point will be at a different position as wind speed increases and when carrying heavy equipment loads.

> **WARNING**
>
> Before attempting any maneuvers or turns, the parachutist must be alert to prevent collisions with other parachutists. This maneuver is especially critical below 500 feet AGL.

CANOPY MANEUVERS

8-48. The various straight-ahead maneuvers are used to affect the glide angle of the canopy. Canopy glide angles can be changed by manipulating either the steering toggles or the risers. Figure 8-16 shows the glide angles for a ram-air canopy at the various toggle settings. As toggles are applied, the angle of attack increases and the glide angle flattens. Airspeed decreases but descent decreases at a greater rate than the airspeed slows, thus increasing the distance covered through the air when the wind is at the jumper's back. Rear risers will offer the best penetration into the wind and with crosswind scenarios.

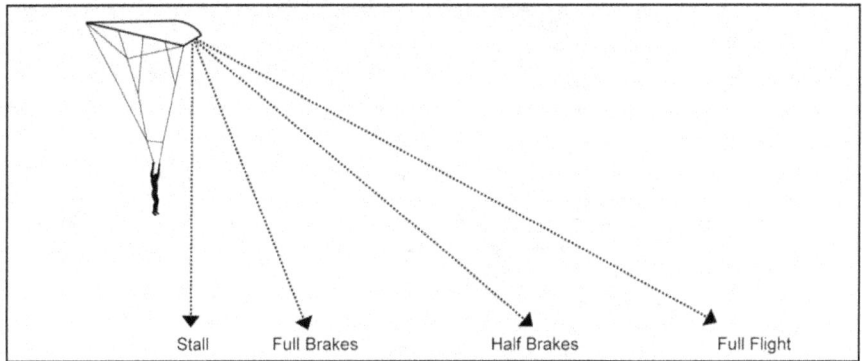

Figure 8-16. Brake-setting glide angles

Chapter 8

FULL FLIGHT (NO BRAKES)

8-49. The maximum canopy flight and penetration for maneuvering are obtained using full flight. The toggles are in the all-up position behind the rear risers (Figure 8-17). Full-flight maneuvering includes the following:
- Toggles are all the way up.
- Greatest forward speed of any toggle setting is 20 to 30 mph.
- Greatest descent rate of any toggle setting aside from sink/stall is 12 to 16 fps.
- It is not an acceptable toggle position for landing.

8-50. Full flight (run) is the quickest way to get from point A to point B when retention of altitude is not a concern. If getting blown backwards on final approach, it will allow a jumper to land closer to the spot than any other toggle setting. The canopy is *least* susceptible to turbulence at full flight because the canopy has the higher airspeed, which gives more pressurization to the canopy making it a more rigid wing. Applying brakes depressurizes the canopy by forcing air out of the wing and slowing the canopy (less airspeed equals less pressure). Also, the higher airspeed gets the canopy out of turbulence faster.

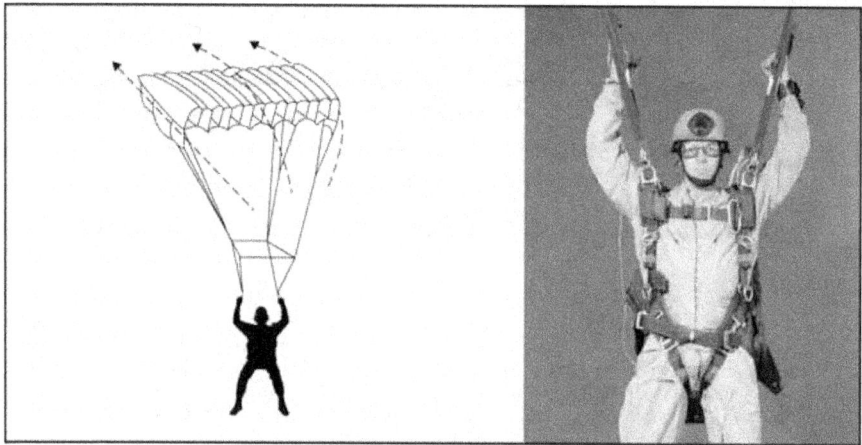

Figure 8-17. Full flight

HALF BRAKES

8-51. The parachutist grasps the toggles and pulls them down to about shoulder or chest level for the half-brakes position (Figure 8-18, page 8-15). The canopy speed will decrease to about a 9- to 12-mph flight, and the rate of descent will increase. This brake setting and both forward speed and descent rates are acceptable for landing. Half brakes also give a jumper the maximum flexibility to adapt to changing wind conditions, which is especially useful on final approach. Higher in the pattern, this brake setting is useful as it allows a jumper more time to make decisions and it allows for higher margins of error than do one-quarter brakes or full run. The half-brakes position includes the following:
- Toggle position is halfway between full run and stall point.
- Forward speed is 9 to 12 mph.
- Descent rate is 9 to 12 fps.
- It is an acceptable toggle position for landing.
- It is the SAFETY POSITION.

Figure 8-18. Half brakes

FULL BRAKES

8-52. The parachutist pulls the toggles to about waist level for full brakes (Figure 8-19, page 8-16). The canopy stops moving forward and the rate of descent increases. In the full-brakes position, the canopy is actually on the verge of a stall. The full-brakes setting is an extremely useful tool for making an accurate jump but is dangerous if used inappropriately. Many jumpers have been injured by using this brake setting at too low of an altitude. All canopies are prone to surging when coming out of this flight mode. The canopy can transition unexpectedly into a stall in the presence of turbulence and sustained use can result in canopy transitioning into a stall. Only experienced jumpers should use full brakes below 200 feet AGL and the canopy should be flying by 100 feet AGL. The full-brakes position includes the following:

- Forward speed is 0 to 5 mph.
- Descent rate is 16 to 24 fps.
- Variance between canopies can be significant.
- It is an extremely unacceptable toggle position for stand-up landings, but safe PLFs can be conducted at deeper brake settings.

STALL

8-53. A stall occurs when the parachutist pulls the toggles below the full-brakes position (Figure 8-20, page 8-16). The angle of attack of the parachute's nose and wing change produce a very great amount of lift for a short time. As the parachute loses forward airspeed and because the parachutist pulled the tail down lower than the nose, the canopy will attempt to fly backward and the rate of descent will increase to a hazardous degree. It is best used prior to entering the pattern or during the downwind leg. The stall maneuver should not be used lower than 300 feet AGL. It is extremely likely that a landing in this mode will result in a serious injury. Transitioning out of the stall maneuver should be done by smoothly raising both toggles and holding until the canopy begins flying again. The canopy should regain flight at the one-half or three-quarters brakes position. Snapping the toggles up or raising them to a very high setting can result in an extreme surge causing a rapid dive and high forward speed. The location of stall points can vary significantly from canopy to canopy. Minor variances in stall point location can exist on the same canopy depending on density altitude and wing loading. Positive identification of the stall point on every jump is critical for jump safety and performance. The stall point is located by slowly and smoothly lowering the

Chapter 8

toggles until the canopy ceases flying. To regain forward airspeed and flight, the parachutist slowly raises the toggles to the half-brakes or three-quarters position to raise the tail. The stall maneuver includes the following:
- Toggles are lower than full-brakes position.
- Forward speed is 0 mph.
- Descent rate is 20 to 26 fps.
- Canopy may have directional instability.
- MC-4 recovers by smoothly raising both toggles to the three-quarters brake position and holding until the canopy begins to fly.
- Toggles may need to be wrapped on the jumper's hands.

Figure 8-19. Full brakes

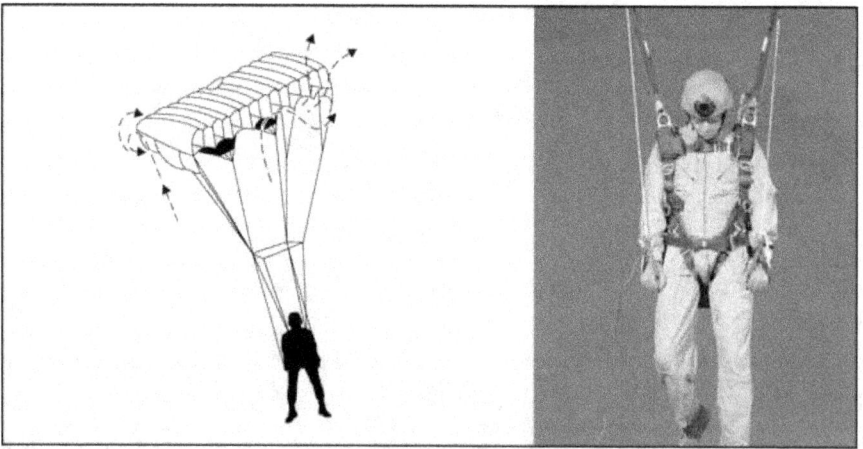

Figure 8-20. Stall

Ram-Air Parachute Flight Characteristics and Canopy Control

> **WARNING**
>
> The parachutist must not move the toggles quickly from the stall to the full-flight position, as the canopy will surge forward with an increased rate of descent. The parachutist must avoid stalling the ram-air parachute below 500 feet AGL.

TOGGLE TURNS

8-54. The parachutist can make turns from the full-flight, half-brakes, and full-brakes positions. Turns from full flight are very responsive, but because of the high forward speed, the turns will cover a wide arc. The parachutist makes these turns by depressing either toggle, leaving the other one at the guide ring. In this type of turn, the parachute will bank and actually dive, causing the parachute to lose altitude quickly. The further the parachutist depresses the toggle, the steeper the bank angle becomes.

SPIRAL TURNS

8-55. Spiral turns (full-glide turns) are basically turns from full flight but maintained (Figure 8-21) for more than 360 degrees of rotation. The parachute will begin diving in a spiral. The first turn will be fairly slow, with shallow bank angles, but the turn speed and bank angle will increase rapidly while the parachutist maintains the spiral. Spiral turns are effective tools for turning but subject the jumper to more banking than flat turns. Sustained spiral turns are an effective maneuver to lose altitude. Common mistakes made while employing the spiral turn include losing track of altitude and coming out of the turn in the wrong direction. This maneuver can often result in a jumper becoming dizzy, disoriented, and off target. It is extremely important that the jumper ensures the airspace is clear below and downwind prior to executing the turn. The spiral turn includes the following:

- Toggle is pulled all the way down on the side the jumper wants to turn while the other toggle is left at full run.
- Canopy banks similar to an aircraft.
- Canopy takes 4 to 6 seconds for the first turn.
- Turn rate, degree of bank, and descent rate will increase with time.
- Turns are held beyond a full revolution.
- The jumper will feel increased pressure in the harness and increasing airspeed.

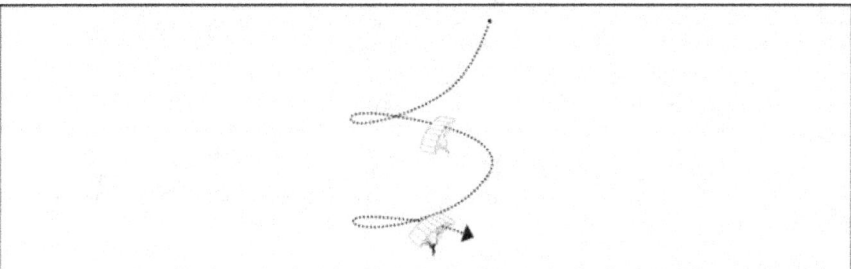

Figure 8-21. Spiral turn

Chapter 8

> **WARNING**
>
> Spiral turns will cause excessively fast diving speed with a rapid loss of canopy control. If the parachutist makes a spiral turn, it should not be conducted in proximity of other canopies or at any point in the traffic pattern. He must NEVER make a spiral turn below 1,000 feet AGL.

FLAT TURNS

8-56. The reduced banking associated with flat turns (off-hand) makes it the preferred type of turn for many parts of the jump. Using flat turns while turning between legs of the pattern helps to minimize altitude loss. Using flat turns for making corrections on final approach will result in a better sight picture. Also, gentle flat turns can be safely used to avoid obstacles near or on the ground. Turns from the half-brakes position result in almost flat turns. Flat turns (Figure 8-22) are generally preferred over spiral turns. Flat turns include the following:

- They are initiated while flying in a partially braked mode.
- Toggle is raised on opposite side from desired turn direction.
- Canopy banks much less compared to a full-glide turn.
- Canopy turns tighter.
- They will consume little altitude.
- Parachutist will remain close to vertical underneath the parachute.

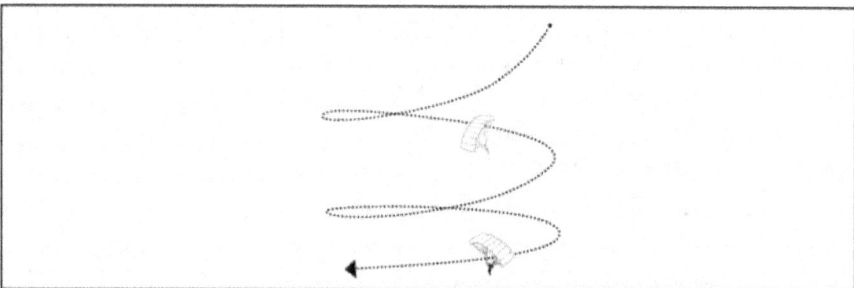

Figure 8-22. Flat turn

Note: The parachutist should use trim tabs located on the front risers to lose altitude, if required. During HAHO operation jumps, the trim tabs can be used to make changes in the angle of attack in order for a lighter jumper to stay in the stack with heavier jumpers. Using the trim tabs eliminates the need for a jumper to pull on the front risers during the entire canopy flight. Steering correction can be made with body weight shifts or minor rear or front riser turns until the toggles are released.

FRONT RISER TURNS

8-57. Sustained front riser turns are an effective tool for gaining vertical separation early in a jump. The same maneuvers listed under spiral turns apply to sustained front riser turns. Front riser turns are an effective maneuver to lose altitude. Common mistakes made while employing the front riser turn include losing track of altitude and coming out of the turn in the wrong direction. This maneuver can often result in a jumper becoming dizzy and disoriented. It is extremely important to ensure that the airspace is clear below and downwind of the jumper prior to executing the turn. Sustained front riser turns put the canopy in a dive with a significant amount of pitch and roll. The canopy will turn to the jumper's blind spot and the jumper will not be able to control the wing until the canopy is midway through the recovery arc. Front riser turns include the following:

- Toggles are to remain in the jumper's hands.
- Toggles should be put in full-flight position.
- Turns are initiated by pulling down either riser on side of desired turn.
- Jumper can grab any upper part of front riser.
- Canopy will bank significantly.
- Turn rate, descent rate, and speed will be very high and increase over time.

REAR RISER TURNS

8-58. Rear riser turns should be used in every jump to orient the canopy to the DZ after opening. If a jumper is faced with an imminent canopy collision after opening, an opposite riser turn will result in a quicker turn than unstowing the brakes and making a toggle turn. The second situation demanding rear riser turns would occur in the case of a broken brake line or detached toggle. The jumper can choose to either steer with rear risers or with the sole functioning steering line and one rear riser. Rear riser turns are similar to toggle turns in that they deflect the rear sections of the canopy. They require more force to execute because the jumper is pulling down a larger area of the canopy. Pulling down a rear riser will pull all of the C and D lines on that side compared with just the trailing edge for a toggle turn. Rear riser turns include the following:

- Jumper leaves toggles stowed if used in case of imminent collision after opening.
- Jumper releases toggles from stows if used in the case of a broken control line.
- Turns are initiated by pulling either rear riser down on side of desired turn.
- Jumper grabs the upper part of rear riser.
- Turns are effective but require more force than pulling down a toggle.

Note: Canopy will stall much quicker using rear risers than when using the toggles. In the event of a broken control line, jumpers should practice this landing technique above 1,000 feet before actually trying to land with it.

LANDING MANEUVERS

8-59. One of the most common sources of injury is during landing maneuvers. The jumper's judgment during this maneuver while flaring his canopy, if incorrect, could result in a high-speed impact with the ground or other hazards on the ground. Additional hazards that could contribute to bad landing maneuvers are changing wind conditions, turbulence during hot days, downdrafts close to the ground, crosswinds, and limited visibility. The following maneuvers will assist the parachutist during his landings.

HALF-BRAKE LANDING

8-60. Half-brake landings are often the best choice for jumps at night, as well as days consisting of low visibility, rough terrain spots that require a high degree of accuracy and/or have turbulence present. Virtually no timing is required so it is an easy maneuver to execute. Accuracy is increased due to the ease of maintaining a consistent sight picture. The canopy is more susceptible to turbulence at half brakes than full flight. If the jumper encounters turbulence close to the ground, he should conduct a full flare and prepare for a PLF. A somewhat lighter landing can be obtained by "punching out" from the half-brakes

setting. About 5 to 10 feet off the ground, the jumper slowly depresses both toggles to the stall point coinciding with touchdown. The jumper should be prepared to perform a PLF. The half-brakes setting—
- Provides for an acceptable landing due to the low descent rate and moderate forward speed.
- Allows for the greatest accuracy.
- Provides less chance of misjudging ground due to poor visibility.

Note: The parachutist can safely land the ram-air parachute in the half-brakes position. This procedure is especially useful during night or limited-visibility operations when he cannot see the ground or if recovering from a stall. He must be prepared to perform a PLF upon ground contact.

FLARE

8-61. The parachutist makes flared landings into the wind. He starts them at an altitude of approximately 10 to 15 feet, with room ahead for the actual touchdown. At 200 feet, he eases both toggles to the full-flight position, allowing airspeed to build. At about 10 feet above the ground (depending on wind conditions), he slowly pulls both toggles downward, timing the movement to coincide with the full-brakes position at touchdown. The flared landing, when properly executed, practically eliminates forward and vertical speed for a short period. If the parachutist slows down the ram-air parachute before the flare point, depressing the toggles will result in a "sink." On high-wind days, the parachutist must be aware that the canopy will react quicker during the flare; therefore, the flare should be conducted slightly lower to the ground. If the flare is conducted too high on a high-wind day, the parachutist may prematurely stall the canopy, falling backward on the ground. On low- or no-wind days, the parachutist must be aware that the canopy will react slower during the flare; therefore, the flare should be conducted slightly higher from the ground. If the flare is conducted too low on a low- or no-wind day, the parachutist may not have slowed the canopy down enough to perform a safe landing.

STAGED FLARE

8-62. Staged flares are a good compromise between half-brake landings and dynamic flares. Landings can be lighter than with a half-brake landing and a staged flare has two significant advantages over a dynamic flare. First, the staged flare does not require as much precision to execute. Second, the staged flare can be terminated and a safe brake setting can be held if timing is off and/or turbulence is encountered. Additionally, staged flares help a jumper develop his timing for executing a dynamic flare. Proper altitudes to initiate are very dependent on density altitude with high-density altitudes requiring the jumper to initiate at a higher altitude than at low-density altitudes. Staged flares—
- Convert forward speed of the canopy into lift.
- Require toggles be moved from full run or one-quarter brake setting to half-brake to full-brake point in incremental steps with pauses at half-brake and three-quarter brake settings.
- Allow jumper to hold safe brake setting if timing is off or turbulence is encountered.

LANDING APPROACHES

8-63. The ram-air parachute landing approach is similar to standard aircraft practice consisting of a downwind leg, a base leg, and a final approach upwind into the target. The pattern can be left or right hand, defined by the direction of turns used in the pattern. For example, a jumper flying a left-hand pattern would be making left-hand turns when turning between the legs of the pattern. The standard pattern offers many advantages to the jumper in the areas of accuracy and safety. The standard pattern allows for a good inspection of the jump spot and makes it easy to monitor changes in wind direction or speed during the jump. It also lends itself well to making adjustments for changing conditions. Most importantly, it provides an orderly landing sequence for multiple jumpers in the air. Components of the pattern include holding area, downwind leg, base leg, and final approach. The parachutist uses his altimeter to assist his visual altitude determination during the pattern for the landing approach.

Holding Area

8-64. The holding area is when the jumper is upwind of the landing area. The holding area—
- Is the point at which the jumper enters the pattern.
- Begins approximately 1,500 to 2,000 feet AGL upwind from the landing area.

Downwind Leg

8-65. The parachutist flies the downwind leg along the wind line, passing the target area at an altitude of 1,000 feet (depending on winds), about 300 feet to the side of the target passing the holding area. He continues the downwind leg about 300 to 400 feet downwind of the target (again, depending on winds). The downwind leg—
- Begins at holding area, approximately 1,000 feet AGL.
- Ends with turn onto base leg, approximately 750 feet AGL.

Base Leg

8-66. When 300 to 400 feet past the target, the parachutist begins a gentle 90-degree turn to fly the base (crosswind) leg across the wind line. He usually flies this leg at 30- to 60-percent brakes, depending on the wind conditions. He may either shorten or extend the base leg to reach the turning altitude. Under low-wind conditions, he flies the base leg to a turning point about 500 feet directly downwind of the target and at an altitude of 500 feet. The base leg—
- Begins at end of downwind leg, approximately 750 feet AGL.
- Ends with turn onto final approach, approximately 500 feet AGL.

Final Approach (Leg)

8-67. Under light-wind conditions (0 to 5 knots) and 500 feet directly downwind of the target, the parachutist makes a braked turn to turn toward the target. He completes the final turn at approximately 500 feet and no lower than 200 feet. On the final approach, braking techniques control descent and flight. The parachutist performs any major control corrections immediately to avoid obstacles only or to follow established landing direction while there is enough altitude and distance to the target. He lowers his equipment at 200 feet. The final approach (leg)—
- Begins at end of base leg (setup point), approximately 300 to 500 feet AGL.
- Lowers equipment.
- Ends with landing.

> **WARNING**
>
> **The parachutist avoids the turbulent air directly behind and above a ram-air parachute by flying offset to a parachute to his front or a minimum of 25 meters to the rear and above. He does not make sharp or hook turns on the final approach or attempt a 360-degree turn.**

8-68. Figure 8-23, page 8-22, shows approximate glide angles for a final approach flown at the half-brakes setting for the MC-4. Figure 8-24, page 8-22, shows an example of landing approaches.

Chapter 8

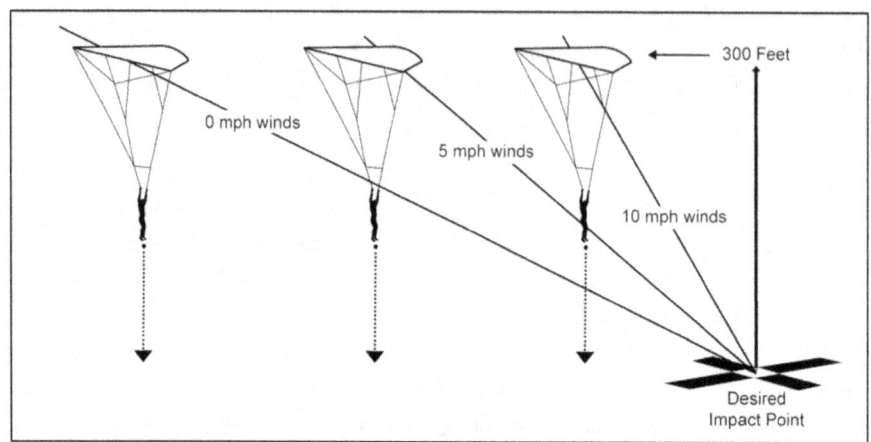

Figure 8-23. Glide angles for a final approach

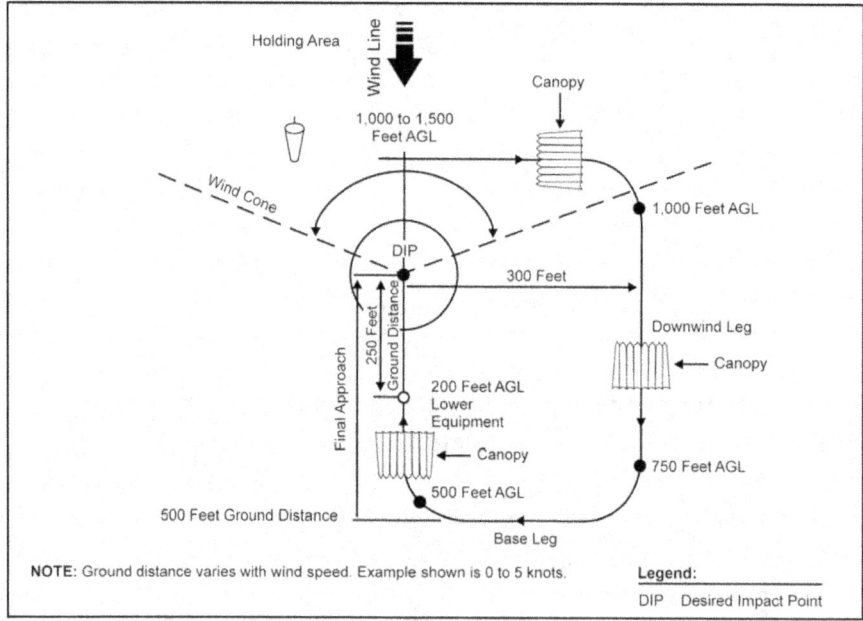

Figure 8-24. Landing approaches

Ram-Air Parachute Flight Characteristics and Canopy Control

> **WARNING**
>
> The preferred direction for landing is into the wind, but it is far more important to have all the canopies land in the same landing direction and follow the established pattern than it is to land into the wind. Any wind that is a quartering headwind is safe for jumpers of any level to land in.

> **WARNING**
>
> The parachutist maintains a sharp lookout for fellow parachutists at 500 feet AGL and below to avoid canopy collisions and entanglements. The lower parachutist has the right-of-way.

HIGH AND LOW WIND PATTERNS

8-69. The parachutist will tighten pattern as wind speed increases; downwind leg should be closer to wind line (Figure 8-25), and the base leg will move closer to the desired impact point.

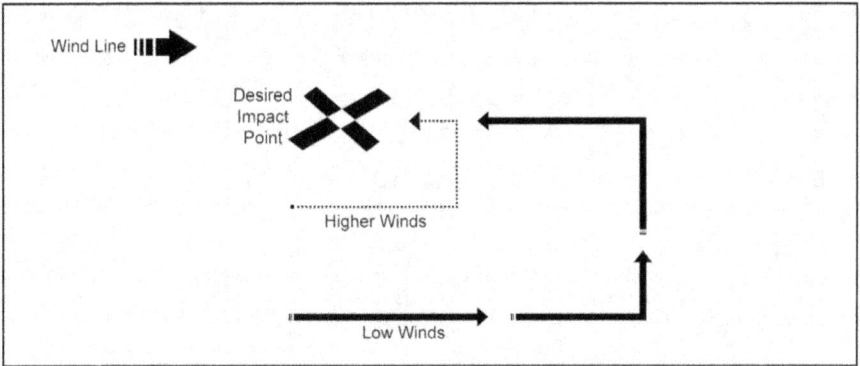

Figure 8-25. High and low wind patterns

ADJUSTING FOR CHANGES IN WIND DIRECTION

8-70. It is not uncommon for the wind to change during the course of jump operations. Most changes are minor and can be corrected for by slight adjustments on final approach. If the change in wind direction is significant (Figure 8-26, page 8-24) and it is recognized prior to entering the pattern, it may be preferable to shift the entire pattern. Shifting the pattern will increase the potential for landing directly into the wind, but jumpers should be cautious to avoid chasing the wind sock with light and variable wind conditions.

Chapter 8

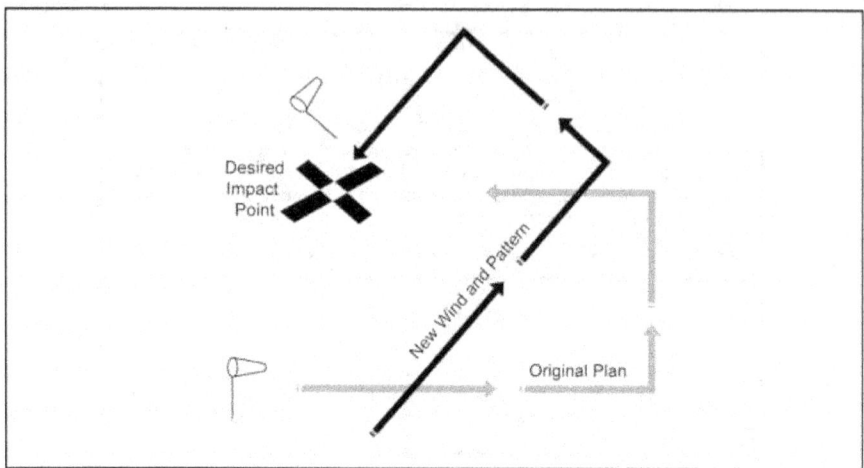

Figure 8-26. Significant change in wind direction

TURBULENCE

8-71. Turbulence is the result of an air mass (wind) flowing over obstructions on the earth's surface. Common obstructions are irregular terrain (bluffs, hills, mountains), man-made features (buildings, elevated roadways, overpasses), or natural features, such as tree lines. A disturbance of the normal horizontal wind flow causes turbulence. As the air mass moves around and over the obstruction, it transforms into a complicated pattern of eddies and other irregular air movements. Turbulence generally affects the flight of the parachute at the most critical time for the parachutist—the last 200 feet of canopy flight.

8-72. In general, with ground wind speeds less than 10 knots, both the windward and leeward sides of an obstruction cause small eddies 10 to 50 feet in depth. When wind speeds are between 10 and 20 knots, obstructions can cause currents that are several hundred feet in depth. Additionally, there will still be eddies on the windward and leeward side near the obstruction. At wind speeds greater than 20 knots, currents formed on the leeward side are carried considerable distances beyond the object that created them. Only minor eddies and currents form over smooth water surfaces. Turbulence is worse over choppy swells closer to the surface of the water because of the wind flow over a constantly changing surface configuration. Over mountains, even light winds (moving air masses) pushed up mountainsides or redirected down valleys can form major eddies and air currents that have violent, abrupt characteristics. Additionally, in HAHO operations in mountains or around hilly terrain, unstable air masses form currents that continue to grow in size and complexity. The resultant turbulence can extend up to thousands of feet AGL. Turbulence is also caused by heat rising off roads, concrete, and urban built-up areas and clearings.

8-73. An example of turbulence is the vortex created by aircraft taking off or landing. The turbulence created by these aircraft can invert smaller aircraft landing too closely behind them. Another example is the turbulence behind another parachutist's canopy. The parachutist who finds himself behind this canopy will feel the turbulence it creates. Turbulence can exist around any cloud mass. Individual clouds probably will not create turbulence. Clouds that mark the leading edge of an air mass probably will contain strong downdrafts. Cloud decks capping mountain ridges will contain very strong downdrafts and abrupt turbulence. Those type cloud formations will contain rapid pressure differentials. Altimeter readings should be suspect because the parachutist could be 1,000 feet lower than the indicated altitude on the altimeter.

Ram-Air Parachute Flight Characteristics and Canopy Control

The parachutist should avoid at all costs clouds that contain thunderhead activity because of the violent turbulence associated with those formations.

ADJUSTING FOR CHANGES IN WIND VELOCITY

8-74. When winds increase during the jump, it is important to recognize the change and make the necessary adjustments (Figures 8-27 through 8-29, pages 8-25 and 8-26). Most adjustments are fairly straightforward and entail reducing the distance traveled on the downwind leg or "cutting corners" of the base leg. Adjusting for a decrease in winds does not always require changing the pattern. Sinking from the original setup point will usually result in an acceptable sight picture. Extending the base leg is acceptable if the parachutist is the last jumper in the stick, but the potential for airspace conflicts increases.

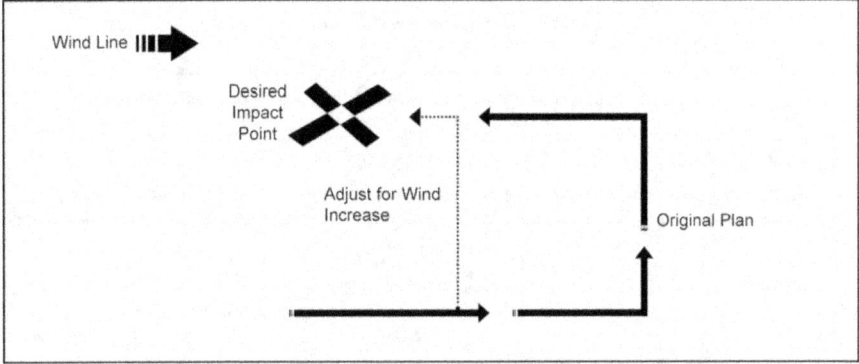

Figure 8-27. Adjusting for increase in winds on downwind leg

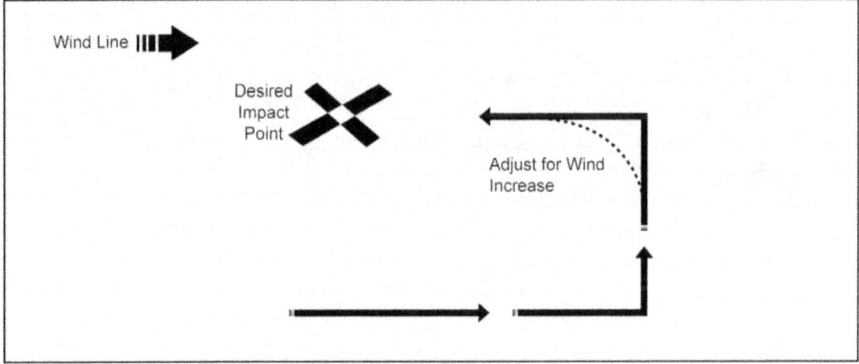

Figure 8-28. Adjusting for increase in winds on base leg

Chapter 8

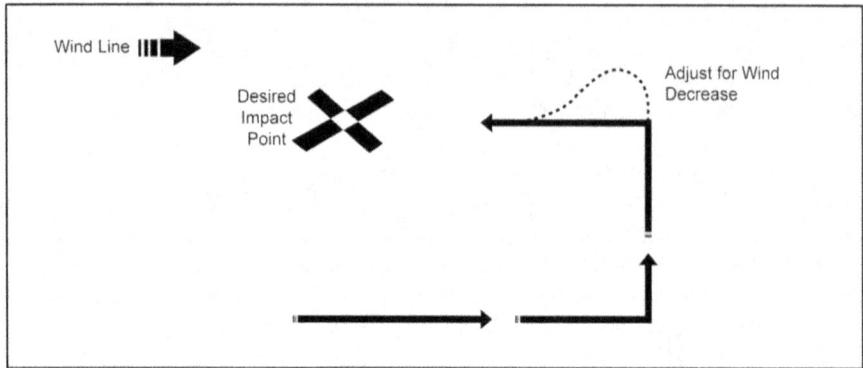

Figure 8-29. Adjusting for decrease in winds on base leg

LAND AND SEA BREEZES

8-75. The thermal differences of air masses associated with the interface along shorelines (oceans, lakes, and rivers) causes land and sea breezes. In the daytime, landmasses warm up faster than water. The air above the land rises, causing a lower air density than over the water. The air flows from the water over the land to replace the lower air density there. This phenomenon creates onshore breezes known as sea breezes (or lake breezes). It is most evident on clear, summer days in lower latitudes. The same phenomenon occurs in reverse in the evening because of the more rapid cooling of the landmass. The reversed process creates land breezes. The airflow over obstacles near shoreline DZs creates turbulence; when farther away from the coast, turbulence might not exist.

Note: If turbulence is encountered at altitude, parachutist should maintain full flight.

VALLEY AND MOUNTAIN BREEZES

8-76. Winds generally flow upslope on warm days in mountainous terrain. They flow downslope in the evening as the air masses cool. During the day, the winds create *valley breezes*; at night, the reverse process creates *mountain breezes*. These breezes, coupled with the airflow over obstacles, can cause strong and unpredictable turbulence.

Chapter 9

Emergency Procedures for Military Free-Fall Operations

Military free-fall airborne operations are inherently dangerous. Emergencies may occur before or during takeoff, during flight, while in free fall, or during canopy descent. Safety considerations require that each parachutist be able to recognize an emergency situation and react accordingly. Any departure from these emergency procedures may interfere with the parachutist's conditioned response. This action can lead to a delay at a critical time with the potential of causing injury or death. This publication strongly recommends that all parachutists follow these established procedures.

REFRESHER TRAINING

9-1. The conditioned response executed as the correct procedure for a particular emergency is a highly perishable skill. Refresher training must include performance-oriented training with special emphasis on emergency procedures and the actions required to respond successfully to any situation. This training must take place before each MFF airborne operation. The duration of the training should be commensurate with the time between airborne operations and, at the very least, until each parachutist is confident in his emergency procedure skills.

EMERGENCY MEASURES

9-2. The procedures established by this publication in response to emergency situations have proven to be the most successful in both MFF training and tactical environments. Figures 9-1 through 9-5 and Tables 9-1 through 9-5, pages 9-1 through 9-10, depict the emergency procedures that will be used with the RAPPS during emergency situations.

Parachutist—
- Learns the location of emergency exits and how to open them.
- Secures all loose items.
- Wears helmet.
- Fastens seat belt securely.

Figure 9-1. Emergency preparations before takeoff

Chapter 9

Table 9-1. In-flight emergency procedures and signals

Situation	Signal	Parachutist's Actions in Fixed-Wing Aircraft	Parachutist's Actions in Rotary-Wing Aircraft
Crash Landing: • During takeoff. • During flight.	• Continuous ringing of alarm bell or verbal warning by aircrew. • Six short rings of alarm bell or verbal warning by aircrew. • One long ring of alarm bell.	• Remains seated until aircraft stops, then exits. • If time and altitude permit, jumps. • If not, secures seat belt. • Braces for impact.	• Follows aircrew instructions. • Pulls legs inside aircraft. • Remains in position. • Covers head with arms. • Clears the aircraft as soon as it stops and moves well away from it. **NOTE:** Jumpmaster ensures all personnel are away from the wreckage.
		NOTE: Parachutists coordinate opening the aircraft exits with the aircrew.	
• Below 1,000 feet AGL.	• Six short rings of alarm bell or verbal warning by aircrew. • One long ring of alarm bell.	• Takes aircraft seats and fastens seat belts. • Prepares for crash landing. • Braces for impact.	• Takes aircraft seats and fastens seat belts. • Prepares for crash landing.
Emergency Bailout: • 1,000 to 3,000 feet AGL.	• Three short rings of alarm bell or verbal warning by aircrew. • Green light. • One long sustained ring of alarm bell.	• Prepares for exit. • Exits at the jumpmaster's command. • Deploys the reserve parachute immediately. • Attempts to land with the other jumpers.	• Exits at the jumpmaster's command. • Deploys the reserve parachute immediately. • Attempts to land with the other jumpers.
• Above 3,000 feet AGL.	• Three short rings of alarm bell or verbal warning by aircrew. • Green light. • One long sustained ring of alarm bell.	• Prepares for exit. • Exits at the jumpmaster's command. • Deploys the main parachute after a maximum 5-second delay. • Attempts to land with the other jumpers.	• Exits at the jumpmaster's command. • Deploys the main parachute after a maximum 5-second delay. • Attempts to land with the other jumpers.
Ditching Over Water With Insufficient Drop Altitude.	• Six short rings of alarm bell. • Verbal warning by aircrew. • One long ring of alarm bell.	• Remains seated. • Secures seat belt.	• Pulls legs inside aircraft. • Remains in position. • Covers head with arms.

Emergency Procedures for Military Free-Fall Operations

Table 9-2. In-flight emergency procedures

Situation	Jumpmaster/Jumper Responsibility
Ramp and Doors Closed: • Main pilot chute deploys. • Reserve pilot chute deploys.	• Shouts **"PILOT CHUTE"** and contains the pilot chute and canopy in the aircraft. • Moves the jumper to the front of the plane. • Disconnects the RSL and cuts away from the main parachute. • Jumper keeps the rig on, sits in the seat, and secures his seat belt. NOTE: In an aircraft emergency, the jumper can still exit the aircraft on his reserve. • Shouts **"PILOT CHUTE"** and contains pilot chute. • Moves the jumper to the front of the plane. • Removes the parachute system from the jumper. • Jumper sits in the seat and fastens his seat belt. • Jumper lands with the aircraft.
Ramp or Door Open: • Main pilot chute deploys. • Reserve pilot chute deploys.	• Same as above; the pilot chute is contained. • Same as above; the pilot chute is contained.
WARNING If parachutist is standing in the vicinity of an open door or ramp and he experiences a premature deployment, he tries to contain it; if any portion of the parachute goes out of the aircraft, he exits immediately to minimize or avoid serious injury.	
• Altimeter Failure.	• Gets the attention of the jumpmaster. • Replaces the altimeter with the spare in the jumpmaster bag. • If an altimeter is not available, the jumper will be moved to the front of the aircraft, seated, with seat belt fastened. NOTE: The jumper can exit the aircraft in an emergency situation.
• Equipment Malfunctions.	• Gets the attention of the jumpmaster. • The jumpmaster will correct the malfunction or make the determination for the jumper to land with the aircraft.

Chapter 9

Table 9-3. Emergencies in free fall

Emergency	Parachutist's Procedures
Collision on Exit.	• Maintains his arch, gently pushes off (with an open hand) the parachutist, regains his stability, checks his altimeter, checks the rip cords, and continues the MFF as planned. **NOTE:** The parachutist does not grab anything.
Instability in Free Fall: • Spinning.	• Counters, relaxes, arches, checks his hands and feet, and maintains altitude awareness. • If unable to gain control of spin, waves off and pulls.
• Tumbling.	• Arches, keeps his head up, checks his hands and feet, and maintains altitude awareness. • If unable to maintain altitude awareness and control tumbling, waves off and pulls.
• Entering a cloud or loss of visibility.	• Stops all movement and returns to a stable, relaxed arch. • Maintains altitude awareness. • Pulls at the prescribed altitude even if he is still in the cloud.
Rucksack Shifts.	• Counters any turns by turning in the opposite direction. **NOTE:** If the rucksack strap moves below his knee, parachutist makes one attempt to replace it while maintaining stability. If unsuccessful, he relaxes and attempts to fly. If the parachutist loses altitude awareness and is unable to gain control, he will wave off and pull.
Premature Opening: • Main parachute.	• Determines which canopy is deployed. ▪ Determined by the three (3) ring assembly, D-bag, and pilot chute. ▪ Conducts a controllability check.
• Reserve parachute.	▪ Determined by **NO** three (3) ring assembly, **NO** D-bag, and pilot chute. ▪ Conducts postopening procedures.
Collision Avoidance During Free Fall.	• Lower jumper has the right-of-way. • Never gets over the top of another jumper. • Uses forward glide, back slide, or side slide to get off a jumper's back.
Lost or Broken Altimeter.	• Immediately clears airspace, waves off, and pulls.
Lost or Broken Goggles.	• Maintains altitude awareness. • Maintains his arch. • Reaches up with both hands symmetrically (keeping elbows high), finds and replaces the goggles. • If unable to find the goggles, squints his eyes and maintains altitude awareness. • If unable to maintain altitude awareness, waves off and pulls.

Emergency Procedures for Military Free-Fall Operations

Table 9-4. Cutaway procedures

Malfunction	Parachutist's Procedures
Total Malfunction: Occurs when the canopy remains in the container assembly after the rip cord has been pulled. **Partial Malfunction:** Occurs when the container assembly opens but the canopy does not fully or properly deploy.	• Throws away the main rip cord. • Looks at the red cutaway pillow, elbow up, and grabs it. • Looks at the reserve rip cord handle, elbow up, and grabs it. • Arches with head up. • Pulls the red cutaway pillow to full-arm extension and throws away the red cutaway pillow. • Pulls the reserve rip cord handle to full-arm extension and throws away the reserve rip cord handle. • Raises right shoulder to disrupt the partial vacuum while continuing to look straight down to ensure reserve deployment. • Performs postopening procedures.

- Upon opening, hands go to rear risers.
- Clears airspace. Turns right to avoid collision, unless left is closer.
- Activates the strobe light, as required.
- Releases the brakes and gains control of the canopy; if controllability is questionable, performs a controllability check.
- If a malfunction cannot be resolved and if the canopy is uncontrollable, the decision to cut away must be made by 2,500 feet AGL and cutaway performed by 2,000 feet AGL.
- Orients himself to the drop zone.
- Locates other jumpers and achieves separation.
- Maintains altitude awareness.
- Checks rate of descent with other parachutists.

Figure 9-2. Parachutist postopening procedures

Parachutist—
- Releases the brakes.
- Looks left, clears airspace, and turns left 90 degrees.
- Looks right, clears airspace, and turns right 90 degrees.
- Determines the stall point.

NOTE: If the canopy requires more than 50-percent opposite toggle input to counter a turn, the canopy is uncontrollable. If the canopy stalls before 50-percent brake setting, the canopy is uncontrollable.

NOTE: If the canopy is uncontrollable, parachutist performs cutaway procedures.

Figure 9-3. Controllability check

Chapter 9

Table 9-5. Malfunction procedures

Malfunction	Parachutist's Procedures
Floating Rip Cord or Unable to See Rip Cord	• Arch, look. • If unable to see the rip cord or if it is floating, locates the cable housing on his right shoulder with his right hand. • Traces the cable housing down to where the rip cord cable protrudes out. • Makes a circle with his index finger and thumb and pulls to full-arm extension. • Makes one attempt; if unsuccessful, performs cutaway procedures.
WARNING Parachutist makes no more than two attempts to locate the rip cord (the initial attempt is the first attempt).	
Hard Pull	• If the pull is unsuccessful, comes across with the left hand in a punching motion and pushes the right hand and rip cord out. • If still unsuccessful, performs cutaway procedures.
Pack Closure	• Raises right shoulder to disrupt the partial vacuum while continuing to look straight down. • If main parachute does not deploy, performs cutaway procedures.
Pilot Chute Hesitation	• Raises right shoulder to disrupt the partial vacuum while continuing to look straight down to clear the burble. • If main parachute does not deploy, performs cutaway procedures.
Horseshoe	• Performs cutaway procedures immediately. • Makes no attempt to clear this malfunction.
Bag Lock	• Performs cutaway procedures immediately. • Makes no attempt to clear this malfunction.
Streamer/Snivel	• Reaches up and releases the brakes. • Pulls the toggles down to full-brakes position for 3 to 4 seconds. • Lets up slowly to 50-percent brake setting. • If the malfunction is not clear, makes one more attempt to pull the toggles down to full-brakes position for 3 to 4 seconds. • If the malfunction still has not cleared, performs cutaway procedures.

Emergency Procedures for Military Free-Fall Operations

Table 9-5. Malfunction procedures (continued)

Malfunction	Parachutist's Procedures
Hung Slider	• Reaches up and releases the brakes. • Pulls the toggles down to the full-brakes position for 3 to 4 seconds. • Lets up slowly to the 50-percent brake setting. • If the slider did not come down, pulls the toggles down to full-brakes position for 3 to 4 seconds. • Lets the toggles all the way up slowly. • If the slider did not come down below the cascade point on the lines, after two attempts performs cutaway procedures. • If the slider came down below the cascade point on the lines, performs controllability check and continues to attempt to get the slider down, maintaining air awareness.
Closed End Cells	• Reaches up and releases the brakes. • Pulls the toggles down to full-brakes position for 3 to 4 seconds. • Lets up slowly to 50-percent brake setting. • If the end cell did not inflate, pulls the toggles down to full-brakes position for 3 to 4 seconds. • Lets the toggles all the way up slowly. • If the end cells have not inflated, performs a controllability check. • If uncontrollable, executes cutaway procedures.
Pilot Chute Over the Nose	• Performs postopening procedures. • Performs a stall and recovery in an attempt to sling the pilot chute to the rear. • If the pilot chute did not go to the rear of the canopy, performs a controllability check. • If uncontrollable, executes cutaway procedures.
Premature Brake Release	• Immediately releases the opposite toggle/brake. • Performs postopening procedures.
Broken Control Lines	• Releases the brakes and steers with the remaining control line and rear riser. • Continues the postopening procedures. • At a safe altitude, determines the stall point with the rear risers. NOTE: Jumper lands using the rear risers; he *DOES NOT* land with one toggle and one riser.
Broken Lines (A, B, C, D)	• Determines which and how many lines are broken. • Performs canopy controllability check. • If uncontrollable, executes cutaway procedures.

Chapter 9

Table 9-5. Malfunction procedures (continued)

Malfunction	Parachutist's Procedures
Line Twists	• Reaches up with both hands and grabs the risers, thumbs down. • Pulls hands apart and kicks in a bicycle motion. • Maintains altitude awareness. • If still twisted by 2,500 feet AGL, executes cutaway procedures. NOTE: Jumper *DOES NOT* release the brakes until all the twists are out.
Holes and Tears (During Postopening Procedures)	• If the hole or tear is in the lower skin of the canopy, performs a controllability check. • If uncontrollable, executes cutaway procedures. • If the hole or tear is in the top skin of the canopy, immediately performs cutaway procedures.
Tension Knots (During Postopening Procedures)	• If a tension knot is noticed in the lines, reaches up and grabs the affected line group and pulls it down to his chest; releases the lines in a snapping motion in an attempt to clear the knot. • Repeats only twice. • If it fails to clear, performs controllability check. • If uncontrollable, executes cutaway procedures.
Dual Canopies: • Main inflated with reserve deployed but not inflated or is still in the D-bag.	• If brakes have not been released, leaves them stowed. • If brakes have been released, lets the toggles all the way up. • Slowly pulls the reserve in and places it between his legs. • Is prepared to perform cutaway procedures should the reserve inflate. NOTE: If the reserve starts to inflate, jumper waits until it is above shoulder level to perform cutaway procedures.

• Both the main and reserve canopies are deployed and inflated.	• First, determines if the canopies are entangled.
	• If the canopies are not entangled, separates the canopies and performs cutaway procedures.
	• If the canopies are entangled or unsure whether they are or not, assumes they are.
	• The canopies will be in one of three configurations:
	  ▪ Biplane, one behind the other.
	  ▪ Side by side.
	  ▪ Down plane.
	• The goal is to keep the canopies together; to do this, steers one canopy with the rear risers and turns that canopy into the other one.
	• Applies minimal input to the canopy to land into the wind safely.
	NOTE: Jumper *DOES NOT* release his brakes in a dual canopy situation.

Emergency Procedures for Military Free-Fall Operations

Table 9-5. Malfunction procedures (continued)

NOTE: There is no special emergency procedure associated with the use of NVGs during MFF operations. If a horseshoe malfunction occurs, the jumper should make no attempt to clear the malfunction and should immediately execute cutaway procedures.

Table 9-6. Canopy entanglement procedures

Situation	Higher Parachutist	Lower Parachutist
Lower parachutist is entangled with higher parachutist, and higher parachutist has a good canopy. Above 2,000 feet AGL.	• Attempts to clear off the lower canopy.	• If canopy cannot be cleared, checks the altitude. • Above 2,000 feet AGL, disconnects RSL and performs cutaway procedures.
	NOTE: If lower canopy is cleared, it should reinflate in 150 to 200 feet.	
1,000 to 2,000 feet AGL.	• Makes every effort to control lower canopy. • Must be prepared to do a PLF.	• Performs cutaway procedures. OR • Jettisons equipment. • Lands with higher parachutist. • Must be prepared to do a PLF.
Below 1,000 feet AGL.	• Makes every effort to maintain control of lower canopy. • Must be prepared to do a PLF.	• Jettisons equipment. • Lands with higher parachutist. • Must be prepared to do a PLF.
	NOTE: The higher parachutist should fly the final approach and land with half brakes.	
Both parachutists are entangled, and neither has a good canopy. At any altitude.	• Gets clear of entangled lines and cuts away (altitude permitting).	• Cuts away after the higher parachutist (altitude permitting).
	DANGER The higher parachutist may be fatally engulfed in the canopies if the lower parachutist performs a cutaway first.	
	• If still unsuccessful at 1,000 feet, both jumpers must deploy reserve parachutes in an attempt to slow the descent. • If only one reserve parachute deploys, the parachutist with the good reserve must bring the other parachutist to the ground. • If both reserves deploy, parachutists cut away from the entanglement. NOTE: Communication between the parachutists and altitude awareness are critical in successful disengagement.	

Chapter 9

Trees	Wires	Water
• Does not lower equipment; jettisons if it was lowered. • Turns canopy into wind. • Brakes as needed (50-percent or more braking position) to achieve vertical descent through the trees. • Prepares for a PLF. • Uses forearms to protect face while passing through trees. • If suspended, signals for assistance. • Attempts to land between smaller trees. NOTE: Goggles and oxygen mask provide additional face and eye protection.	• Avoids wires at all costs, even if a downwind landing is required. • Throws away rip cord. • Turns off oxygen. • Slows canopy down. • Streamlines body while passing through the wires. • If entangled, remains motionless until power is disconnected. • Prepares to do a PLF after passing through the wires. • If the parachute is entangled in the wires and contact with the ground is made, cuts away from the main chute immediately and moves away. NOTE: If time and altitude permit, parachutist unhooks the RSL and jettisons equipment.	• Jettisons oxygen mask and equipment. • Unhooks RSL. • Unfastens chest strap and waist strap. • Inflates flotation device, if available. • Turns canopy into the wind. • Uses brakes to slow airspeed. • After entering water, releases leg straps (as feet contact the water) and swims free of the harness and upstream from the canopy. • If being dragged in the water, cuts away the main canopy. • If trapped under the canopy, follows a seam to the edge. • Signals for assistance using emergency devices. NOTE: On entering water, parachutist must be prepared for a normal landing or a PLF.

Figure 9-4. Parachutist emergency landing procedures

Parachutist—
- After landing, releases one toggle and pivots in direction of retained toggle.
- Pulls the toggle hand over hand until either the canopy collapses or he has canopy fabric in hand.
- Attempts to run behind the canopy or downwind of the canopy.
- If unable to recover from a drag, ensures the RSL has been disconnected and pulls the cutaway pillow to release the main canopy.

Figure 9-5. High-wind landing procedures

ACTIONS FOR DUST DEVILS AND TURBULENT AIR

9-3. Parachutists should stay alert under canopy for signs of swirling or erratic wind conditions. The DZSO may use red smoke or flares to warn of visible turbulence, such as dust devils. Parachutists avoid turbulence at all costs by maneuvering away under canopy. If the parachutist is unable to avoid the turbulence, he should maintain full flight and remove all slack from the brake lines to prepare for a possible canopy collapse. If the canopy does begin to collapse, the parachutist should quickly conduct a 12- to 24-inch strike on the toggles to prevent collapse. Depending on the altitude, the parachutist should reattempt this procedure until the canopy reinflates or landing is imminent. As the parachutist approaches the ground, he should flare the canopy fully and be prepared to conduct a PLF.

9-4. If the parachutist lands and is overtaken by a dust devil, he should—
- Try to gather up the canopy.
- Lay down on top of the canopy.
- If unable to control the canopy, disconnect the RSL and cut away.

This page intentionally left blank.

Chapter 10

High-Altitude High-Opening and Limited-Visibility Operations

Standoff delivery techniques offer the commander a unique method for infiltrating trained operational elements. The RAPPS gives the commander a tactical capability to infiltrate these elements by parachute without requiring the aircraft to overfly the intended DZ. These elements can be released at an offset release point and navigate long distances under canopy. The flight characteristics of the reserve parachutes of the RAPPSs are identical to the main parachutes. This fact increases the chance of a successful infiltration should a cutaway from the main parachute take place because of a malfunction.

Note: For parachute systems that have a smaller reserve canopy than the main canopy, the mission commander planning the operation must plan for contingencies that address the reduced glide capability should a cutaway from the main parachute take place. Canopy openings at 6,000 feet AGL or above are considered HAHO jumps.

TECHNIQUES AND REQUIREMENTS

10-1. The parachutist uses a combination of delayed free-fall and HAHO techniques if making exits at an altitude above 25,000 feet MSL. He can also deploy his parachute at intermediate altitudes to minimize the chance of parachute damage or injury to himself upon canopy deployment, while using the glide advantage of the RAPPS.

WARNING

The maximum deployment altitude of the MC-4 RAPPS is 25,000 feet MSL.

10-2. The commander should consider altitude requirements when conducting training at altitudes. It is recommended that routine HAHO training be conducted at or below 19,999 feet MSL. Conducting training at lower altitudes eliminates the need for oxygen prebreathing and additional support personnel, and minimizes the chance of parachute damage and injury to the parachutist due to opening forces. The parachutist is also less likely to encounter physiological problems and cold-weather injuries.

10-3. HAHO standoff parachuting requires extensive airspace clearance. Additionally, this training must take place in areas having alternate DZs should the parachutist (or element) not be able to reach the primary DZ.

10-4. Accurate weather data is essential. Wind directions and speeds are critical for route planning. Air temperatures are important for preparing against exposure injuries. An excellent source of real-time accurate weather comes from artillery unit support. If the meteorological exploitation team weather section is deployed near the DZ, this organization has the ability to supply wind direction and speed every 1,000 feet up to 25,000 feet MSL or higher, depending on the conditions.

Chapter 10

> **WARNING**
>
> Icing conditions may occur at high altitude or during adverse weather conditions. Ice formation on the parachute canopy adversely affects its flight characteristics by increasing the rate of descent and decreasing its responsiveness. This condition results in less distance traveled under canopy with a decrease in canopy control.

SPECIAL EQUIPMENT

10-5. Individual body armor should be rigged and worn with all pouches in a manner that allows for a proper fit of the MC-4 RAPPS (Figure 10-1). The added bulk of the body armor will not affect the performance of the MC-4; however, jumpers and jumpmasters must ensure that all handles are free and clear of any equipment and easily manipulated during any emergency procedures.

10-6. Special precautions must be taken to prevent exposure injuries to the parachutist at high altitude. Gloves are necessary to protect the hands. The gloves, however, must not interfere with the manual activation of the main parachute or the performance of emergency procedures. The following paragraphs discuss special equipment that the parachutist should use. Future MFF individual clothing/protective equipment, like the USMC parachutist individual equipment kit in Chapter 2, should be looked at to determine the needs of the force when jumping in these extreme conditions to keep the parachutist comfortable and safe at any altitude or in extreme conditions.

Figure 10-1. Jumper with individual body armor

Toggle Extensions

10-7. Toggle extensions permit the parachutist to keep his hands at waist level during extended flights. They also allow for improved blood circulation to the hands and arms and lessen fatigue. Another technique is to leave the brakes stowed and simply steer the parachute using the risers to make needed corrections. If necessary, extreme cold weather gloves can be put on after canopy deployment to keep the parachutist's hands from numbing and to increase circulation.

> **WARNING**
>
> Jumper must not use the toggle extensions during landing for flaring. Toggle extensions affect the control range of toggle input to the canopy, and will not allow the toggles to be extended far enough for flaring the canopy during landing. Jumper should put wrist all the way through toggle extension loops. Extensions cannot be restowed once used; letting them hang beneath toggles can cause interference with jumper's equipment, which could inhibit the jumper's ability to flare.

Navigational Tools

10-8. HAHO parachute operations require the use of navigational tools to assist the parachutist with navigating from the predetermined "opening point" to the "point of impact." Oftentimes, tactical parachute operations are conducted at extreme high altitudes in order to maximize the parachutist standoff distance. Extreme high-altitude operations place a higher reliance on the GPS for navigating from the opening point to the intended point of impact. Temperatures at these extreme high altitudes have an adverse effect on the electronic and LED screen of the attached GPS unit and require additional thermal protection.

Compass

10-9. Each parachutist needs a compass to determine direction should he separate from the group or during limited visibility, such as when passing through cloud layers. A marine-type, oil-dampened compass that is unaffected by pressure changes or cold weather is recommended. The compass (Figure 10-2) must show direction regardless of its mounted attitude on the parachutist. The parachutist takes care when mounting the compass to avoid erroneous readings caused by interference from radios or other electronic navigation aids. He adjusts the declination of his compass while wearing all his accompanying equipment. This action will account for all magnetic variances caused by accompanying metal objects.

Figure 10-2. Compass mounted to high-altitude high-opening navigation board

Chapter 10

Electronic Navigation Devices

10-10. The parachutist mounts the electronic navigation or guidance devices by a multi-adjustable, articulating front plate for mounting the board to the parachutist by means of a variety of harnesses and Velcro to secure the front plate (Figure 10-3). The electronic navigation board is enclosed in a container and folded up on the chest during free fall, so as not to interfere with the manual activation of the main parachute or the performance of emergency procedures. The container with the electronic navigation device can be opened once the main canopy has deployed to assist the parachutist with the heading and other information when flying the canopy to the desired impact point. Parachutists should take caution that they not become fixated on the navigation device and to stay alert of their surroundings during flight. Considerations should also be taken into account that the use of electronic navigation devices may increase the likelihood of detection during infiltration.

Note: Due to the extremely cold temperatures at these altitudes, regular commercial hand warmers may be placed inside the electronic navigation devices for additional thermal protection for navigation devices that do not have their own heated storage box to keep the device from freezing.

Figure 10-3. Navigation aid attaching point

WARNING

The use of GPS-assisted parachute and aerial delivery operations is on the rise. Soldiers involved in these operations routinely bring GPS receiver/repeaters onto the aircraft to provide remote wireless in-flight updates to their jumper and/or cargo bundle guidance systems. Army Regulation (AR) 70-62, *Airworthiness Qualification of Aircraft Systems*, requires that carry-on equipment with an in-flight mission requirement be assessed by the appropriate Army-level Airworthiness Release (AWR) Authority to ensure the equipment will not negatively impact the aircraft or its subsystems. That assessment will determine the extent of airworthiness qualification/testing and documentation required for in-flight use. The failure to do so can lead to aircraft system problems ranging from inaccurate drop points to aircraft system shut-downs—particularly on aircraft equipped with more modern digitalized aircraft systems. Civilian contract air operators may experience similar aircraft system problems. Units are prohibited from using nonapproved systems in Army aircraft until they have been assessed and approved for use. The devices listed below have been issued an AWR and are currently approved for use on the CASA 212-200; however, their use requires disabling the aircraft's onboard GPS navigation system:

- Joint Precision Airdrop System (JPADS), Inc. Block III Kit and Receiver.
- GPSRKL1M-AXX-PM/5-TF Military Mobile L1 Repeater.

Requests to use equipment other than that listed above should be directed to: Commander, USASOC, ATTN: Special Operations Aviation, Special Programs Office, AOAO-SP, Fort Bragg, NC 28310-9610.

10-11. One type of navigation board that is being used is the Wilcox Parachutist Navigation Board described below and in Figure 10-4, page 10-6. Figure 10-5, page 10-7, shows the Wilcox Parachutist Navigation Board going from the closed (jump) position to the open (under canopy) position. Figure 10-6, page 10-7, shows the navigation aid attached to the parachutist. The red LED backlight serves as a visual indicator of heater function. When the LED backlight turns off, the heater will also turn off shortly thereafter. All components can be easily manipulated with gloved hands.

Chapter 10

10-12. The components (Figure 10-4) consist of the following:
- (1) A multi-adjustable, articulating front plate for mounting the board to the operator by means of a variety of harnesses. A Velcro adhesive assists in securing the front plate to the harness.
- (2) A release lever allows for easy access to the GPS unit by means of a hinged cover.
- (3) A heated GPS storage box protects the GPS unit from freezing at extremely cold temperatures and is capable of accommodating handheld GPS units up to 3.25 inches wide x 6.25 inches high.
- (4) An MA-230 standard military altimeter, included with the navigation board (heated version), features a red backlight for night operations.
- (5) A three-position power switch provides three modes of operation: Off, Low, and High.
- (6) The battery compartment houses a 123 3-volt lithium battery.
- (7) A standard ball compass features a white face and bold black lettering for ease of viewing during night operations. The ball compass features a red LED backlight that can be operated at high or low power and can accommodate a backup chemlight.
- (8) An MA-230 housing adapter provides a means for storing and using a backup mini chemlight for the ball compass in an emergency situation.
- (9) Up to four mini chemlights can be housed at 3 o'clock, 6 o'clock, 9 o'clock, and 12 o'clock, to provide a means of illuminating the altimeter in an emergency situation.

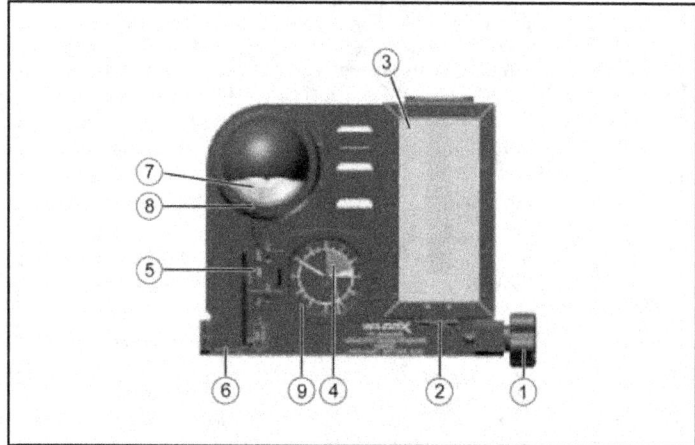

Figure 10-4. Wilcox Parachutist Navigation Board

High-Altitude High-Opening and Limited-Visibility Operations

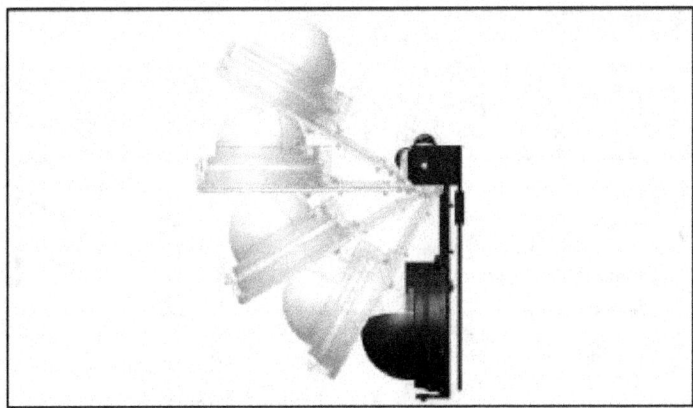

Figure 10-5. Parachutist navigation board closed and open position

Figure 10-6. Navigation aid attached to parachutist

COMMUNICATION EQUIPMENT

10-13. The parachutist can use radios for air-to-air or air-to-ground communications. He mounts the radio so that it does not interfere with the manual activation of the main parachute or the performance of emergency procedures. The radio can be mounted into a modified container that the waistband runs through to attach it to the parachutist. This procedure is approved for operations inside the aircraft or during canopy flight. The use of radios may increase the likelihood of detection during infiltration. The MICH interfaced with the oxygen mask (POM), AN/PRC-148 (multiband inter/intrateam radio [MBITR]), and MBITR interface is all the parachutist should need while on the aircraft and under canopy. The MICH/Peltor can also be plugged directly into the aircraft to communicate with the aircrew.

FREE-FALL DELAYS

10-14. As an aircraft increases altitude, the aircraft's true airspeed must increase to maintain a constant indicated airspeed due to decreased air density. True airspeed is the actual speed of the aircraft through the air mass. When true airspeed exceeds terminal velocity, the parachutist must allow for longer delays to decelerate to a safe speed for parachute deployment (Table 10-1).

> **WARNING**
>
> Failure to take the minimum required delay can result in serious injury to the parachutist and parachute damage.

Note: Jumpmasters must take into consideration the DZ (in feet AGL) for any delays in parachute opening during MFF operations.

Table 10-1. Required free-fall delays

Exit Altitude (in Feet MSL)	Delay
Below 20,000	4 seconds
Above 20,000	Pull altitude will be predetermined; pull altitude will be no less than 1,500 feet below drop altitude rather than a set time delay.

PARACHUTE JUMP PHASES

10-15. The HAHO standoff parachute jump has four phases. Each of these phases is discussed in the following paragraphs.

EXIT, DELAY, AND DEPLOYMENT

10-16. On the command GO, the group leader exits the aircraft. The remainder of the element exits the aircraft at designated intervals using the same exit technique as the group leader:

- Each parachutist free-falls for the required delay or until reaching the predetermined pull altitude.
- The exit interval will be established to ensure canopy separation between parachutists at opening. The exit interval will be based on the type of aircraft, its speed, and the mission requirements.
- A parachutist experiencing a malfunction must immediately start emergency procedures to minimize loss of altitude.
- Upon deployment, the group leader checks with the element for malfunctions, then assumes the initial flight heading. Should a member of the element be beneath the group, the element must execute the rehearsed tactical plan (lose altitude to reform the group or follow the low parachutist).

ASSEMBLY UNDER CANOPY

10-17. The opening altitude should be a minimum of 1,000 feet above any cloud layer to allow enough altitude for the element to assemble under canopy. Each parachutist flies his canopy to his rehearsed position within the formation. Each parachutist assumes the group leader's heading.

High-Altitude High-Opening and Limited-Visibility Operations

FLIGHT IN FORMATION

10-18. The "wedge" and the "trail" formations are the easiest to control and to maintain in flight (Figure 10-7 and Figure 10-8, page 10-10). The group leader (low parachutist) has the primary responsibility for navigation. All parachutists should have navigation aids when they jump.

10-19. During HAHO operation jumps, element members in the formation maintain relative airspeed and position with the group leader. They do this maneuver by trimming their canopies using the trim tabs (Figure 10-9, page 10-10) on the front risers and by braking. Doing so changes the angle of attack in order for a lighter jumper to stay in the stack with heavier jumpers, thus eliminating the need for a jumper to pull on the front risers during the entire canopy flight. Steering correction can be made with body-weight shifts or minor rear or front riser turns until the toggles are released.

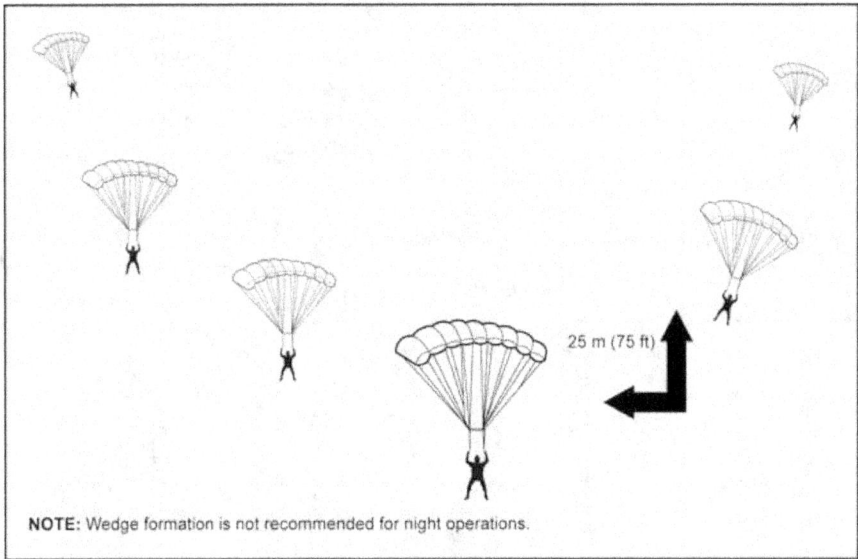

Figure 10-7. Wedge formation

Chapter 10

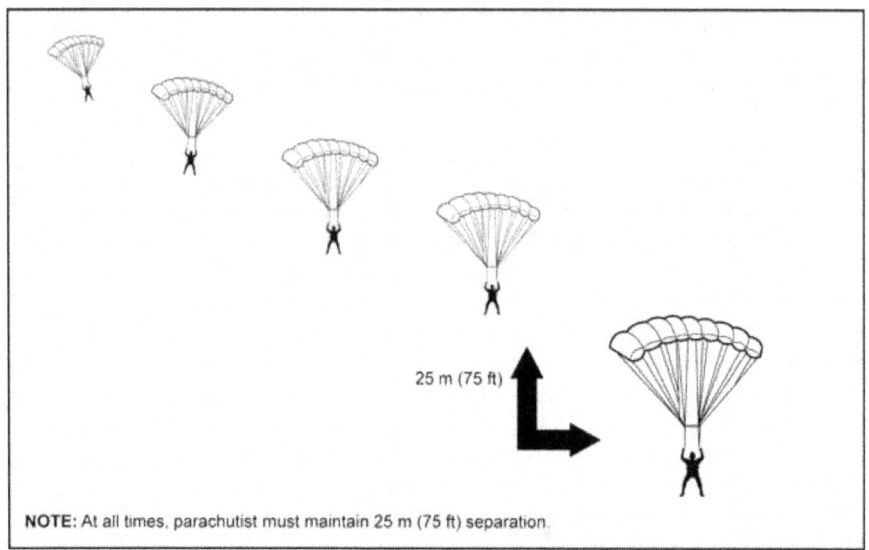

NOTE: At all times, parachutist must maintain 25 m (75 ft) separation.

Figure 10-8. Trail formation

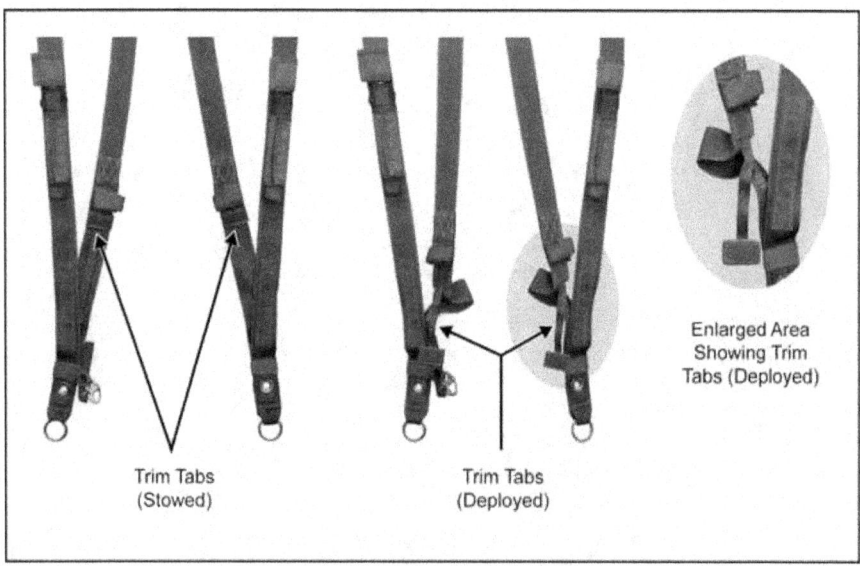

Figure 10-9. Trim tab locations

10-20. Under limited visibility conditions, such as when passing through a cloud layer, each parachutist goes to half brakes and maintains the compass heading until he regains visual contact with the formation or as stated in unit SOP. Each parachutist must maintain altitude awareness and keep a sharp lookout for other parachutists.

FINAL APPROACH AND LANDING

10-21. The group leader initiates the landing pattern at about 1,000 feet AGL in the landing area. Each parachutist removes any trim tab settings prior to 1,000 feet AGL to prevent injury on landing from the increased forward speed.

10-22. The landings are staggered to avoid the turbulence directly above and to the rear of the other ram-air canopies. Each parachutist prepares to do a PLF should visibility prevent him from seeing the ground.

LIMITED-VISIBILITY OPERATIONS

10-23. MFF infiltrations during periods of limited visibility (adverse weather or darkness) have a higher chance of success than strictly daylight operations. Limited-visibility infiltrations offer surprise and increased security due to reduced enemy observation capability. Limited-visibility operations require a high degree of skill and individual discipline. A well-rehearsed tactical plan executed by personnel proficient in MFF skills is critical to success.

ADVERSE WEATHER

10-24. Foggy, overcast, or mostly cloudy conditions effectively prevent observation from the ground. However, adverse weather conditions present special problems for the MFF parachutist. (Chapter 14 discusses weather factors.) High winds and precipitation can degrade canopy performance and make control difficult. Entering clouds may cause disorientation and lead to detachment separation under canopy, free-fall collisions, or canopy entanglements. The loss of depth perception due to ground fog, smoke, or haze may prevent the parachutist from executing a proper landing.

10-25. In free fall, the parachutist stops all maneuvering upon entering a cloud. He activates the main parachute at the designated altitude, even if he has not passed through the cloud layer. In clouds under canopy, he flies the canopy at the half-brakes position to help prevent a mid-air collision during limited visibility.

NIGHT OPERATIONS

10-26. Night MFF parachuting offers the same advantages as parachuting during adverse weather, especially during the first quarter, new moon, and last quarter moon phases. Night free-fall parachuting is the most psychologically demanding of parachute operations. Extensive training must take place at night. During this training, the parachutist develops confidence in the equipment and his abilities.

10-27. Commanders must weigh the tactical situation when placing lighting devices on the parachutist and on the parachute canopy for safety and control during free fall and canopy flight. At a minimum, illumination devices are used for altimeters and other instruments.

10-28. The use of oxygen dramatically improves night vision. Wearing the oxygen mask until the landing is a recommended procedure. The commander may consider using oxygen for all night free-fall operations, even if the jumping altitude does not require it.

10-29. The lack of depth perception at night may prevent the parachutist from executing a proper landing. The parachutist flies the parachute at the half-brakes position and performs a PLF on contact with the ground. Various night illumination techniques exist to identify parachutists, group leaders, or subunit elements while under canopy. Some techniques involve attaching the devices in the aircraft and some must be activated and placed on the canopy before packing the parachute. Some of these techniques include rheostatic electroluminescent riser lights, chemlights on the parachutist's body and on the risers, strobe light on the back of the helmet, and other electrical systems placed in pockets on the canopy's top skin.

MILITARY FREE FALL WITH NIGHT VISION GOGGLES

10-30. Operational areas frequently have little or no cultural lighting to illuminate DZs and objective areas during night parachute operations. Helmet-mounted NVGs improve the margin of safety for SOF MFF parachutists performing night MFF missions by providing visual cues to the DZ terrain features, and the ability to clearly see other jumpers and obstacles while under the canopy. Better vision translates into increased situational awareness during low-illumination deployments. NVGs are worn during MFF operations to reduce the risk of injury and improve the capability of MFF-coded elements by enhancing visual situational awareness during limited visibility. The jumpmaster can use NVGs to help him while spotting from the aircraft. The parachutist should also use them during canopy flight as an aid to navigation and formation flying. NVGs may be worn for all MFF operations; however, if they are not worn in the down-and-locked position during HALO operations, they should be in the up-and-locked position until after postopening procedures (USASOC Policy Number 20-10, *Wear of Night Vision Devices During Military Free-Fall Operations*).

Note: Airborne commanders/jumpmasters will verify that only jumpers who have completed NVG training participate in NVG-supported MFF operations. The jumpmaster will ensure that only helmets and NVGs listed in the USASOC Personnel Airdrop Systems (PAS)/Approved and Authorized for Use List (AAUL) are utilized during all MFF operations, as well as verifying that NVG rigging is done IAW the approved training support package.

10-31. The following is the minimum recommended qualifications prior to conducting MFF with NVGs:
- Experienced jumpmaster that has performed MFF jumps with NVGs within the past 120 days to train and determine if all jumpers are to the standards needed for this type of training.
- Four hours of ground training with hanging harness for riser manipulation and emergency procedures with NVGs (to include rigging and attachment procedures for NVGs).
- 15 to 30 minutes of wind tunnel flying with NVGs (recommended).
- Three day-familiarization jumps with NVGs (turned off).
- Two night jumps (no equipment, weapon, or oxygen) (NVGs powered on).
- Minimum 5,500 feet AGL training altitude with maximum 5-second delay before main canopy deployment.

Note: It is recommended that MFF NVG task-certified personnel perform this task a minimum of once every 120 days for currency.

WARNING

Jumpers must ensure the NVG mount remains in the LOCKED position. Jumpmasters will verify the lock is engaged and bungee cords are attached to helmet and NVGs during jumper inspection and at the 4-minute window before the jump.

NIGHT VISION GOGGLES AUTHORIZED

10-32. NVGs authorized for MFF NVG operations are the AN/AVS-6(V)3 (Figure 10-10, page 10-13), AN/PVS-14 (Figure 10-11, page 10-13), AN/PVS-15 (Figure 10-12, page 10-13), AN/PVS-23, and AN/VIS-9. (Reference USAF Warfare Center Air Combat Command Project 07-169A Final Report, *Free-Fall Night Vision Devices Tactics Development and Evaluation*, dated November 2008, and the MFF School, Yuma Proving Ground, Arizona, for the latest information.)

High-Altitude High-Opening and Limited-Visibility Operations

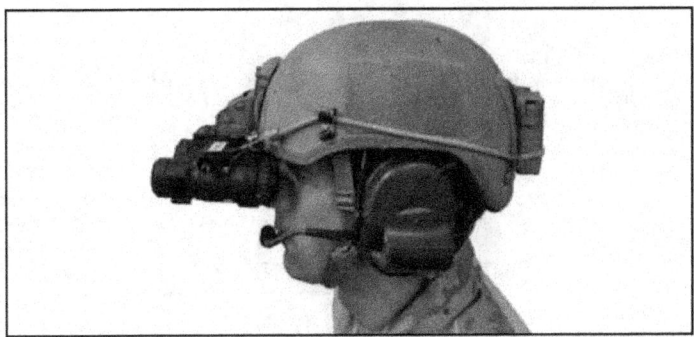

Figure 10-10. AN/AVS-6(V)3

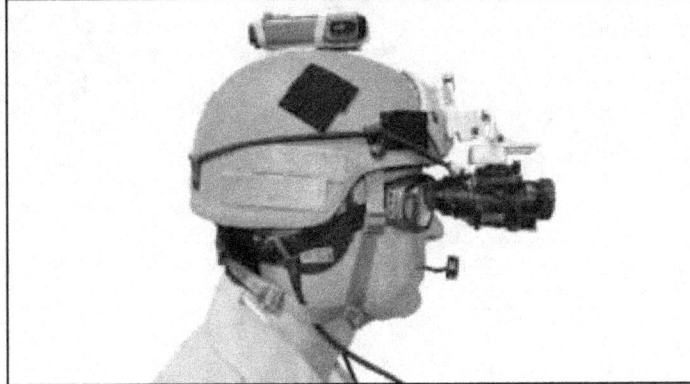

Figure 10-11. AN/PVS-14

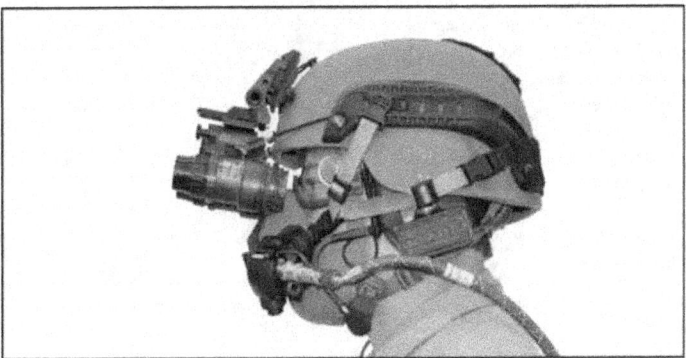

Figure 10-12. AN/PVS-15

Chapter 10

MOUNTS

10-33. Mounts (Figure 10-13) should provide a strong attaching point with a low profile from the helmet that is least obtrusive, snag free, and allows for the best ergonomics/adjustment with all goggles and is "permanent." The mount should have a break-away feature when exposed to harsh environmental and combat conditions to reduce injuries to the parachutist's head or neck if risers or parachute lines come into contact with the NVG or mount. Mounts should also have the capability and compatibility to switch from one type of NVG to another by only changing the mount arm.

Note: All manufacturer mounting installation requirements should be followed to reduce injury to the parachutist.

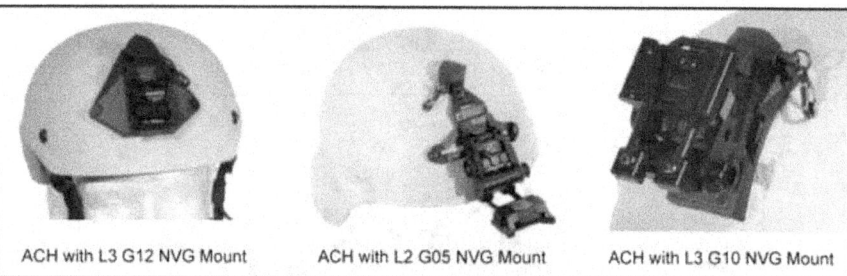

| ACH with L3 G12 NVG Mount | ACH with L2 G05 NVG Mount | ACH with L3 G10 NVG Mount |

Figure 10-13. Night vision goggle mounts

Note: Additional testing should continue to keep up with new NVGs and mounts to maintain safety for the SOF parachutist when conducting night MFF operations.

NIGHT VISION GOGGLE PREPARATION

10-34. Before each jump the parachutist should inspect the entire NVG system to ensure that all components are serviceable and free from any defects caused from prior training, combat, or airborne operations. This inspection includes the following:
- Check components for serviceability.
- Install new batteries.
- Clean lenses.
- Check visual acuity and focus to infinity, preferably using the NVG lane tester.
- Close the open end of the bungee cord hook fasteners by using paper tape and rigger's tape.
- Use 550 cord to provide a loop connection point to the NVGs.
- Ensure bungee strap connector is serviceable and attached to helmet and NVGs (Figure 10-14, page 10-15) and that the NVGs are pulled down and fit securely in place at the jumper's eye or eyes.

Note: The securing lanyard/bungee cord should be short enough so that if NVGs become dislodged from the mount, there is minimal slack, thereby reducing risk of horseshoe malfunction.

High-Altitude High-Opening and Limited-Visibility Operations

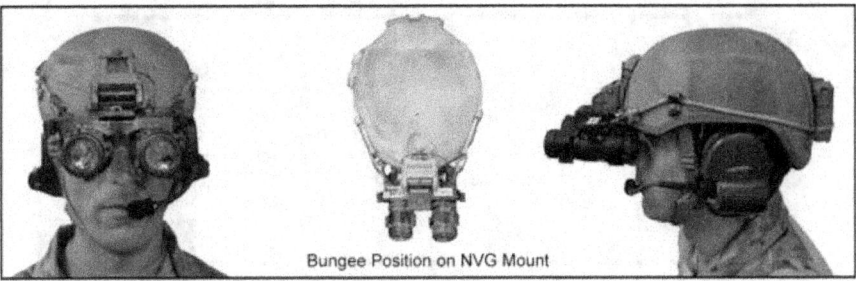
Bungee Position on NVG Mount

Figure 10-14. Bungee position on night vision goggle mount

Note: Bungee retainers will be a minimum of 5/32 inches (4 mm) and not larger than 1/4 inch (6 mm) with a hook not to exceed 2 inches in length. The bungee will be secured to the helmet with appropriate-sized cable clamps or a rail-mounted retention system. Cable clamps/ties will not have a loop larger than 3/8 inches.

HELMETS

10-35. Helmets authorized for MFF NVG operations are the MICH, 3/4 MICH, Maritime MICH, and the Protec. The MICH is preferred because of the ballistic protection it provides the parachutist during landing and follow-on combat operations once on the ground. Regardless of which helmet is used, it should fit snuggly to help prevent shifting on exit and while in free fall. Jumpers should expect the added weight of the NVGs to cause shifting and must be prepared to correct their helmet position as needed.

Note: Only helmets and NVGs listed in the USASOC PAS/AAUL will be used during all MFF operations.

Note: Some parachutists have found that the ACH/MICH helmet insert (padding) at high altitudes expands, freezes, and causes the helmet to get too tight causing pressure on the parachutist's head. Parachutists should not overtighten the helmet when conducting HAHO operations. When wearing the ACH/MICH with or without oxygen, the chin strap is routed underneath the chin.

Note: The ACH becomes the MICH when worn with communications equipment.

10-36. For helmet preparation, the parachutist should—
- Check components for serviceability.
- Install NVG mount as required and ensure proper helmet fit.
- Attach NVG to mount system and LOCK in position.
- Connect the bungee strap to the NVG (use paper tape to close the open ends of the bungee cord hook fasteners).
- Ensure the infrared (IR) strobe is operational and attach the chemlight.

Chapter 10

> **WARNING**
>
> Jumpers wearing NVGs have an increased chance of horseshoe malfunctions due to the additional helmet fixtures. The modified pull technique drastically reduces the chance for a horseshoe malfunction. All horseshoe malfunctions should be treated as such and jumpers should immediately execute cutaway procedures.

10-37. Per AFI 11-410, *Personnel Parachute Operations*, only oxygen masks certified and approved for use may be used. Oxygen masks will be fitted and inspected by a qualified jumpmaster with experience in jumping with NVGs. When jumping oxygen with NVGs, it is recommended that the oxygen mask be kept on until landing unless otherwise required by emergency procedures. The oxygen mask helps support the NVGs and releasing/lowering it may cause the NVGs to shift and restrict visibility of other jumpers and terrain.

COMMUNICATIONS

10-38. All MFF NVG jumpers should use radio communication to increase situational awareness of the team. The finger push-to-talk (PTT) device is recommended because it allows jumpers to maintain canopy control and communicate while flying.

GENERAL CONSIDERATIONS

10-39. MFF with NVGs can increase situational awareness and safety during reduced lighting operations, but there are several issues jumpers should consider. Proper training and rehearsal will help minimize these issues:

- Jumpers should know the pros and cons of NVG use, restrictions, and alternatives; for example, exiting with NVGs down and powered on, exiting with NVGs locked in the up position and turned off, or exiting with the NVGs in a pouch and putting them on after the jumper is under a good canopy.
- Jumpers must have an understanding of all equipment and materials needed to complete rigging procedures of the NVGs for MFF operations that include the approved types of NVGs, authorized mounting brackets, and a review of the USASOC PAS/AAUL authorized helmets available for use during NVG operations.
- NVGs should be focused to infinity to provide the best clarity while under canopy.
- If the helmet is not properly adjusted, it may shift after exit. The jumper may need to readjust the helmet after postopening procedures are complete.
- The normal sight picture for checking the altimeter will change when wearing NVGs. The jumper should practice checking the altitude on the ground and in the aircraft prior to exit. The jumper can look beneath the NVG to identify and read the altimeter.
- When beginning the pull sequence (Chapter 7), the jumper must be sure to look forward after identifying his main rip cord and before pulling it because the risers may hit the NVGs—potentially knocking them off.
- Turning his head to see jumpers behind him while under canopy is difficult. Lack of training and rehearsal of this can increase the potential for canopy collisions. A useful technique to look behind him without turning his head excessively is to grasp and push one side of the risers and "kick and twist" in the harness to rotate his body in that direction. The jumper should avoid unpredictable turns during canopy manipulation.

High-Altitude High-Opening and Limited-Visibility Operations

- NVGs limit the jumper's peripheral vision; therefore, he should fly accurate, predictable, and briefed patterns. Other jumpers are flying with the same limitations and extra attention must be given to situational awareness throughout the jump.

> **CAUTION**
> NVGs provide greater situational awareness during night MFF operations; however, jumpers should always be prepared to land with half brakes and to conduct a PLF.

JUMPMASTER CONSIDERATIONS WITH NIGHT VISION GOGGLES

10-40. Jumpmaster personnel inspections remain essentially the same for jumping with NVGs. The following additional procedures should be completed on each parachutist:

- Check all NVG components for serviceability.
- Ensure helmet is snug (nape and chin straps tight).
- Check NVG dovetail mount for proper attachment.
- Check that NVG mount is in LOCKED position.
- Turn ON and lower NVGs; verify ON and in proper position on the parachutist (if not, stop the jumpmaster personnel inspection and correct).
- Turn OFF and raise NVGs.
- Ensure straps are secured and taped.
- Ensure battery pack and power cables are secured to the helmet and properly stowed.
- Ensure bungee connector is serviceable and connected to the NVGs.
- Check IR strobe light for serviceability.
- Verify briefed lighting attachments and placement.

Note: Jumpmaster will instruct jumpers to lower NVGs and turn them ON at the STAND-UP call (2 minutes). The decision to jump with NVGs in the up or down position and turned ON or OFF will be made during rehearsals for the operation being conducted.

Note: While performing outside-the-aircraft spotting duties, the jumpmaster should hold the NVGs securely in place with one hand.

This page intentionally left blank.

Chapter 11

Military Free-Fall Drop Zone Operations

A DZ is any designated area where personnel and equipment may be delivered by means of parachute or free drop. DZs for MFF operations are selected during premission planning using all available intelligence sources. DZs are selected by the ground unit commander and are located where they can best support the ground tactical plan. The air mission commander recommends approach headings and selects initial and subsequent timing points based upon the routes to the DZ, terrain obstructions, ease of DZ identification, and enemy defenses. Final approval of selected DZs is a joint decision made by the ground unit commander and the supporting air unit. This chapter outlines the basic selection criteria, markings, and procedures used in support of MFF operations, as well as the qualifications and responsibilities of key DZ support personnel.

RESPONSIBILITIES

11-1. DZ size and selection are the joint responsibility of the air component commander or Commander, Air Force Special Operations Command (COMAFSOC), and the supported force commander. The supporting air unit is responsible for airdrop accuracy and safety of flight. The supported ground unit is responsible for establishment, operation, safety on the DZ, and the elimination or acceptance of ground hazards associated with the DZ. The jumpmaster is responsible for accuracy when jumpmaster-directed release procedures are used. AFI 13-217, *Drop Zone and Landing Zone Operations*, has additional information.

Note: For an MFF DZ using MC-4 or approved equivalent parachutes deployed in free fall or by static line, the jumpmaster will determine the minimum size DZ based on the number of personnel to be dropped, jumper proficiency, and the prevailing winds.

11-2. If a DZ is selected that does not appear in the Assault Zone Availability Report (AZAR), the unit must complete the survey request in full and it must state whether the unit has obtained permission to conduct the exercise. Any other information relating to the area being used as a DZ should also be stated, such as—

- Nearest facility capable of landing type of aircraft being used for mission (must include name, title, and phone number of the individual contacted for authority to land).
- Medical facility for medical evacuation and hospital support.
- Communications capabilities (for FLASH/priority of report).
- If the drop is to be made on civilian-owned land or on a non-Department of Defense government reservation, written permission from the owner/agency must be attached to the request.
- Airspace clearance from the Federal Aviation Administration or the local range control agency.
- Facilities available for storing and/or repacking of air items.
- Wind historical data for time and date of drop.
- Any other pertinent information.

11-3. The supporting air unit is responsible for airdrop accuracy and safety of flight. The supported ground unit is responsible for the establishment of a DZ, DZ operations, safety measures on the DZ, and the elimination or acceptance of ground hazards associated with the DZ. The jumpmaster is responsible for accuracy when jumpmaster-directed release procedures are used. AFI 13-217 has additional information.

Chapter 11

DROP ZONE SELECTION CRITERIA

11-4. The joint force commander gives guidance on DZ size in operation plans and operation orders. The ground unit commander selects the general area of the DZ where it will best support the ground tactical plan. DZ selection should be based on the following criteria:

- *Mission supporting.* Some of the main considerations when selecting a DZ that supports the mission are—
  - Method of insertion (HALO or HAHO).
  - Elevation and drop altitude.
  - Location and capability of enemy forces.
  - Recognizability during limited visibility.
  - Distance from the objective area.
  - Terrain between the DZ and the objective area.
  - Built-up areas.
  - Time available for movement to the objective area.
  - Amount of equipment being carried.
  - Physical characteristics of available DZs and surrounding areas.
  - Relative number of obstacles in the area.
  - Proximity to alternate and contingency DZs.
- *Supporting aircraft.* When considering the capabilities of the supporting aircraft, parachutists take the following into account:
  - Type of aircraft.
  - Capabilities of the aircraft.
  - Skill level of the aircrew.
  - Availability of backup aircraft if the primary aircraft has mechanical problems.
- *Infiltration route.* The primary, alternate, and contingency DZs should be selected so that the aircraft can overfly them in order without making major course corrections. Air routes to and from the DZ should not conflict with other air operations, restrictive terrain, restrictive airspace, or fall within the enemy's air defense umbrella.
- *Security.* The DZ must provide security from the enemy threat. The DZ should be located away from enemy positions and built-up areas.
- *Safety.*
- *Weather and astronomical conditions.* Seasonal weather and astronomical conditions in the area must be considered. If conducting a water jump, the tides, waves, currents, and sea state must be considered.
- *Size.* There is no minimum size for MFF DZs according to STANAG 3570 and AFI 13-217. The jumpmaster will determine the minimum size of an MFF DZ based upon the experience and capabilities of the parachutists. An area 50 meters by 100 meters is the recommended minimum DZ size for training.
- *Undesired landing areas (DZ hazards).* Some considerations include the following:
  - *Rising terrain*: Landing into the hill could cause injury and thermal updrafts could keep jumpers in the air longer.
  - *Tall timber*: Falling out of a tree might cause serious injuries; there is a high probability of a lengthy letdown of the reserve from the top of a tree. Snags have been known to fall over if landed in, which could also snap tops, increasing likelihood of injury. Hardwoods are brittle and can cause possible injury when landing in them. Turbulence near treetops could make it difficult to land safely at the desired impact point. Parachutes could get hung in trees disclosing the infiltration location.
  - *Side hill landings*: The hill's steepness could be a safety problem if a jumper does not contour the hill.

- *Power lines*: These could be hard to see, especially in fading light, and there is a greater risk for serious injury.
- *Fences*: These blend in with the landscape and present a hazard.
- *Deadfall*: High risk of extremities catching as the parachute carries jumper forward on landing.
- *Ice*: There is a higher probability of injury occurring if a jumper busts through ice. The situation will likely be more serious if bodies of water went undetected due to snowy landscape. There is a possibility of more jumpers being needed to help assist with recovery and medical attention, taking away from conducting mission.
- *Water*: Landing in water could require additional equipment and personnel to recover lost and damaged equipment. Recovery time could take away from mission.
- *Rocky ground*: Large outcroppings can be notorious for blocking wind. Landings could be rough. There is a possibility of more jumpers needed to assist with medical attention.

- *Aerial power line restrictions*. For the purpose of this publication, all restrictions apply to aerial power lines operating at 50 volts or greater. Power lines present a significant hazard to jumpers. Jumpers can sustain life-threatening injuries from electric shock and/or falls from a collapsed canopy. To reduce this hazard, power lines should not be located within 1,000 meters of any DZ boundary. If power lines are located within 1,000 meters of any boundary, coordination with the power company must be made to shut off power not later than (NLT) 15 minutes prior to TOT. If power cannot be interrupted, the flying mission commander, aircrew, and jumpmaster must conduct a risk assessment of the mission. Included, as a minimum, are the type of jump, jumper experience, aircrew experience, ceiling, and surface/altitude wind limits required to approve, suspend, or cancel the operation. To further minimize risks, consideration should be given to altering the mission profile to raise or lower drop altitudes, change DZ run-in/escape headings, or remove inexperienced jumpers from the stick. If possible, power lines should be marked with visual markings (lights, smoke, or VS-17 panels).

WARNING

At no time will military personnel attempt to climb power line poles to position or affix markings to wires or poles.

Note: During USAF MFF operations, aircrews should ensure the jumpmaster/team leader is aware when aerial power lines are within 1,000 meters of the intended point of impact. Non-USAF personnel will comply with their Service guidance for power line procedures and restrictions.

DROP ZONE SURVEYS

11-5. A DZ survey is required for all airdrop training missions involving U.S. personnel and equipment. Completing the DZ survey process involves a physical inspection of the DZ and documenting the DZ information on AF IMT Form 3823 (Drop Zone Survey). The using unit completes the DZ survey and forwards it through appropriate channels for review and approval. The using unit is defined as the unit whose personnel or equipment are being airdropped. The DZ survey review process involves the following steps:

- *Step 1*: The surveyor/MFF jumpmaster (AF IMT Form 3823, item 4a) physically surveys the DZ and completes the ground portion of AF IMT Form 3823. Once completed, AF IMT Form 3823 is forwarded to the ground operations review authority for approval (AF IMT Form 3823, item 4c). The ground operations review authority is normally the surveyor's commander or designated representative. This review ensures the AF IMT Form 3823 is complete, accurate, and meets the criteria for planned airborne operations.

- *Step 2*: Using unit forwards the survey to the USAF regional/wings tactic office for a safety-of-flight review (AF IMT Form 3823, item 4d). A safety-of-flight review is completed by an airdrop-qualified pilot or navigator on all DZ surveys. The purpose of a safety-of-flight review is to ensure an aircraft can safely ingress and egress the DZ.
- *Step 3*: Regional/wings tactic office forwards the survey to the appropriate operations group commander for review and final approval (AF IMT Form 3823, item 4e). This approval assures that the safety-of-flight review has been conducted and the DZ is considered safe for specified airdrop operations.
- *Step 4*: Once AF IMT Form 3823, item 4e, has been completed the survey is approved for use. Copies of the survey are forwarded to Headquarters, Air Mobility Command/DOKT, 402 Scott Drive, Scott Air Force Base, Illinois 62225-5320, for inclusion into the Zone Availability Report (ZAR) database.

11-6. The ZAR is a comprehensive listing of approved assault zones available for use by the Department of Defense. Use of the ZAR will expedite mission planning, enhance safety, and avoid duplication of surveys. Information contained in the ZAR does not replace the need for a completed DZ survey before conducting airdrop operations. Completed surveys are available via facsimile (FAX) on-demand system (also located at Scott Air Force Base, Illinois, at DSN 576-2899 or commercial [618] 256-2899).

DROP ZONE PERSONNEL QUALIFICATIONS AND RESPONSIBILITIES

11-7. The airborne commander designates key personnel for each airborne operation. These key personnel are the primary jumpmaster, assistant jumpmaster, safety personnel, oxygen safety personnel (when required), departure airfield control officer (DACO), DZSO/drop zone support team leader (DZSTL), and the malfunction officer (MO). The qualifications and responsibilities of DZ support personnel are listed in the paragraphs below. FM 3-21.220, *Static-Line Parachuting Techniques and Training*, includes further discussion of responsibilities during airborne operations.

DROP ZONE SAFETY OFFICER/DROP ZONE SUPPORT TEAM LEADER

11-8. The DZSO/DZSTL must be a commissioned officer, warrant officer, or noncommissioned officer (NCO) (E-5 or above for proficiency jumps; E-6 for tactical jumps). The airborne commander makes sure the DZSO/DZSTL is a current qualified static-line or MFF jumpmaster, has performed the duties of assistant DZSO/DZSTL in support of an airborne operation involving personnel or heavy equipment at least once, and is familiar with MFF operations IAW this manual. The MFF jumpmaster briefs the DZSO/DZSTL on the DZ markings, communications, and operating procedures that will be used.

11-9. The DZSO/DZSTL has overall operational responsibility for the DZ. He conducts a ground or aerial reconnaissance of the DZ before the drop to make sure there are no safety hazards. Other responsibilities include—

- Establishing personal liaison with the USAF drop zone control officer (DZCO) and STT, and discussing drop procedures (USAF troop carrier aircraft).
- Clearing the DZ of unauthorized personnel and vehicles.
- Briefing and posting road guards, if required.
- Ensuring medical personnel are in position.
- Ensuring that the DZ is operational 1 hour before TOT.
- Establishing communications with the DACO NLT 1 hour before TOT.
- Maintaining continuous surface wind readings NLT 12 minutes before TOT. (Peacetime ground wind training limits will not exceed 18 knots.) There are no winds aloft restrictions.
- Giving the pilot the ground winds and the CLEAR TO DROP or NO DROP signal 2 minutes prior to the scheduled TOT.

Military Free-Fall Drop Zone Operations

Note: The CLEAR TO DROP or NO DROP signal that is relayed to the pilot 2 minutes prior to TOT does not indicate the final wind reading. A NO DROP signal can be relayed to the pilot, any time afterwards, if surface winds increase beyond the authorized limit.

- Receiving from the pilot the number of parachutists that have exited the aircraft after each pass.
- Relaying strike reports to the aircraft pilot.
- During night drops, ensuring that all lights on or next to the DZ (except for DZ markings) are turned off 15 minutes before drop time and remain off during the jump.
- Directing the recovery crew to assist parachutists and to retrieve equipment in trees.
- Assisting in medical evacuation of injured personnel from the DZ.
- Immediately after the completion of the jump, asking the pilot if any personnel or equipment did not drop, and then relaying this information to the airborne commander on the DZ.
- In the event a malfunction occurs, securing the equipment and allowing no one to disturb it until the MO has completed his on-site investigation. If an MO or an NCO is not physically located on the DZ, the DZSO/DZSTL turns it over to an appropriate parachute maintenance facility.
- Recording the necessary information for the parachute operation report.
- Closing the DZ.

UNITED STATES AIR FORCE DROP ZONE CONTROL OFFICER

11-10. The USAF DZCO represents the airlift commander. He supervises all USAF personnel on the DZ. He also observes drop operations. Other responsibilities include—

- Evaluating all factors that might adversely affect safety.
- If conditions make drop operations unsafe, directing the STT to relay that information to the appropriate USAF commander as soon as possible and to display the established NO DROP signal on the DZ.
- Directing the use of STT equipment.
- Canceling drops when requested to do so by the Army DZSO.
- Keeping the Army DZSO advised on ground wind speed on the DZ.
- Preparing the necessary log and reports for submission to the airlift control element or the appropriate USAF commander.

SPECIAL TACTICS TEAM

11-11. The STT marks the DZs with proper navigational and identification aids. The STT establishes ground-to-air communications at DZs, as well as communications with designated control agencies. Other responsibilities include—

- Providing the U.S. Army DZSO with surface weather and low-level (up to 1,500 feet) winds aloft observations.
- Exercising air traffic control over aircraft in the vicinity of a specific DZ, as directed.

MALFUNCTION OFFICER

11-12. The investigation of personnel, parachutes, and equipment malfunctions receives the highest priority and is secondary in priority only to medical aid for the injured. This investigation supersedes all other aspects of the operation, to include ground tactical play. Prompt and accurate investigations and reporting could save lives and equipment. The report provides data to determine if a system or procedural training change is necessary to prevent future occurrences. The MO is subordinate to the DZSO/DZSTL and is a member of the DZ support team. Any assistance required by the MO must pass through the DZSO/DZSTL, who controls the DZ.

Chapter 11

11-13. The MO must be a commissioned officer, warrant officer, or NCO (minimum grade of E-5). The MO must be a trained parachute rigger who is familiar with airdrop, parachute recovery, and aircraft personnel parachute escape systems IAW AR 59-4, *Joint Airdrop Inspection Records, Malfunction/Incident Investigations, and Activity Reporting.*

11-14. The organization that provides the air items will provide the MO. He will be present on the DZ during all personnel and equipment drops and will be familiar with requirements. The MO must have the following minimum equipment in his possession during duty performance:

- A communication capability with the DZ control party.
- A good-quality camera to take photos of malfunctions or incidents (video camera preferred). Photographic equipment is essential for the proper performance of MO duties. Pictures of malfunctions greatly assist in investigations.
- The forms and clerical supplies necessary to tag equipment and initiate reports.
- Binoculars or NVGs.
- Transportation to move around the DZ.

11-15. If a malfunction occurs, the MO immediately conducts an on-site investigation of the causes of the malfunction. The MO photographs the malfunctioned equipment, or the malfunction as it happens, and the malfunction site that shows possible causes of the malfunction. The MO secures, identifies, tags, and numbers airdrop equipment involved in the malfunction incident. The MO then prepares and submits DD Form 1748-2 (Joint Airdrop Malfunction Report [Personnel Cargo]), to report all airdrop malfunctions (IAW AR 59-4/OPNAVINST 4630.24D/MCO 13480.1D/AFJ 13-210[1]), as well as any other required reports.

Note: MOs must prepare complete and accurate MFF accident reports. The fielding of new MFF equipment and the introduction of new MFF procedures depends on the feedback of the reporting process to detect accident patterns. There are several forms used by the Services in addition to DD Form 1748-2. Appendix E contains an example of the amount of detail that should be included in an accident report.

MILITARY FREE-FALL DROP ZONE MARKINGS

11-16. MFF infiltrations usually take place on blind DZs because of the general ineffectiveness of visual markings when viewed from high altitudes (HALO) and extended distances (HAHO). DZ identification is normally by location in relation to major terrain features.

11-17. DZ markings are sometimes used when the tactical situation permits, and it is desirable to indicate wind direction to the descending parachutists (Figure 11-1, page 11-7). FM 3-05.210, *Special Forces Air Operations*; FM 3-21.38, *Pathfinder Operations*; and AFI 13-217 outline approved marking techniques. Markers that can be used with the approved markers are the wind sock, wind streamer, wind blade, wind arrow, smoke, two-light method (one red and one green), wind "T," and IR light source.

11-18. There are several types of wind socks (Figure 11-2, page 11-7) that are used on airfields and DZs for MFF and civilian skydiving. Normally, a wind sock is what is seen on airfields for aircraft but can assist jumpers as well in determining the direction and velocity (somewhat) of the wind. Wind socks come in 5-, 10-, and 15-knot categories determined by the "erectness" of the wind sock to let the user know an estimate of the velocity. Naturally the jumper will have to know the type of wind sock in use in order for it to be of benefit for velocity approximation. For MFF, the 15-knot version is used since this upward end of the scale is closest to the maximum landing conditions. On most civilian DZs, the wind blade or tetrahedron is used. These do not give an approximation on the wind speed; however, they are much easier to be seen by jumpers in the air. On MFF DZs, the wind "V" or other wind direction device is required, and, depending on the DZ, there may be permanent wind socks, such as on Philips DZ at the MFF School in Yuma, Arizona.

Military Free-Fall Drop Zone Operations

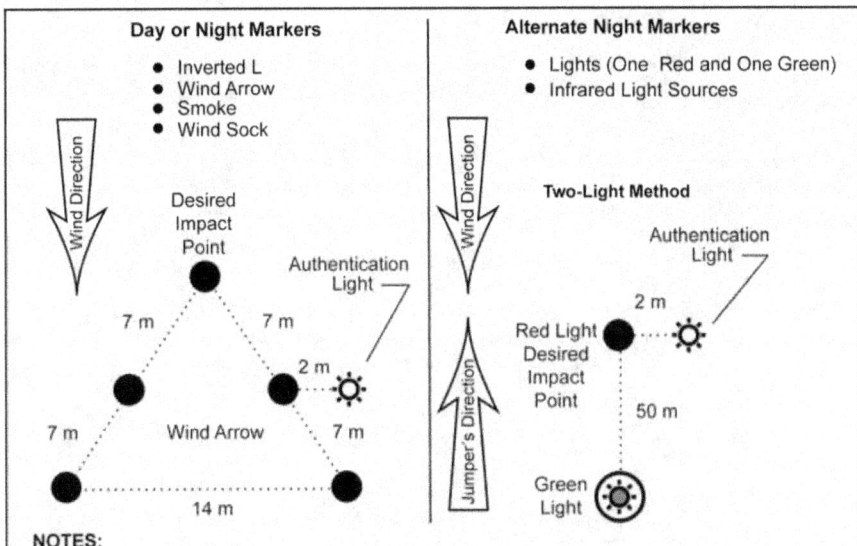

Figure 11-1. Military free-fall drop zone markings

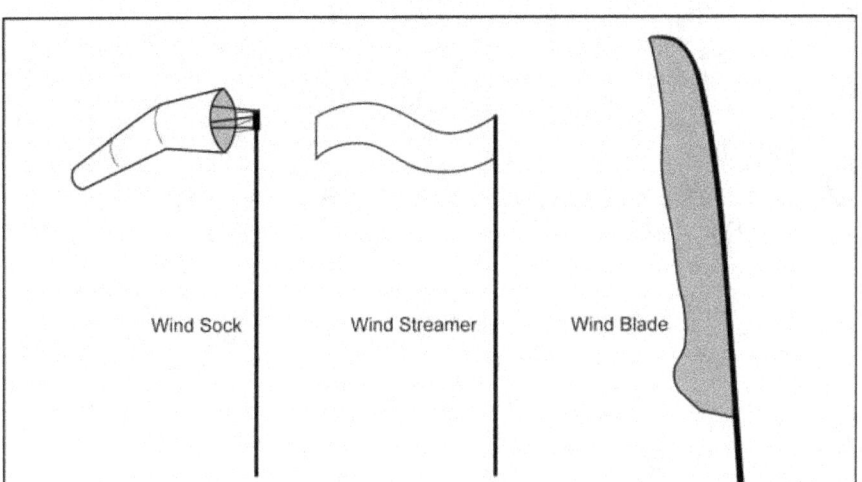

Figure 11-2. Examples of wind socks

Chapter 11

NONSTANDARD DROP ZONE MARKINGS

11-19. The tactical situation may dictate the use of nonstandard DZ markings. When nonstandard markings or identification procedures are used, it is imperative that all appropriate participants be thoroughly briefed.

11-20. The unmarked DZ is not authenticated with any type of visual or electronic marking. Unmarked DZs are normally used for contingency operations and may not have a DZ party present. Air Force Special Tactics personnel; combat rescue officers; pararescue; rescue squadron assigned or supporting survival, evasion, resistance, and escape (SERE) specialists; and USSOCOM-assigned forces are authorized to drop on unmarked DZs. During training missions, a DZ control party must be on site for safety.

11-21. The two DZ marking systems commonly used during MFF operations are the wind arrow and the two-light system (Figure 11-1, page 11-7):
- The wind arrow is formed by placing visual markers on the ground in the shape of an arrowhead. The arrow is aligned pointing into the wind. The arrow tip marker is placed on the desired impact point. Jumpers fly their approach to land facing the direction of the arrow.
- The two-light system consists of one red light and one green light. The red light is placed on the desired impact point and the green light is placed between 15 and 50 meters downwind. Jumpers will be briefed on the actual separation of lights. Jumpers fly their approach to landing from green light to red light.

WATER DROP ZONE MARKINGS

11-22. Water drops can be conducted on marked or unmarked DZs. Marked DZs will have mutually agreed-upon markings (visual or electronic). Markings that do not mimic local maritime navigational aids (buoys, channel markers, and so on) should be selected.

Note: Ground parties and aircrews must coordinate and brief NO-DROP markings for all types of DZs, to include water DZs.

EN ROUTE AND TERMINAL NAVIGATIONAL AIDS

11-23. A variety of electronic navigational aids are available to support DZ operations, including the tactical air navigation system (TACAN), zone marker, or radar beacons. These navigational aids are used at the discretion of the joint force air component commander, joint force special operations component commander, or mission commander. For MFF airdrops, the beacons will be placed on the point of impact (AFI 13-217).

HIGH-ALTITUDE RELEASE POINT AND MILITARY FREE-FALL DROP ZONE DETECTION

11-24. Location in relation to major terrain features identifies the HARP. Appendix F contains methods of computing the HARP. The HARP may be marked, if known, when the tactical situation permits. In heavily vegetated, mountainous, or urban terrain and during conditions of restricted visibility, DZs and HARPs may be difficult to detect. Electronic beacons or radar transponders and appropriate tracking devices help aircraft personnel and parachutists locate DZs or HARPs. Expedient methods, such as balloons and pyrotechnics, may also help aircraft personnel and parachutists locate DZs or HARPs. In situations where secrecy is important, aircraft and parachutists equipped with automatic direction-finding equipment may conduct drops using only the radio homing beacon. Parachutists may also use the GPS with portable terminals.

AIRCRAFT OR HIGH-ALTITUDE HIGH-OPENING TEAM IDENTIFICATION

11-25. In air-to-ground identification, the aircraft or HAHO team identifies itself to the reception committee by arriving in the objective area within the specified time limit. The aircraft or HAHO team also identifies itself by approaching at the designated drop altitude and track (aircraft).

11-26. In ground-to-air identification, the reception committee identifies itself to the aircraft or team by displaying the correct marking pattern within the specified time limit and using the proper authentication code signal.

AUTHENTICATION SYSTEM

11-27. There is no standard authentication system for unconventional warfare reception operations. During mission planning, the commanders concerned agree on the authentication system they will use. Signal operation instructions prescribe the authentication procedures.

11-28. Authentication may take the form of a coded light source, panel signal, radio contact, homing beacon, or combinations thereof. Authentication may be used individually or with the marking pattern. When using a homing beacon or radar transponder for authentication, the commanders concerned will jointly agree upon positioning and turn-on and turn-off times during mission planning.

11-29. Detachments conducting MFF operations during special reconnaissance missions are not going to have the opportunity in most cases of being assisted by a reception committee or lighted DZ during the infiltration. This type of operation will take additional planning and rehearsals on the detachment's part. Training with NVGs, electronic navigation devices, compasses, communications while under canopy, rough terrain landings, caching of equipment, thrall map and terrain analysis of desired impact point, and all contingency plans for lost jumpers or medical situations must be taken into consideration. This type of mission is not limited to just special reconnaissance; it can also be used to get into position for direct action by giving the detachment the element of surprise by not having large numbers of vehicles, hovering aircraft flying into the area, or foot traffic moving toward the objective causing the target to be empty upon arrival. Additional forces in vehicles or by helicopters (from staging areas) should assist the detachment as soon as the detachment begins to move on the objective. This timing is critical to the MFF detachment and the success of the mission.

This page intentionally left blank.

Chapter 12
Deliberate Water Military Free-Fall Operations

This chapter outlines the policies, procedures, and restrictions for conducting deliberate MFF operations into water DZs. Individual Services will use their applicable regulations and SOPs when conducting Service-pure MFF operations into water DZs. The procedures outlined in this chapter are different from the emergency water-landing procedures discussed in Chapter 9.

Note: When conducting joint deliberate water MFF operations, Services must determine if a waiver requirement for minimum exit altitudes with or without the use of an EAAD is required for joint operations and which regulation is to be followed due to the differences within the Services.

Note: The minimum exit altitude for Navy MFF parachutists conducting water jumps, to include following cargo over the ramp, is 2,500 feet AGL without EAADs. The minimum exit altitude for the Maritime Craft Aerial Delivery System (MCADS) with MFF parachutists following the boat over the ramp is 3,000 feet AGL without EAADs.

ADDITIONAL SUPPORT REQUIREMENTS

12-1. All basic parachute support operations outlined in Chapter 11 must be used when conducting deliberate water parachute operations. Listed below is the additional support needed for parachute operations using water DZs. Parachutists should refer to individual Service regulations for additional restrictions.

PARACHUTIST RECOVERY BOATS

12-2. A minimum of one power-driven parachutist recovery boat is required for every parachutist being dropped on the same pass if parachutists are not combat swimmer, combat diver, waterborne infiltration course, scout swimmer, or second-class swimmer certified. If the parachutists are combat swimmer, combat diver, waterborne infiltration course, scout swimmer, or second-class swimmer certified, then the requirement is one parachutist recovery boat for every four parachutists on the same pass. At 2 minutes from TOT, all engines must be running and the recovery boats must be circling the command and control boat before the CLEAR TO DROP signal is relayed to the aircraft. If conducting low-altitude drops and no ground-to-air communication is established, this formation will indicate a CLEAR TO DROP signal to the aircraft commander. To indicate a visual no-drop situation, all recovery boats will scramble from formation.

12-3. The number of parachutists exiting the aircraft per pass will be limited to the number of parachutist recovery boats available. Parachutist recovery boats must have an inflatable boat or ladder rigged alongside if they have a freeboard of more than 3 feet or if the boats do not provide an easy platform for recovery of personnel. Boats assigned as parachutist recovery platforms may only be used to assist in the recovery of equipment after all parachutists have been recovered. The boat coxswain cannot act as the DZSO/DZSTL, MO, safety swimmer, or medic.

Chapter 12

EQUIPMENT RECOVERY BOATS

12-4. A minimum of one power-driven boat is required for every two equipment platforms dropped on the same pass. Equipment recovery boats are to be used in the recovery of equipment parachutes and platforms.

12-5. Recovery boats assigned to recover personnel do not meet this requirement when parachutists and equipment are on the same pass. Equipment recovery boats must be large enough to recover cargo parachutes and platforms. The boat coxswain cannot act as the DZSO/DZSTL, MO, safety swimmer, or medic.

SAFETY SWIMMERS

12-6. Safety swimmers must be qualified swimmer/divers IAW Service publications. A minimum of one safety swimmer is required to be onboard each recovery boat. The safety swimmer must have fins, a facemask, a knife, a flare, and an inflatable life preserver. For night drops, safety swimmers should have a light that is visible for 1 mile (for example, a chemlight) and an emergency light visible for 3 miles (for example, a strobe light).

12-7. The safety swimmer will be used to recover personnel and equipment and assist parachutists, as needed. The safety swimmer cannot be assigned additional duties, such as the DZSTL, MO, boat coxswain, or medic.

PARACHUTIST REQUIREMENTS

12-8. Currency requirements for conducting deliberate MFF water jumps include the following:
- *Training before jump*. Commanders must ensure individuals meet the qualifications as specified in USASOC Regulation 350-2, *Airborne Operations*, Service regulation, and the unit air special operation procedures.
- *Parachutist swimmer qualification*. Parachutists must be qualified swimmers IAW Service regulations before making a water parachute drop.
- *First water jump*. Personnel must be current parachutists to conduct their first water jump. Their first water jump must be made during the day and without combat equipment.
- *First night water jump*. Parachutist training requirements for conducting night water jumps will be IAW Service publications.
- *Jumper currency*. Personnel who are not current can use a water jump for refresher provided it is done during the day and without combat equipment.

Note: The final decision for water jump training while deployed will be forwarded to the first O-6 in the chain of command for approval.

EQUIPMENT REQUIREMENTS

12-9. Equipment requirements for conducting deliberate MFF water jumps include the following:
- *Minimum equipment*. Each parachutist must have the following minimum equipment for a water jump:
    - Life preserver (Chapter 5).
    - Long-sleeved top or wet suit.
    - Booties, coral shoes, jungle boots, or equivalent.
    - Fins.
    - Helmet (equipment waiver).
    - Knife and approved day/night flare.
    - Chemlight (night operations only).

Deliberate Water Military Free-Fall Operations

- *Equipment waivers.* Helmets can be waived by the commanding officer based on operational requirements and a risk assessment (for example, wet suit hoods or cold weather hoods).
- *Flotation.* Parachutists must ensure they wear enough flotation to enable them to be positively buoyant in the water. If an injury occurs to the parachutist, he must be able to float without swimming.
- *Inflatable life jacket.* When using a UDT life preserver, parachutists must route the parachute harness chest strap (Chapter 5, Figure 5-71, page 5-59) underneath the life jacket to allow proper inflation in an emergency and not interfere with any emergency procedures.

> **CAUTION**
> Routing the chest strap over the UDT vest will prevent the life vest from inflating properly and may cause injury to the parachutist.

- *Altimeters.* Altimeters are required for every jump except water jumps with delays less than 10 seconds. Units should coordinate for waivers when conducting deliberate water MFF parachute operations without an altimeter IAW their Service regulations.
- *Automatic rip cord releases (ARRs).* ARRs are required for all MFF parachute operations. The Military CYPRES 2 can be used during water operations. The Military CYPRES 2 is waterproof to a depth of 15 feet (5 meters) for a duration of 15 minutes. Procedures as outlined in the user's guide for water operations must be followed in order to retain serviceability of the Military CYPRES 2. The supporting parachute rigger activity will identify by serial number and track by annotating in shop records the following information: date, DZ, salt or fresh water, and estimated depth and duration of submersion for all Military CYPRES 2s that have been used for water operations. Additionally, they will perform post-water-operation procedures as outlined in the user's guide. Commanders shall be advised of the probable cost involved to replace the Military CYPRES 2 in the event that guidelines for water operations are infringed.
- *Safety lanyards.* Only 80-pound cotton tape is authorized as the safety lanyard for swim fins. The safety lanyards must be short enough not to catch or snag on anything during exit.
- *Reserve static line.* When making a deliberate water jump with the military RAPPS, parachutists must disconnect the RSL once they have a good canopy over their heads. This action will prevent the reserve from being deployed if the main is cut away while in the water.
- *Placement of fins.* During an exit for a water parachute drop, the jumper may wear his fins as described in one of the three methods listed below. From each configuration, the parachutist must be able to put the fins on either under canopy or in the water. The fins may be—
  - Worn on feet as normal with 80-pound safety lanyards. This method may be used if the parachutist does not have to walk far to exit. Short fins are recommended if the parachutist must walk in the aircraft to exit.
  - Taped vertically to shins with foot through strap and 80-pound safety line. Holding the fin vertically with the strap down, the parachutist places his foot through the fin strap. He tapes the top of the fin to the front of his leg, folding the end of the tape over to make a quick-release tab. He then secures the fin to his ankle with a short piece of 80-pound cotton tape.
  - Attached or fastened to a separate belt. The fins must be worn in front on the parachutist's thigh or in the back under the pack tray. Fins must be placed so as not to interfere with parachute deployment or the parachutist's ability to remain stable during free fall.

12-10. Whenever possible, the parachutist should wear his fins on exit. If the parachutist does not have his fins on during exit, then he should wait to put them on until **after** entering the water. Doing so allows the parachutist to concentrate on canopy grouping at low altitudes. Aircraft configuration and SOP will determine the proper location.

DROP ZONE REQUIREMENTS AND MARKINGS

12-11. DZ requirements and markings for conducting deliberate MFF water jumps include the following:
- *Establishment of the DZ.* The DZ must be established not less than 60 minutes before the TOT to allow time for the DZSO to monitor DZ conditions.
- *Surface winds.* Surface winds (Table 12-1) shall not exceed 18 knots.
- *Sea state.* Sea state (Table 12-1) shall not exceed limits IAW Service publications.
- *Water depth.* The depth of the water must be at least 10 feet.
- *Water temperature.* Minimum safe water temperature for personnel drops is 50 degrees Fahrenheit (10 degrees Celsius) unless an appropriate exposure suit is worn. Partial or full exposure suits should be considered whenever water temperatures are below 72 degrees Fahrenheit.
- *Air-to-ground communications.* Personnel must establish a positive visual or electronic signal for DZ identification before the drop for water parachute operations. Only a positive visual or electronic signal for DZ identification is required; however, radio communications are highly recommended to assist in verifying the DZ. (USASOC units require radio communications.) Parachutists must use positive night visual signals (for example, beacons or strobes) for night drops to avoid confusion and to aid in positive identification.
- *DZ communication.* All DZ safety craft must be equipped with boat-to-boat radio communications.
- *DZ configuration.* The DZ is configured IAW Service regulations.

Table 12-1. Wind/sea state observation chart

Wind Velocity	International Description	Wind Force (Beaufort)	Average Wave Height (Feet)	Sea Indications	Sea State
<1	Calm	0	0	Like mirror.	0
1–3	Light Air	1	0.05	Ripples with appearance of scales.	0
4–6	Light Breeze	2	0.18	Small wavelets; crests have glassy appearance but do not break.	1
7–10	Gentle Breeze	3	0.6	Large wavelets; crests begin to break; scattered whitecaps.	2
11–16	Moderate	4	2.0	Small waves, becoming longer. Fairly frequent whitecaps.	3
17–21	Fresh	5	4.3	Moderate waves, taking a pronounced long form; many whitecaps.	4
22–27	Strong	6	8.2	Large waves begin to form; white foam crests more extensive; some spray.	5
28–33	Near Gale	7	14	Sea heaps up, white foam from breaking waves blown in streaks along direction of waves.	6
34–40	Gale	8	30	Moderately high waves of greater length; crests break into spindrift; foam blown in well-marked streaks in direction of wind.	7
41–47	Strong Gale	9	36	High waves. Dense streaks of foam; sea begins to roll; spray affects visibility.	8

PARACHUTIST PROCEDURES FOR WATER JUMPS

12-12. Parachutist procedures for conducting MFF water jumps include the following:
- *Water parachute jump.* Procedures for a premeditated water parachute jump after exiting the aircraft are described below. Parachutists—
  - Check parachute and locate other parachutists. Parachutists turn canopy toward the DZ.
  - Disconnect RSL and release waistband.
  - Continue to steer and group with other parachutists to the target.
  - At no lower than 200 feet above the water, turn into the wind and release the chest strap (500 feet is recommended with combat equipment).
  - Confirm leg strap snap hook locations.
  - Flare canopy to land (land with half brakes for night jumps).
  - After entering the water, release leg straps and crawl out of the harness.
  - Put fins on, if required.
  - Swim to the center of trailing edge (tail).
  - Hand the center of the trailing edge (tail) and harness to recovery boat.
- *Reserve static line.* When making an MFF water jump with the MC-4, parachutists must ensure they disconnect the RSL once under a good canopy. This action will prevent the reserve from being activated if the main is cut away while in the water.
- *Life preserver use.* If the parachutist is unable to stay above the water, he must either add air using the oral inflation tube or inflate his life preserver with the CO_2.
- *High winds.* If a parachutist is being dragged in high winds, he must roll over on his back and attempt to collapse the canopy by pulling in on one steering toggle. If this is not possible, he then performs a cutaway on the RAPPS. He must ensure the RSL system is disconnected before cutaway of the main.
- *No-wind landings.* In a no-wind landing condition, the canopy may possibly land on top of the parachutist. If this occurs, parachutist must remain calm and avoid getting tangled in the suspension lines. He should create an air pocket by splashing the water and lifting the canopy above the water. Then he finds a seam and follows it to the edge of the canopy. In an emergency, the parachutist uses his knife to cut through the canopy.
- *Equipment flotation.* The reserve parachute will float for a short time; however, if the parachute starts to sink, parachutist should make no attempt to hang on or recover it.

DROP ZONE PROCEDURES FOR PICKUP OF PARACHUTISTS AND EQUIPMENT

12-13. DZ procedures for pickup of parachutists and equipment include the following:
- *Recovery boat assignments.* Recovery boats must have assigned duties by the DZSTL so as to minimize confusion during the recovery procedure. These assignments must be briefed by the DZSTL/DZSO before setting up the DZ.
- *Recovery priority.* Recovery boats will first pick up any parachutist who signals he is in trouble or has deployed his reserve parachute. Parachutists always have priority for pickup over cargo chutes or equipment.
- *Approaching parachutists in the water.* Boat coxswains must approach the parachutist perpendicular to the wind to avoid drifting or being blown over the parachutist or the parachute. Caution must always be taken not to operate the propeller (screws) while the parachutist is alongside in the water. The engine should be placed in neutral. If the parachute gets entangled in the propeller (screws), the boat coxswain turns the motor off while the safety swimmer frees it.
- *Recovery of RAPPSs.* The parachutist must hand the center of the trailing edge (tail) and then the harness to the boat crewman. The suspension lines should be daisy-chained starting from the

harness end. After the lines are daisy-chained, the canopy will be pulled in from the trailing edge (tail) first to allow the water to drain out the leading edge (nose).
- *Recovery of parachutes and platforms.* Recovery of equipment after a water parachute jump is only administrative. Combat conditions will call for the sinking of parachutes and platforms. All swimmers except one should be in the combat rubber raiding craft or move away from it before sinking the platform. Parachutes and platforms may be intentionally sunk on training jumps as long as procedures are used to prevent the equipment from resurfacing and becoming a navigation hazard.

NIGHT WATER PARACHUTE OPERATIONS

12-14. For night water MFF parachute training, parachutists are required to be equipped with a light visible for 1 mile (chemlight), an emergency light visible for 3 miles (strobe), and a flare for emergencies in the water. During free fall and under canopy, parachutists display a light (for example, a chemlight) visible for 1 mile as a safety measure to prevent mid-air collisions or entanglements. Parachutists are not required to be marked for combat situations.

WATER JUMPS WITH COMBAT EQUIPMENT

12-15. Requirements for water jumps with combat equipment include the following:
- *Combat equipment limitations.* Jumping with combat equipment is authorized for water parachute jumps. Parachutists should minimize the amount of equipment they jump with in the water for safety reasons. Parachutists are not authorized to jump with rifles rigged on themselves. They must place rifles and other weapons in buoyant weapons bags and will be lowered with combat equipment. Rifles rigged on the parachutists may easily entangle with suspension lines in the water. Whenever possible, parachutists place as much equipment as possible in the combat rubber raiding craft load except for individual survival gear.
- *Jumper currency.* Parachutists conducting water parachute operations with combat equipment must be current and have previously made at least one noncombat equipment water parachute jump.
- *Equipment rigging.* Equipment packs jumped on the individual must be rigged to be positively buoyant in water. Equipment should be dip-tested for buoyancy before the jump. The equipment is rigged and attached as described in Chapter 5.
- *Parachutist procedures.* When jumping equipment, it is recommended to make the turn on final approach at 500 feet, but no lower than 200 feet, to allow additional time to unfasten the chest strap and lower the equipment. After the parachutist enters the water, he must disconnect the equipment after getting out of the harness.

Chapter 13
Jumpmaster Responsibilities and Currency Qualifications

This chapter establishes the procedures and techniques that jumpmasters use in MFF parachute operations. It delineates duties and responsibilities, regardless of unit, location, and mission. Units may have to supplement this guidance with SOPs to perform certain missions. FMs 3-21.220 and 3-05.210 include further discussion on responsibilities during airborne operations.

RESPONSIBILITIES

13-1. The airborne commander designates the key personnel for each airborne operation. These key personnel are the primary jumpmaster, assistant jumpmaster, oxygen safety personnel (when required), DACO, DZSO/DZSTL, and MO. Each aircraft has a designated primary jumpmaster, an assistant jumpmaster, and oxygen safety personnel (when required). The airborne commander gives the designated primary jumpmaster command authority over, and responsibility for, all airborne personnel and their associated equipment onboard a jump aircraft. The primary jumpmaster assigns tasks to the assistant jumpmasters and oxygen safety personnel appointed to help him. The primary jumpmaster can delegate authority but cannot delegate responsibility. Table 13-1, pages 13-1 and 13-2, lists jumpmaster responsibilities.

Table 13-1. Jumpmaster responsibilities

Location	Responsibilities/Actions
At the Unit Area	Receive operation officer's briefing.
	Receive weather-decision or mission-abort criteria from airborne troop commander.
	Check manifest (DA Form 1306 [Statement of Jump and Loading Manifest]).
	Organize planeload.
	Appoint assistant(s) and/or safety personnel.
	Brief personnel.
	Inspect personnel and equipment.
	Conduct prejump training.
At the Departure Airfield	Coordinate with departure airfield commander.
	Make weather decision.
	Authorize issue of the parachutes.
	Inspect personnel (Appendix G).
	Inspect equipment.
	Inspect aircraft (Appendix H).
	Attend jumpmaster crew briefing (Appendix I).
	Give planeside briefing, as appropriate.
	Announce station time to personnel.

Chapter 13

Table 13-1. Jumpmaster responsibilities (continued)

Location	Responsibilities/Actions
In Flight	Remain ground-oriented.
	Constantly check personnel.
	Enforce flight rules and regulations.
	Issue time warnings.
	Oversee preparation, placement, and drop of free-fall bundles.
	Give heading corrections to flight crew (when using jumpmaster release).
	Perform outside safety checks of the aircraft and DZ before personnel jump.
	Issue jump commands.
On the Drop Zone	Account for personnel and equipment.
	Oversee care and evacuation of injured personnel.
	Ensure jumpers turn in air items/equipment.
	Report to DZSO (peacetime).

QUALIFICATIONS

13-2. For appointment by the airborne commander as either a jumpmaster or assistant jumpmaster for an airborne operation, the individual must be a graduate of the MFF Jumpmaster Course (note below includes further information). He must have performed jumpmaster duties within the previous 6 months or attended MFF jumpmaster refresher training. An assistant jumpmaster must have performed assistant jumpmaster duties at least twice before being designated as a jumpmaster.

Note: The Commandant, USAJFKSWCS, is the proponent for the conduct of MFF courses of instruction. Only graduates of a USAJFKSWCS-recognized MFF jumpmaster course may perform duties as an MFF jumpmaster. The only recognized Navy MFF jumpmasters are those who hold a Navy MFF jumpmaster graduation certificate dated before 16 June 1989 or those who have graduated from the USAJFKSWCS MFF Jumpmaster Course. The only recognized Air Force MFF jumpmasters are those who have graduated from the USAJFKSWCS MFF Jumpmaster Course and those previously qualified Air Force free-fall jumpmasters who have undergone an MFF jumpmaster upgrade certification using USAJFKSWCS criteria. The only recognized Marine Corps MFF jumpmasters are those who have graduated from the USAJFKSWCS MFF Jumpmasters Course.

CARDINAL RULES FOR THE JUMPMASTER

13-3. General rules stress that the jumpmaster must—
- Never sacrifice safety for any reason.
- Rehearse jumpmaster procedures on the ground.
- Face the open jump door when in flight.
- Maintain a firm handhold on the aircraft when working in or close to an open jump door or ramp.
- Never allow anyone in or near an open jump door or ramp who is not wearing a helmet and safety harness connected to the aircraft or who is not wearing a parachute. The helmet requirement may be waived for deliberate water jumps.

CURRENCY AND REQUALIFICATION REQUIREMENTS

13-4. An MFF jumpmaster must be USAJFKSWCS-trained or have formally undergone transitional training in a proponent-recognized school environment from the MC-3 system to the RAPPS. He must have performed primary or assistant jumpmaster duties within the last 6 months where parachutists actually exited the aircraft while using a jumpmaster-directed release.

13-5. Previously qualified MFF jumpmasters who do not meet proficiency and currency requirements will meet the following requalification requirements:

- Undergo MFF parachutist refresher training outlined in Appendix B.
- Receive JMPI training for the primary MFF parachute system used in his parent unit.
- Receive refresher training in wind drift (HARP) calculation for MFF mission profiles.
- Receive oxygen equipment refresher training.
- Perform assistant jumpmaster duties for one MFF jump.
- Execute under-canopy navigation techniques specific to the navigation aids unique to the parent unit.

An MFF jumpmaster who meets the currency criteria will conduct the requalification and refresher training.

Note: Whenever possible, a jumpmaster-directed release should be used to enhance MFF jumpmaster skills.

This page intentionally left blank.

Chapter 14

Weather Factors for the Military Free-Fall Jumpmaster

The MFF jumpmaster depends on knowledge of and ability to interpret weather phenomena to make critical decisions concerning the infiltration portion of the mission. As the commander's advisor in MFF-unique requirements, the jumpmaster must maximize knowledge of weather subtleties and know the effects of terrain on parachuting conditions in order to make informed recommendations and decisions concerning the application of MFF parachute operations.

Note: The JPADS (Appendix J) uses mission-planning and weather forecasting software and can receive en route mission changes and weather updates via satellite links before and during deployment from the aircraft.

CRITICALITY OF WEATHER KNOWLEDGE

14-1. Weather information derives from a variety of sources. These include fixed forecasting and observation facilities, manned and unmanned aerial reconnaissance platforms, civilian and military satellite imagery, en route air transport readings (pilot report), the global communications network, and intelligence reports from the operational area. This data is exploited by the MFF jumpmaster and his supporting headquarters to enhance the accuracy of his MFF infiltration planning. Rarely are all of these resources available. The MFF jumpmaster then relies upon an in-depth area analysis during the premission planning isolation. Effective interpretation of weather information products is supported by the understanding of the physical principles that drive weather events. The MFF jumpmaster applies this knowledge to the terrain and probable conditions in the AO for that mission. The end product from this analysis is a preplanned release point determination of where to exit the aircraft to land in the vicinity of the target area.

14-2. Operational weather services are provided to the U.S. Army by the USAF Air Force Weather Agency (AFWA) and Joint Air Force-Army Weather Information Network (https://weather.afwa.af.mil/jaawin). Air Force weather technicians (forecasters) assimilate data from a variety of sources—balloons, satellites, radar, and computer forecast models available from the AFWA—to provide readings or products that are useful to the MFF jumpmaster. These forecasts are specific to a certain time frame or locale. It is important that a datum parameter or particular weather product is applied to the specific operational need for which that product was intended. This is to preclude decisions being made on outdated or unsuitable data. Once assessed for utility, the information is then used by the MFF jumpmaster in the computation of a release point for a HAHO or HALO MFF parachute operation.

14-3. Weather trend knowledge is important in situations where conditions deteriorate en route to the mission area or when the infiltration is a blind drop. Understanding weather trends assists in the formulation of mission abort criteria and quantifiable contingency parameters. Knowledge of terrain along the planned canopy line of flight gives indications of potential air mass turbulence and other daily or seasonal canopy flight hazards. The winds aloft data help identify what can be significant environmental or atmospheric planning factors that affect MFF parachute operations. These include the location of the jet stream or low-level jets, the existence and extent of cloud formations, or extreme temperature conditions.

Chapter 14

MISSION PLANNING TOOLS

14-4. Planning is conducted in two stages. Detailed area and terrain analysis is accomplished before the mission is directed to be executed. It is conducted as part of contingency planning or as part of an operation plan formulation while reaction time is plentiful. It addresses specified, as well as implied, planning tasks pertinent to the infiltration as found in the mission tasking directive. An operational area update is conducted during the planning isolation immediately before the mission's launch. It is used to validate previous planning parameters, update the intelligence, and incorporate current weather data. The following are only some of the meteorological and terrain planning products that can be employed in planning for an MFF infiltration (formats vary and include printouts, two-dimensional computer graphics, matrix charts, and imagery):

- Forecast charts of winds aloft can be tailored to the jumpmaster's needs.
- Satellite or other platform imagery of the operational area, including primary DZ, alternate DZs, abort rally points, and exfiltration points.
- Weather trend (24, 48, and 72 hour) in the mission area.
- Computer-enhanced digitized terrain analysis for a HAHO flight route (line-of-sight analysis from various points under canopy). The same products are especially useful for night infiltrations.
- Seasonal analysis of prevailing winds, air density altitude, and clouds.
- Area topography within 10 nautical miles of DZ center of mass. Derived from CD-ROM National Geospatial-Intelligence Agency (NGA) databases for ground movement planning. Terrain elevations in the vicinity of the planned opening point (OP).
- Temperature influence of large water masses affecting canopy performance K factors and subsequent wind drift calculations.
- Moon phase, percent illumination, moonrise, moonset, sunrise, sunset, end of evening nautical twilight, end of evening civil twilight, beginning morning nautical twilight (BMNT), beginning morning civil twilight, and the location (azimuth and elevation) of the sun or moon in the sky at a given time.
- HAHO/HALO exit and landing windows (for example, exit in light at altitude, land in darkness on the DZ).
- Incorporation of above pertinent factors into the survival, evasion resistance, escape, and recovery plan.

14-5. Weather aids and products must be employed in mission planning for regularly scheduled proficiency training. The MFF jumpmaster becomes familiar not only with what the products provide informationally, but more importantly, with what they do not contain. Information gaps drive information requirements and specific requests for intelligence information (RFIIs). Weather-specific information requirement taskings are filled by the supporting staff weather officer (SWO), who is usually in the USAF. Regardless, weather products are useful only if the MFF jumpmaster knows their meteorological content. It is the job of the weather staff to establish the information conduit, provide the products, and assist in their interpretation. It is the responsibility of the MFF jumpmaster to be able to concisely articulate his weather planning requirements. It is also the MFF jumpmaster's responsibility to be the primary interpreter of the data provided by the SWO. These skills should be exercised during all peacetime MFF proficiency training. It is critical that they are employed during any tactical exercise. The following are some of the tools and products accessible to the MFF jumpmaster:

- *Low-level prognostic chart*: A chart that gives the significant weather forecast of conditions at 12- and 24-hour intervals from the surface to 24,000 feet MSL. By comparing prognostic charts, the MFF jumpmaster determines expected surface weather, its direction and speed of movement, degree of cloud cover, degree of turbulence induced by frontal air masses, expected wind patterns, and subsequent directional changes.
- *Flight hazards forecast*: A forecast that shows areas of icing conditions, clear air turbulence, and thunderstorms above 10,000 feet MSL. These are expected worst-case conditions not associated with thunderstorm activity during the time indicated on the product.

- *Radar summary chart*: A collection of radar weather reports showing precipitation echo locations, altitudes of cloud/cell tops, direction of storm cell movement, and type of precipitation. Because this type of convective activity changes extremely rapidly, charts more than 2 hours old must be carefully considered.
- *Hazard chart*: Graphics depicting aerial extent and timing of hazardous weather, including tornadoes, thunderstorms, heavy rain/snow, icing, strong winds, and/or other required parameters.
- *Winds aloft prognostic chart*: Reflects winds and temperatures from the surface to 18,000 feet MSL. Temperatures are read in degrees Celsius (°C). The MFF jumpmaster should use the Standard Lapse Rate (SLR) (2°C or 3 degrees Fahrenheit [°F] per 1,000 feet) to adjust for temperatures at altitude and calculate windchill. Weather forecast models and applications now provide winds and temperature forecasts for all altitudes at 1,000-foot intervals. Jumpmasters should ensure they have a precision airdrop system forecast of winds, temperatures, and pressure over their target, which is available via the AFWA Joint Air Force-Army Weather Information Network.
- *Weather advisories/significant meteorological information*: Contains weather phenomena potentially hazardous to light aircraft (or parachutists under canopy during MFF operations). It includes sand and dust storms.
- *Sky cover/ceiling*: Includes the cloud bases and multiple layers or decks. It is reported as the amount of sky actually covered by clouds. For example, it can technically be overcast with areas still open between cloud cells. The cumulative amount of sky cover determines if it is scattered, broken, overcast, partially obscured, or totally obscured. Total obscurity is defined as no vertical visibility upward or downward from the MFF jumpmaster's perspective. Partial obscurity is usually manifested in MFF parachute operations when the pilot cannot see the slant view from the cockpit, but the MFF jumpmaster can see straight down with no difficulty. Current satellite imagery from either geostationary or polar orbiting satellites provide near-real-time ground-truth assessments of cloud cover.

14-6. Weather support technology and military information transfer systems have improved exponentially over the last few years. Development of forecasting capabilities has improved for meso- and microscale forecasting. The ultimate standard using artificial intelligence techniques is to provide mesoscale descriptions of weather phenomena from 2,000 kilometers to 2 kilometers of the target area and microscale information within 2 kilometers of the target site. Mesoscale weather data affects release point calculations. Microscale information is pertinent for DZ landing patterns and tandem bundle release point calculations. This degree of detail is enhanced by a digitized database to accommodate the nonstandardized products required. Computer analysis software and the integration of products from a variety of surface, air breathing, and satellite platforms provide a multitude of tools capable of enhancing an MFF jumpmaster's release point calculation. Tasking the supporting weather staff's systems in regularly scheduled training establishes the mechanism by which the desired wartime products are developed and transmitted. The establishment of the need for a particular or unique product and its regular use during recurring training decreases future operations request response time. It establishes the user's requirements database. The first time the staff hears of a desired MFF-unique weather product should not be in an outside the continental United States (OCONUS) forward launch site a few hours from mission launch.

14-7. U.S. civilian, military, and foreign satellites can produce a variety of weather products useful to the MFF jumpmaster. However, the SWO must be made aware of the need to establish the information requirements before the fact, thus formalizing the pipeline to create, deliver, customize, or enhance a desired product. The products may be printouts, graphics, or digitized cross-sectional representations and consist of a wide spectral range of imagery. Some examples are preprocessed vertical profile products of cloud formations, wind speed, direction, temperature, humidity, or barometric pressure differentials. The product display is packaged in user-established increments, parameters, or layers of the atmosphere. It could be nothing more than an IR-enhanced satellite observation of the snowfall depth on the intended DZ or ice thickness on the river the team must cross after infiltrating. Once identified, the datum will be color enhanced to show bands or intensity (average fog thickness on the intended DZ at 0430 hours) with a specified magnitude range of the measured element. For example, the IR color enhancement can also

Chapter 14

differentiate obstacles, such as height of vegetation or obscured ditches on the intended DZ. These parameters are user defined to meet the MFF jumpmaster's tactical planning information requirements.

14-8. Software programs can overlay NGA geographical boundaries on intended imagery products. Many of the terrain products are digitized, providing a wide range of computer manipulation possibilities for planning purposes. They can plot proposed ground movement routes from DZ to the objective or HAHO canopy flight routes in mountainous terrain from OP to the DZ. The superimposed data can be to any scale the user requires. Tailored packages of data, such as vertical profiles of cloud formations, altitude AGL of the cloud bases, and associated wind speeds, are easily developed specific to a unique terrain feature or target locale. Cloud animation techniques and color enhancement can depict future cloud movement during seasonal weather activity or for a specific, anticipated set of mission conditions. Once in the database, the user-specified algorithm parameters can be integrated into a variety of other associated weather products.

14-9. In concert with the SWO and the supporting staff, MFF-capable units should develop MFF-unique weather effects' critical threshold values to supplement information requirements. These values are derived from the experience base of the unit's most skilled parachutists and the assumptions paragraph of the mission-tasking directive. Some are specified and others are implied. Several sets of quantifiable values should be developed. These value sets accommodate varying factors, such as overall skill level of the infiltrating team (for example, accounting for the most inexperienced man). The sets might include infiltration under ideal conditions; less than tactically ideal conditions (for example, 17-knot winds and no illumination); marginal conditions (for example, 22-knot winds and 1,500-foot ceiling); conditions that dictate a shift from HAHO to HALO infiltration (for example, presence of the jet stream); and conditions that might dictate cancellation of the MFF infiltration entirely.

14-10. The assigned decision matrix parameters assist in deciding whether the operation is HAHO or HALO, will employ an intermediate deployment altitude (high-altitude medium-opening [HAMO]), are conducted at oxygen altitudes, incorporate a low-level exit, or are cancelled. Essentially, the MFF jumpmaster and team's senior leadership currently undergo the same mental decision process, although with fewer weather tools to aid the factual basis for their decision. It should always be remembered that weather forecasting is not an exact science. Nothing, including weather products, replaces the experience that can only come by actually executing training jumps in unique and tactically realistic conditions.

14-11. Once validated, decision matrix threshold values are used as a trigger mechanism for any specified weather category or product of interest. When these critical values are compared to observed or forecasted conditions, computer-enhanced products are more easily produced. When the product is predefined, vice a nebulous, general RFII which begs clarification, the collection echelons can respond immediately without having to create a product from scratch. Essentially, the MFF jumpmaster, in concert with the directing commander and supporting staff, identifies parameters which can affect the execution of that operation's MFF infiltration. These are cataloged as an intelligence preparation of the battlefield (IPB) matrix. This matrix indicates that when winds aloft from a cardinal direction reach a certain speed, a HAHO infiltration onto a waterside DZ should be cancelled due to the now-increased probability of the team landing offshore in the lake. This same matrix, however, might also indicate that a HALO infiltration has a better chance to succeed, as the wind conditions negatively affecting a HAHO operation will mask a HALO operation's canopy opening noise.

14-12. These parameters become a set of definable, quantified criteria around which a weather product is designed. Details such as wind speeds at critical altitudes or any other parameter envisioned in the concept of the operation or assumptions paragraph are listed so that they can be graphically depicted and incorporated into the product requested. A unit or developer of the RFII (usually the MFF jumpmaster) who cannot adequately articulate the requirement encumbers the system. This causes time delays by creating an obligatory request for clarification because the original request is unreasonable (too broad or not specific enough). Without specificity, the information the requestor receives is extraneous and largely useless to operational planning.

14-13. Establishment of a solid IPB matrix of measurable criteria additionally permits validation during normally scheduled training. The exercise sensitizes supporting staff echelons to their existence and familiarizes them with their purpose. More importantly, it establishes a taskable database at the collection echelon which increases responsiveness at the most critical time (usually during preinfiltration mission planning). A digital terrain database for an envisioned contingency target takes time to program, but once established it can be used as a database tool from which further products can be modeled.

14-14. For example, a computer-generated view of a valley depicts the visual field of the baseman of a HAHO stack wearing NVGs looking down at a 60-degree angle from 4,000 feet above the valley floor. This view is also used to model potential turbulence as the stack flies over the valley's western ridgeline when the wind is from the west. When modeling of weather phenomena is matched with canopy flight algorithms (K factors), clouds with bases matching the forecast at infiltration (low-level nephanalysis) can be plugged into the database. Each parachutist has a chance to see the view of the infiltration DZ as it will look from under canopy where it is calculated that the stack will descend through the cloud base along the planned canopy flight route. This view is modified to reflect the illumination conditions which match the forecast for the infiltration.

14-15. If forecasted winds are strong for the infiltration, the same program's database is used to identify aborted or alternate DZs short of (upwind) and beyond (downwind) the ridgeline directly to the east of the primary DZ. If winds are even stronger, the IPB weather product matrix might indicate to abort the MFF infiltration for those conditions and to execute alternative infiltration plans. Other MFF-unique taskings can be driven by the database and IPB matrix. An example is cloud-free line of sight (CFLOS) which is the altitude and degree of clarity with which the DZ can be seen from under canopy while wearing a specified version of night vision equipment or with the naked eye. A second parameter potentially derived from CFLOS is a horizon profile. This is the ambient light terrain feature horizon line generated from user-specified altitudes AGL when facing a given azimuth orientation. It is very similar to the maritime planning reference, *List of Lights, Radio Aids, and Fog Signals*, which is the view of a coastal horizon looking inland as depicted by its light marker buoys, urban skyline, and terrain horizon line. For example, when sitting under canopy at 6,500 feet, facing 121 degrees magnetic at 0240 hours and the moon is at 36-percent illumination and 47 degrees off the horizon, the team's leader wants to know what a parachutist will see of the hill to the northwest of the DZ. Knowing that the target area is off the northeast shoulder of the hill, this product gives the parachutist another tool by which to visually reference the canopy line of flight during the hours of darkness. This product is valuable in target areas where there is little commercial electrical power, no nighttime urban landmarks, and few visual references.

14-16. When the planned DZ location is extremely dark and difficult to differentiate from the surrounding area, the database can be manipulated to provide perspective detail. This detail might be the angle of the electric lights of the town 2.5 kilometers away in the parachutist's field of view when under canopy directly over the DZ at 2,000 feet when facing into the wind. This detail can provide visual clues of the direction to set up the desired landing pattern to help ensure upwind landings for tandem bundle delivery. This database is also useful in contingency planning where a parachutist might get separated from a HAHO stack or have to divert to an alternate DZ with no landing wind direction indicators after executing emergency cutaway procedures. Mission preparation studies using observation line-of-sight products (for example, from known guard posts on the target, given the light conditions forecasted at infiltration, the distance the team under canopy can be visually identified) will suggest how close to the target the team can land and remain undiscovered. It might also indicate if a distraction is warranted, perhaps necessitating that the team split up under canopy and land at separate locations to approach the target from different directions.

ATMOSPHERE

14-17. It is important that the MFF jumpmaster be familiar with the characteristics of the atmosphere. This aids in interpreting weather data so the MFF jumpmaster can apply existing conditions against contingency and immediate operational requirements. Fifty percent of the earth's air by weight is found below 18,000 feet MSL. Therefore, air density decreases as one travels farther from the earth's surface. It is this attribute which affects parachute performance, barometric device function, and human physiology (supplemental oxygen requirements). Atmospheric oxygen content drives "time of useful consciousness" at high altitudes. Air density thins to the point that at 43,000 feet above MSL, partial pressure suits are required. For comparison, outer space starts at approximately 64,000 feet MSL, requiring fully pressurized suits (Figure 14-1, page 14-6).

Chapter 14

Event	Altitude (Feet)	Altitude (Meters)	Pressure (Inches of Mercury)	SLR Ambient Air Temperature Celsius)	SLR Ambient Air Temperature (Fahrenheit)
Top of some thunder clouds	70,000	21,350	1.3	-125.0	-151.0
"Outer Space" starts	64,000	19,520			
	50,000	15,250	3.4	-85.0	-91.0
Use of partial pressure suits required	43,000	13,115		-56.7	-69.7
Commercial airline cruising altitude	35,000	10,675	7.0	-49.4	-57.5
MC-4 maximum opening altitude	25,000	7,625	11.8	-35.0	-16.0
One half the earth's atmosphere by weight below this altitude	18,000	5,490	14.9	-21.0	-5.1
Supplemental breathing oxygen required	13,000	3,965	18.5	-10.0	20.0
USAF aircrews on oxygen	10,000	3,050	20.6	-5.0	23.4
Average MFF opening altitude	4,000	1,220	24.9	5.1	41.0
Mean sea level (standard atmosphere)	Mean sea level	0	29.92	15.0	59.0

Figure 14-1. Atmosphere

14-18. Water vapor is found mostly below 30,000 feet MSL. The water vapor content of an air mass is temperature dependent: as the temperature of the air mass increases, its ability to hold water vapor increases. Therefore, hot air can hold more water vapor than cold air. Air also has weight: one square inch of air at sea level exerts a pressure of 14.7 psi. This is equivalent to a 29.92-inch tall column of mercury at sea level or 1013.25 millibars of pressure. The bar and millibar are the metric standards used to calculate the activation settings for most U.S. and foreign military ARRs currently in use (Figure 14-2, page 14-7).

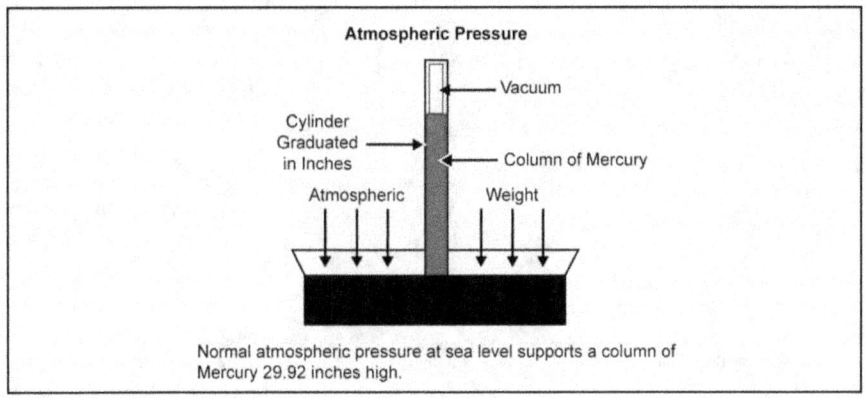

Figure 14-2. Mercury barometer

WEATHER

14-19. Weather is defined as the local state of the atmosphere with respect to its temperature, turbulence, moisture content, and resultant degree of cloudiness. These factors interact in combinations which have a variety of militarily significant effects on MFF parachute operations. Weather is characterized by five meteorological elements: air temperature, humidity, clouds/precipitation, atmospheric pressure, and wind.

14-20. Above the upper limits of the troposphere (35,000 feet MSL), only air temperature, atmospheric pressure, and wind velocity are key factors. The remaining phenomena are factors below the troposphere because they can only occur when water vapor (atmospheric moisture) is present. Moisture also affects icing, turbulence, and environmental stress on the parachutist.

14-21. Weather is caused by the heating and cooling cycles of the earth's surface. This cycle is the evaporation of moisture into the atmosphere, its condensation, and the subsequent precipitation of those water droplets. The presence of moisture causes the militarily significant environmental, temperature, and atmospheric pressure changes. It is the atmospheric pressure variances that cause wind and generate clouds. These pressure-variance phenomena are driven by daily and seasonal temperature differentials over the face of the earth's surface. Understanding these relationships permits the MFF jumpmaster to identify when these factors will adversely affect MFF parachute operations.

14-22. World terrain for potential MFF operations varies from extreme mountain ranges to valleys and flat deserts, some even below sea level. Each type of terrain is affected differently by localized wind flow patterns (prevailing winds and their seasonal variations) and specific, seasonal weather phenomena. The MFF jumpmaster is responsible for identifying these factors during the area analysis and incorporating them into the infiltration planning process. This lessens their adverse impact and maximizes their potential tactical value. When properly considered, weather factors contribute as much as any other factor to mission accomplishment. For example, the dust storms encountered on the infiltration helicopter flight into Desert One had a serious operational effect on mission accomplishment.

14-23. An example of a common, localized weather phenomenon that most MFF parachutists have encountered is called virga. Virga occurs because all precipitation does not necessarily reach the earth's surface. It is characterized by precipitation that evaporates before hitting the ground. This normally happens when dry air is at or just below a moisture-bearing cloud base. The phenomena is characterized by the presence of snow, hail, or rain at or below exit altitude, but no evidence of precipitation is observed or experienced on the ground. An MFF parachutist might fall through several thousand feet of rain after exit, but experience no precipitation under canopy. In fact, virga conditions contribute to masking the aural and visual signature of an element infiltrating an operational area.

Chapter 14

TEMPERATURE

14-24. One key environmental factor affecting atmospheric phenomena is temperature. The earth's surface is heated by incoming solar radiation (insolation). At night, the reverse effect is in thermal cooling (terrestrial radiation). These two basic mechanisms, coupled with the presence of water vapor, are what fuel nearly all the local weather conditions that affect MFF parachute operations.

14-25. Daily variations in terrestrial temperatures affect the airspace from the earth's surface to approximately 4,000 feet AGL. Daily ground-induced temperatures in the free air above 4,000 feet do not vary significantly. When conditions are still, MFF parachutists on night jumps will often feel a drastic temperature change while in free fall or under canopy when passing through this approximate altitude. Land surfaces have a greater daily temperature variance than water. The range of temperature varies with locale, season, latitude, and type of ground surface (for example, sand, rock, urban, plowed earth, runway complex, or type of vegetation).

14-26. Larger daily temperature variations occur over DZs at higher elevations MSL. The same effect occurs over DZs that absorb more thermal radiation. This phenomenon also induces and fuels thermal updrafts. For MFF jumpmaster planning purposes, daily temperature variations and resultant convective/thermal uplifting are much smaller over thick (dark, therefore cooler) vegetation and deep water. For these two sets of conditions, canopy glide slope is significantly reduced, requiring a shortened OP calculation in wind drift formulas.

14-27. The mechanism by which temperature affects air masses varies with factors such as humidity content and rate of temperature change. Generally, air cools (lapses) as it rises. Close to the earth (below 4,000 feet), the rate at which this occurs depends on the local conditions; however, usually an SLR is used. The SLR is 2°C or 3°F for each 1,000-foot-increment increase in altitude. This rule of thumb is used to determine ambient air temperatures at altitude. For example, if the temperature on the ground is 50°F, at 10,000 feet AGL, the ambient air temperature is 20°F [50 - (10 x 3)]. For terminal free fall or for MFF jumpmasters exposed outside the aircraft, the velocity of the slipstream (knots indicated air speed [KIAS]) is factored against a standard windchill chart using the ambient SLR temperature to determine total windchill effect (Table 14-1, page 14-9). This indicates the degree of environmental protection that the jumpmaster or parachutists will require on the operation. Figure 14-3, page 14-10, depicts a temperature scale comparing Celsius (Centigrade) and Fahrenheit.

14-28. The SLR for atmospheric temperature effects assumes a standard temperature of 15°C or 59°F at sea level. Atmospheric temperature decreases at the SLR as one climbs in altitude. This progression continues linearly until 35,332 feet MSL, which is the altitude where air theoretically becomes isothermal (constant temperature). Because of low atmospheric moisture, a balance is created between sun heat absorbed from space and that reradiated to space. This results in an ambient air temperature of approximately -65°F. Air temperatures (before factoring windchill) above this altitude can drop as low as -121°F. Seasonal, latitudinal, and jet stream factors serve to moderate this extreme temperature differential between the ground surface and MFF altitudes.

14-29. A temperature inversion is a condition where the air mass temperature increases instead of decreases with a gain in altitude. This situation is most frequently found close to the ground (below 500 feet AGL) on a still, clear, relatively cool night. When the ground loses its heat from the day (radiationally cools), the air layer directly above it is heated. With the ground now cooler than the air mass directly above it and when no significant air mass movement or mixing action occurs, the warm air mass acts like a blanket. The temperature differential can be drastic, sometimes exceeding 20°F.

14-30. When enough moisture is trapped under an inversion blanket of warm air (perhaps from rain the previous day or heavy morning dew), ground fog occurs. The greater the temperature difference and the more moisture present, the thicker and denser the fog layer. This is quite common on open areas that characterize potential tactical DZs, where post-sundown and pre-dawn conditions are ideal for inversions that will blanket the DZ. This is significant for parachutists on final approach under canopy and during the landing flare. Additionally, sound carries much farther under these conditions.

Weather Factors for the Military Free-Fall Jumpmaster

Table 14-1. Military free-fall operations windchill determination

Cooling Power of Wind Expressed as Equivalent Chill Temperature

Wind Speed		Ambient Air Temperature (Degrees F)													
Knots	MPH	40	35	30	25	20	15	10	5	0	-5	-10	-15	-20	-25
Calm	Calm	\multicolumn{14}{l}{Equivalent Chill Temperature}												
3-6	5	35	30	25	20	15	10	5	0	-5	-10	-15	-20	-25	-30
7-10	10	30	20	15	10	5	0	-10	-15	-20	-25	-35	-40	-45	-50
11-13	15	25	15	10	0	-5	-10	-20	-25	-30	-40	-45	-50	-60	-65
16-19	20	20	10	5	0	-10	-15	-25	-30	-35	-45	-50	-60	-65	-75
20-23	25	15	10	0	-5	-15	-20	-30	-35	-45	-50	-60	-65	-75	-80
24-28	30	10	5	0	-10	-20	-25	-30	-40	-50	-55	-65	-70	-80	-85
29-32	35	10	5	-5	-10	-20	-30	-35	-40	-50	-60	-65	-75	-80	-90
33-36	40	10	0	-5	-15	-20	-30	-35	-45	-50	-60	-70	-75	-85	-95
Winds above 40 have little additional effect		\multicolumn{5}{l}{LITTLE DANGER}					\multicolumn{3}{l}{INCREASING DANGER (Flesh may freeze within one minute)}			\multicolumn{2}{l}{GREAT DANGER (Flesh may freeze within 30 seconds)}	

Note: Determine ambient air temperature at altitude using SLR formula and then factor in wind chill.

(Continued columns: -30, -35, -40, -45, -50, -55, -60)

Wind Speed	-30	-35	-40	-45	-50	-55	-60
5	-35	-40	-45	-50	-55	-65	-70
10	-50	-60	-65	-75	-80	-90	-95
15	-70	-80	-85	-90	-100	-105	-110
20	-80	-85	-95	-100	-110	-115	-120
25	-85	-95	-105	-110	-115	-125	-135
30	-90	-100	-110	-115	-120	-130	-140
35	-95	-105	-115	-120	-125	-135	-145
40	-100	-110	-115	-125	-130	-140	-150

Chapter 14

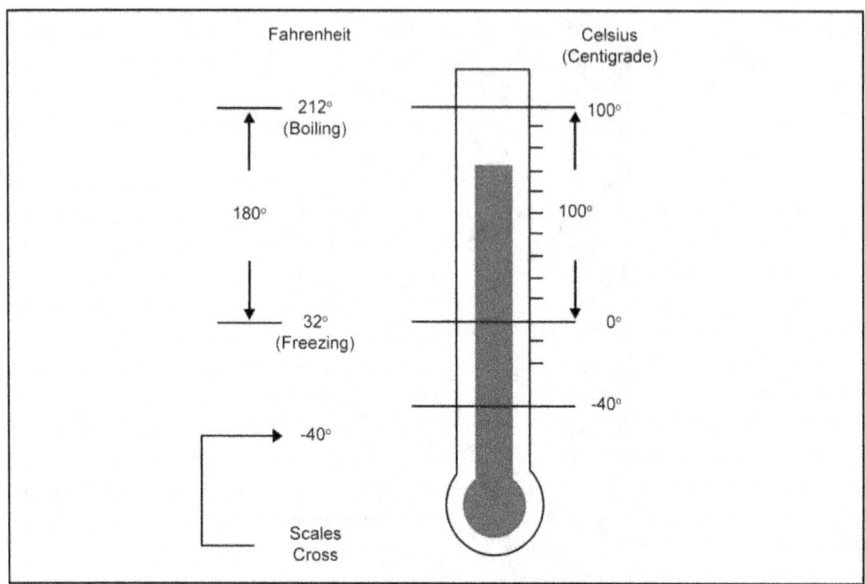

Figure 14-3. Temperature scales

14-31. Over water, where temperatures are generally cooler closer to the surface than at higher altitudes, inversion conditions are usually consistent over the last 4,000 feet of altitude of canopy flight. Because of the lack of thermal convection experienced over land, the canopy lift over water is reduced over the lower portion of the canopy flight. This is the same effect experienced on cool nights with no wind conditions where the dynamic stall ability of the parachute is further reduced due to the lack of convective lift. The canopy glide slope significantly decreases. Most often found over water, a shortened OP is required to offset the reduced lift and must be planned for by the MFF jumpmaster. The rule of thumb for inversion conditions over large cold water bodies is that canopy glide performance can degrade by up to 30 percent once under 4,000 feet AGL.

TEMPERATURE EFFECTS ON ATMOSPHERIC PRESSURE

14-32. In general, air pressure decreases with altitude or in colder air. Environmentally, both of these conditions are met as one climbs in altitude. Therefore, a temperature stabilized altimeter displays a higher altitude barometrically (MSL) than its actual altitude AGL. In other words, a parachutist is closer to the ground than the altitude reflected on his altimeter. This is especially true after an altimeter has been exposed to colder temperatures, such as those found during MFF HAHO operations. For DZs at higher elevations, altimeters can read as much as 500 feet higher than actual jumper altitude AGL. With potential altimeter mechanical lag of up to 250 feet, a parachutist can be seriously mistaken in determining his altitude above the ground, especially on a night descent. The temperature-affected altitude indicated on a HAHO parachutist's altimeter will always be several hundred feet different than a three-dimensional satellite navigation or Global Orbiting Navigation Satellite System (GLONASS) GPS satellite triangulated altitude (Figure 14-4, page 14-11).

14-33. In warmer air masses, air pressure decreases slightly less rapidly as one climbs in altitude. A parachutist's true altitude AGL is higher than the indicated altitude in hotter weather conditions or after the altimeter has been exposed to direct heat (for example, after lying on the concrete flight line in direct sunlight on a 95°F day just prior to the jump).

14-34. Warm air has less density than colder air; therefore, less lift is generated by lifting surfaces such as ram-air parachutes. Design performance and work capacity is decreased. The same effect occurs in the lower density air found at DZs at higher altitudes MSL. Hot air conditions at high altitudes must be considered by the MFF jumpmaster since parachute performance can be seriously degraded, especially when under combat-equipment-induced higher-suspended weights. Other examples are the increased runway length and reduced allowable cargo load for aircraft operating at higher altitudes during hot conditions. The same principle applies to any parachute, whether for a tandem bundle's cargo round or a square personnel canopy. The lifting capacity is reduced, the glide slope decreases requiring an adjusted K factor, and the overall rate of descent increases. Landings are generally harder due to a ram-air parachute's decreased ability to generate transitional lift during the landing flare.

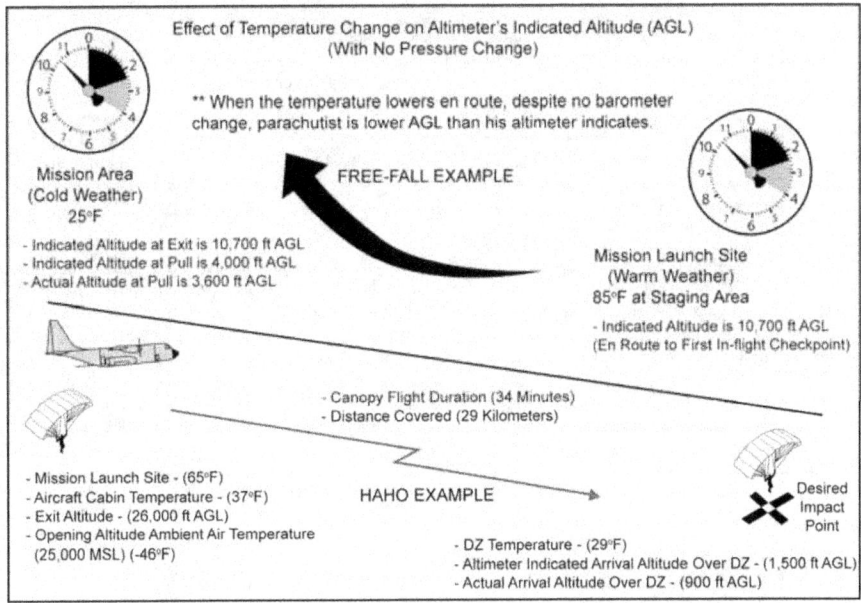

Figure 14-4. Effect of temperature change on altimeter's indicated altitude (AGL)

MEASURING ATMOSPHERIC PRESSURE

14-35. Atmospheric temperature tends to even out (increased entropy) over the earth's surface. Daily, regional, and seasonal differences in heating and cooling patterns cause air density variations. Horizontal variations are very small compared to the magnitude of vertical pressure change that occurs with an increase or decrease in altitude. The scale of horizontal pressure change compared with vertical change is on the scale of one-thousandth of an inch of mercury. Regardless, it is these small horizontal pressure differentials that cause all atmospheric circulation (winds) and fuel most other weather phenomena.

14-36. Atmospheric pressure is the result of gravity acting on the mass of air that comprises the earth's atmosphere. This pressure is measured by instruments which are all variations of two types of barometric devices. The first type is a mercury barometer which is impractical in MFF applications. The second type of atmospheric pressure measuring instrument, most commonly used in MFF operations, is a variation of the aneroid barometer. Generally, an aneroid barometer is a sealed metal bellows containing a partial vacuum, such as the military MA2-30 wrist altimeter and other analog display altimeters. The bellows expands and contracts in response to the changes in air mass density surrounding it. When the device goes higher into lesser atmospheric pressure, the bellows expands. When it goes lower into higher air pressure, the bellows is compressed. A pointer linked to the bellows moves around a calibrated dial, which for MFF applications reads in 100-, 250-, and 1,000-foot increments.

14-37. Other altimeter variations, regardless of the type of readout (such as a digital display), are still aneroid sensing devices. Because they are mechanical devices, altimeters are never 100-percent accurate. They are all environmentally sensitive to some degree. Additionally, some electronic types of barometric sensors, such as EAADs/ARRs, are affected by electrical fields. This includes the potential fields induced by intrateam radios or ambient environmental static electricity found in the immediate vicinity of storm cloud formations. These conditions potentially affect all electrical instruments (GPS receivers, compass board-mounted electronics, radio-controlled bundle systems), as well as electrically fired EAADs/ARRs (for example, Sentinel, FXC, and Military CYPRES 2). For example, analog altimeter needles have an acceptable manufacturer's tolerance of up to ±250 feet, simply due to mechanical lag. They are, however, useful because of their compactness and general consistency. The effects of most minor and some potentially major environmentally induced inaccuracies can be negated through prior planning by the MFF jumpmaster.

14-38. A common MFF example that demonstrates aneroid altimeter false readings is a parachutist who is falling at terminal velocity in a back-to-earth or inverted body position. As any object falls, it creates a zone of low pressure above it. If a parachutist falls with his back down and attempts to read the altimeter, the altimeter's aneroid is sensing a lower atmospheric pressure than the parachutist's actual altitude AGL. Consequently, the altitude reflected on the altimeter face is reading a false altitude AGL. The altimeter reads a higher altitude than the parachutist's actual altitude.

14-39. Additionally, the altimeter needle fluctuates rapidly due to the inconsistent airflow through this burble. If the parachutist is wearing combat equipment, the size of the low pressure zone is larger, the effect of the burble is greater, and the altimeter reading has more extreme fluctuations. This situation is dangerous for a parachutist approaching a designated pull or key decision altitude, as the parachutist could be up to 700 feet lower (500 feet lower without rucksack) than what is reflected on the altimeter. This could result in loss of adequate time to react to a parachute malfunction. It could place the parachutist low under canopy relative to accompanying parachutists. In general, it contributes to a loss of altitude awareness. At night and with oxygen equipment, this situation has the potential to be catastrophic. Figure 14-5 depicts atmospheric pressure change resulting in false readings.

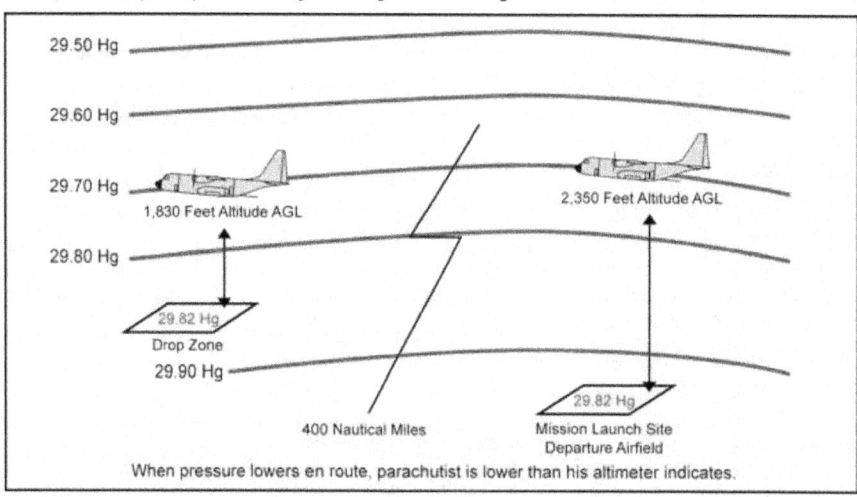

Figure 14-5. Atmospheric pressure change over large distance resulting in false altitude (AGL) readings

**This Page Intentionally Left Blank
As Part of Change 1**

Chapter 14

**This Page Intentionally Left Blank
As Part of Change 1**

**This Page Intentionally Left Blank
As Part of Change 1**

ENVIRONMENTAL EFFECTS ON ALTIMETERS

14-40. An altimeter is simply an air pressure measuring device. It only senses air pressure above MSL. It does not know the intention of a parachutist or its own altitude relative to an intended DZ's ground elevation. For example, an altimeter indicates 10,000 feet MSL when all environmental factors affecting its aneroid chamber equate to 698 millibars, whether or not it is 10,000 feet above a ground surface. It is the MFF jumpmaster's responsibility to understand, identify, and, through prior planning, apply corrections for the environmental factors which change a linear, standard pressure reading MSL in relation to altitude AGL.

14-41. MFF parachutists only have one altimeter to reference. Because altitude pressure deviations occur on the infiltration flight, EAAD/ARR and parachutist altimeter advances or back offs for intended DZs should be set on altimeters before takeoff, not en route. Sometimes it is necessary to adjust altimeters en route (diversion to an alternate DZ), but it should be recognized that some atmospheric deviation occurs over long distances; barometric devices will not reflect or fire quite as accurately. Setting before launch provides a known starting point and later in-flight reference point for adjustments that could be significantly incorrect otherwise. This point of reference is most often used in adjusting instruments when aborting from the primary to an alternate DZ. When referencing aircraft altimeters in flight to effect EAAD/ARR and altimeter adjustments, it is critical that the correct cockpit altimeter is used. The barometric pressure in inches of mercury, when reflected on an altimeter that has been adjusted to the departure airfield elevation MSL, is called the aircraft altimeter setting. When in doubt, the parachutist asks the pilot which altimeter is reflecting altitude AGL or MSL. Some adjustment rules of thumb are—

- 1 millibar = 32.25 feet of altitude (always rounded up to 33 feet).
- 1,000 feet of elevation = 30.3 millibars.

Note: The above two equivalencies are true only up to approximately 15,000 feet MSL, after which reduced air density changes the relationship.

14-42. Altimeter error induced by temperature variation can be significant, even when the altimeter is properly set for surface conditions (Figure 14-9, page 14-17). Essentially, the relationship is this: if warmer air is at altitude, the altimeter will read lower than its actual altitude AGL. This means that the parachutist is actually higher AGL than the altitude reflected on the altimeter. If the temperature aloft is cooler than the surface, the altimeter will read higher than its actual altitude AGL. This means that the parachutist is actually lower than the altimeter's indicated altitude.

14-43. A rule of thumb that can be used to identify the maximum potential discrepancy of altimeters in cold weather environments is that for a given altitude, allow 2 percent for every 5°C (or 9°F) temperature increment making up the differential between the surface and that specified altitude.

EXAMPLE:

Surface temperature is 12°C.
Temperature at 16,000 feet AGL is -18°C.
Temperature differential is 30°C.
30°C = 6 each 5°C increments.
2 percent times 6 increments = 12 percent of 16,000 feet = 1920 feet.

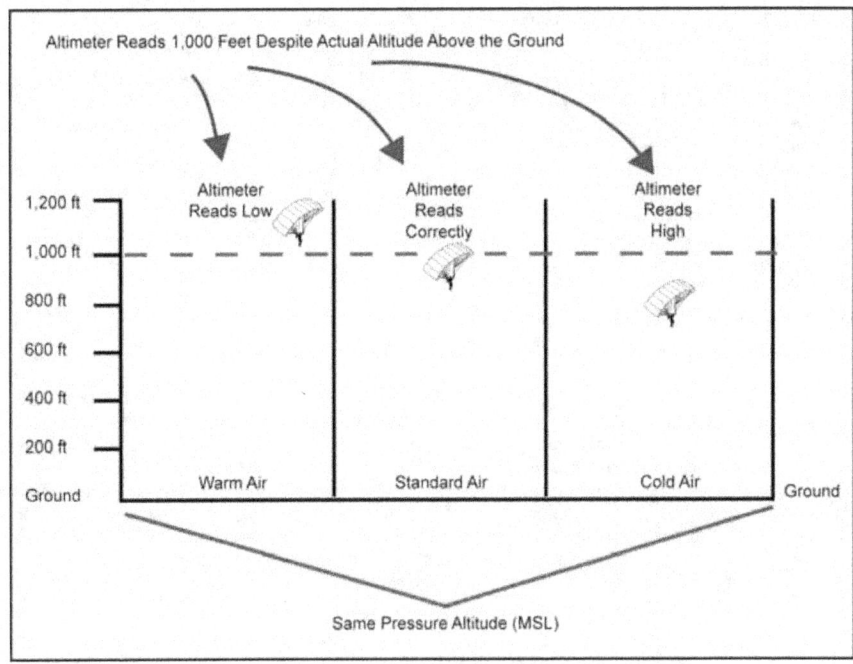

Figure 14-9. Air density variation with temperature change

AIR DENSITY ALTITUDE

14-51. Air density is the factor that has the biggest effect on parachute flight characteristics. Standard or textbook atmospheric conditions rarely exist. Therefore, the MFF jumpmaster must understand and plan to accommodate the changes in canopy flight performance caused by atmospheric pressure, temperature, humidity, or their combined effects. For example, a DZ can have an effective air density altitude that varies several thousand feet from the DZ's actual altitude MSL. It is possible for a DZ physically located at 9,000 feet MSL to have an air density altitude that causes parachute flight characteristics comparable to those at 14,000 feet MSL.

14-52. Parachute performance characteristics that are degraded by air density altitude are opening shock (such as in HAHOs), glide slope, and landing flare effectiveness (amount of required run out). Canopies simply become less efficient as lifting surfaces, especially with the higher suspended weights which accompany tandem bundles, combat equipment configured parachutists, or tandem pairs. Overall, parachute lift is degraded. Essentially, air density decreases with an increase in altitude, temperature, or humidity. When any or all of these factors are combined (high-altitude DZ on a hot, humid day), parachute performance is degraded to the point that despite using a ram-air canopy, a statistically significant number of parachutists can expect to have seriously hard (potentially injury-producing) landings.

14-53. This effect is tangible and is a key planning factor as low as 4,000 feet MSL. For round canopies (tandem bundles), operations above 4,000 feet MSL must carefully consider the effects of air density altitude. When these effects are combined with no-wind conditions (no wind velocity to create transitional lift at landing flare), regardless of the type of canopy being employed, the effects are again magnified. An increase in wind velocity helps offset canopy performance degradation caused by higher density altitudes.

Chapter 14

The MFF jumpmaster plans infiltration times to capitalize on the local weather conditions that best counter the effects of air density altitude.

14-54. Air density altitude effect is also valid, but to a lesser degree, for oxygen altitudes which can contribute to the effects of hypoxia. For example, it is possible below 10,000 feet MSL to have an air density altitude of 13,000 feet (+) MSL in terms of symptoms of hypoxia. This is based on the air temperature locally reducing air density. This marginally less dense air contains a lower volume of oxygen molecules, which means there is less oxygen available to the lungs. Its effect is magnified when the parachutist or aircraft crew member is physically exerting himself. This condition is usually manifested only after prolonged exposure (30 minutes or more) at this effective altitude. It is another reason to use oxygen all the way to landing for MFF operations at higher DZ elevations MSL and night operations. Doing so will minimize the effect and enhance night vision.

MAPPING OF PRESSURE SYSTEMS

14-55. On weather maps, atmospheric pressure MSL is plotted in millibars based on observations by reporting stations. Lines, called isobars, are drawn connecting equal values of reported pressure in the same way that contour lines depict elevation on topographic maps (Figure 14-10, page 14-19). The pressure differential mapping standard for weather map graphics (atmospheric pressure contour interval) in North America is an isobar for every four millibars pressure. When isobars are graphically displayed depicting varying pressures, they create pressure patterns. Patterns that depict low-pressure systems have a counterclockwise ground wind flow. In the Northern Hemisphere, hurricanes, typhoons, tropical storms, tornadoes, and waterspouts are all examples of low-pressure phenomena. If MFF parachute operations must be conducted in the vicinity of low-pressure systems, the following rules of thumb are useful in blind drop planning. Relative to its direction of movement, the strongest winds and most intense part of the storm are found on the right front quarter of the cell. The weakest portion of the storm cell is the left rear quarter in relation to its movement path.

14-56. High-pressure systems have a clockwise ground wind flow. Compared to a low-pressure system, high-pressure air masses have fewer clouds and lighter, calmer winds. Turbulent areas that exist in a high-pressure system are less concentrated. High-pressure areas are predominantly found over cold or cooler ground surfaces. These air masses are generally denser and barometrically will have higher millibar readings than low-pressure systems. High-pressure systems generally have conditions more conducive to MFF parachute operations.

Weather Factors for the Military Free-Fall Jumpmaster

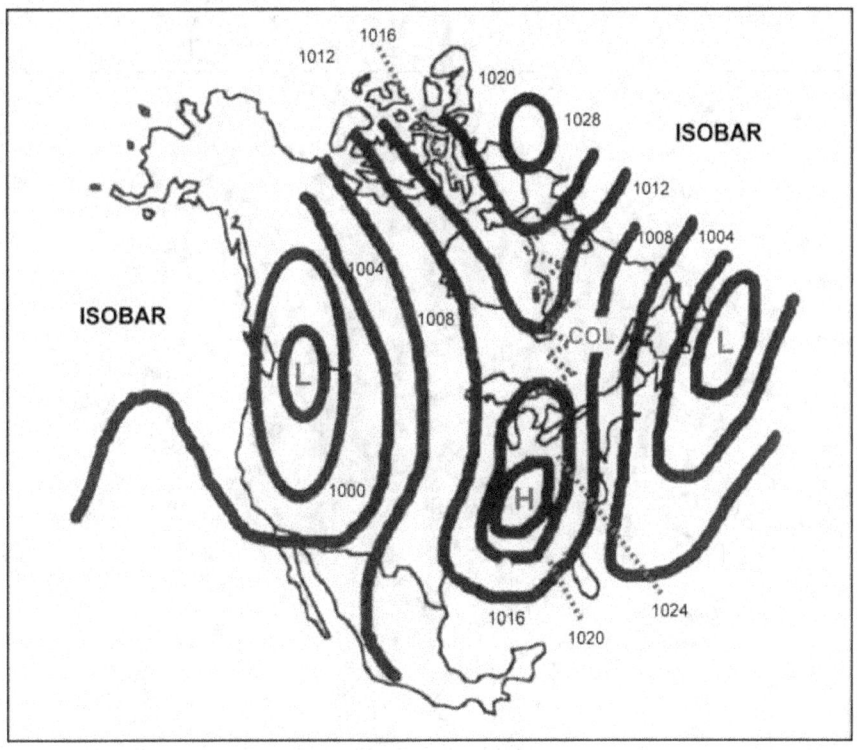

Figure 14-10. Pressure systems

ATMOSPHERIC CIRCULATION

14-57. Knowledge of atmospheric circulation patterns is needed to fully apply mission-planning parameters for long-range MFF infiltrations. As tactical missions will not be conducted on known DZs and long-range infiltrations are the norm, the potential negative impact of weather considerations is significant for the MFF jumpmaster and preparatory planning. The efforts and ability of the MFF jumpmaster affects the entire mission's contingency-planning requirements.

14-58. In general, the unequal heating of the earth's surface is the basis for atmospheric circulation on the macro scale. Regional variations in surface temperatures and topography complicate the movement of these air masses on the micro scale. The resultant airflow sets up irregular circulation patterns. Air mass patterns are additionally affected by the earth's rotation (Coriolis effect).

14-59. Secondary circulation is a description of the movement of the air masses found closest to the surface of the earth. For example, prevailing westerlies are regionally caused by the earth's rotation. This type of phenomenon can move air masses as much as 500 miles in 24 hours. Some sections of an air mass may move more rapidly or flow differently due to the localized effects of topographic irregularities.

14-60. There are two types of air masses. They are differentiated by their atmospheric pressure. They are also characterized as moving or stationary. In general, low-pressure systems and their associated frontal activity can adversely affect parachuting conditions. High-pressure systems are, in general, relatively free of bad weather and are more conducive to MFF parachute operations (Figure 14-11).

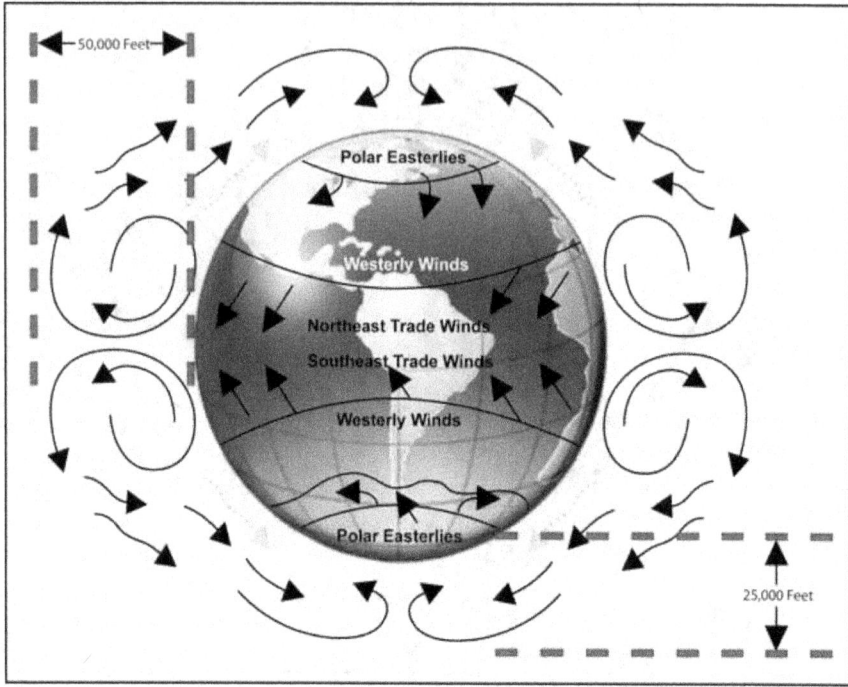

Figure 14-11. General pattern of atmospheric circulation

WIND FLOW MECHANICS

14-61. These systems are categorized into two types of winds—surface and aloft. There are two types of winds aloft. One is the jet stream, which is a band of maximum winds found along the earth's middle latitudes. This feature is concentrated in the troposphere and generally flows from the west. Dipping as low as 15,000 feet MSL, it can dominate planning factors affecting HAHO and, to a lesser degree, HALO MFF operations. The normal average altitude for the jet stream is 30,000 feet MSL. When the air acquires velocity, the Coriolis effect deflects it to the right (with respect to the flow) in the Northern Hemisphere and the reverse in the south. The term *gradient wind* refers to the theoretical wind that blows parallel to the isobars; in fact, in the lower parts of the atmosphere near the ground, friction with the surface deflects the gradient flow. The angle of deflection varies from approximately 10 degrees over the ocean to approximately 45 degrees over irregular land surfaces. Thus, analysis of the general wind patterns from upper air weather maps can be done; winds closer to the surface or where terrain is rugged or high will have more terrain-induced effects on the flow.

14-62. Surface winds are affected by factors caused by air mass contact with the surface of the earth. Friction between an air mass and the earth reduces surface wind speed to about 40 percent of the velocity of gradient winds aloft (Figure 14-12, page 14-21). It is this same friction which, in conjunction with the Coriolis effect, causes surface wind to flow across (perpendicular to) the isobars instead of parallel to them. The velocity of the wind is indicated by the spacing of the isobars. The closer the isobars, the higher the pressure gradient and, consequently, the higher the wind velocity. In the Northern Hemisphere, surface winds flow clockwise around and away from a high-pressure center. They flow counterclockwise and in toward a low-pressure center. In the Southern Hemisphere, the winds still blow around and away from highs and in toward lows, but the direction of the flow is reversed; for example, counterclockwise around a high and clockwise around a Southern Hemisphere low.

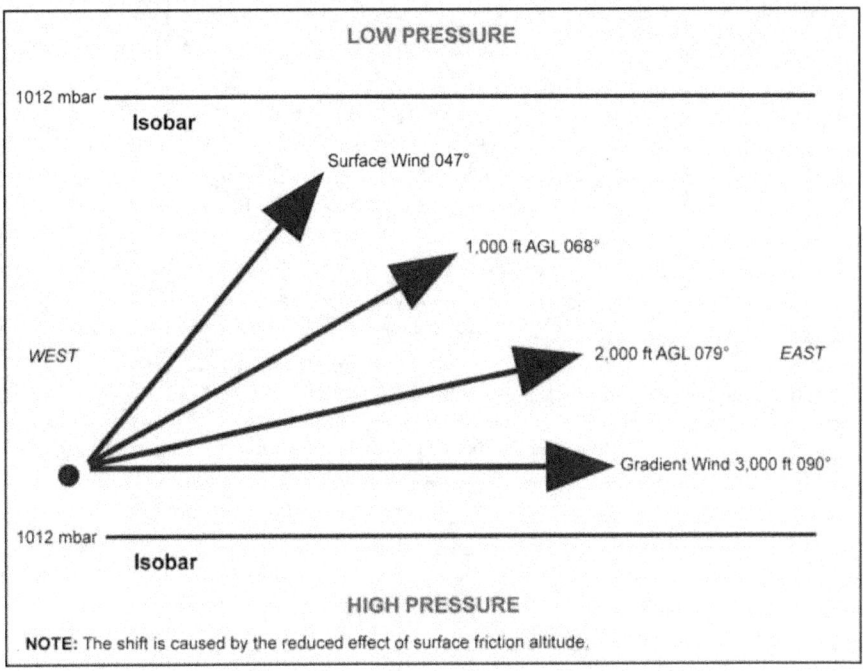

Figure 14-12. Wind shift with altitude increase

14-63. Surface wind flows across isobars from high pressure toward low pressure. Friction with the surface generally only affects air masses up to an approximate maximum altitude of 3,000 feet AGL. The degree of effect will depend on the local contour of the surface. It is least over water bodies or flat deserts and greatest over irregular mountains. The effect of friction also gradually decreases with altitude. The ground surface effect is turbulence, which is a localized result of air mass friction and obstacles on the earth's surface. Since surface friction decreases with altitude, air mass wind velocity generally increases with altitude AGL. Above the friction layer (3,000 feet AGL), wind speed is determined by the dimension of the pressure gradient (closeness of the isobars) and is fairly constant or slightly increases until reaching altitudes affected by the jet stream (Figure 14-13, page 14-22).

14-64. In HAHO operations, topography influences parachute flight-path planning because of its ability to channelize and deflect prevailing winds. Additionally, the jet stream (with seasonal variations) is a significant factor in release point determination. For example, the jet stream can dip as low as 15,000 feet MSL and achieve wind velocities of more than 270 knots. The jet stream is most often found where a cold front boundary intersects the 500-millibar level aloft. It is found in segments 1,000 to 3,000 miles long, is 100 to 400 miles wide, and 3,000 to 7,000 feet thick. A thorough premission area study will identify many of these factors that are unique to the intended AO. Later, during mission planning isolation, just prior to infiltration, an operational area intelligence update with current weather provides the data used to determine detailed prevailing wind effects on the MFF infiltration. It is the responsibility of the MFF jumpmaster to assess these factors and advise the commander of their potential combined impact on that portion of the mission (Figure 14-14).

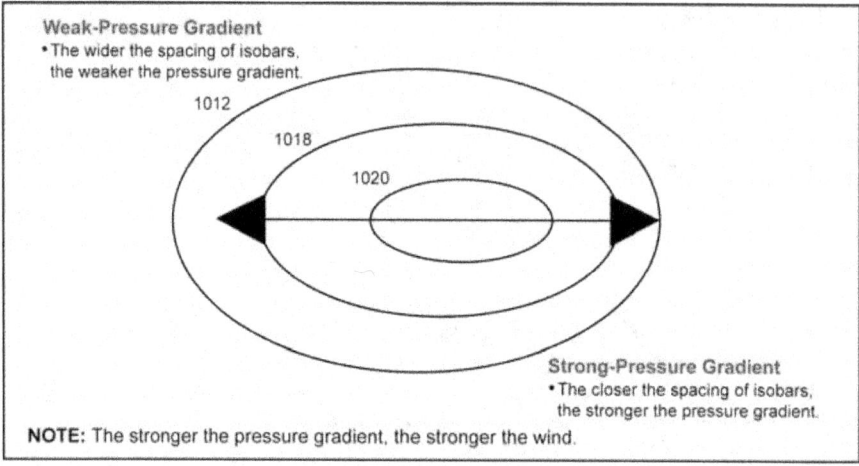

Figure 14-13. Pressure gradient principles

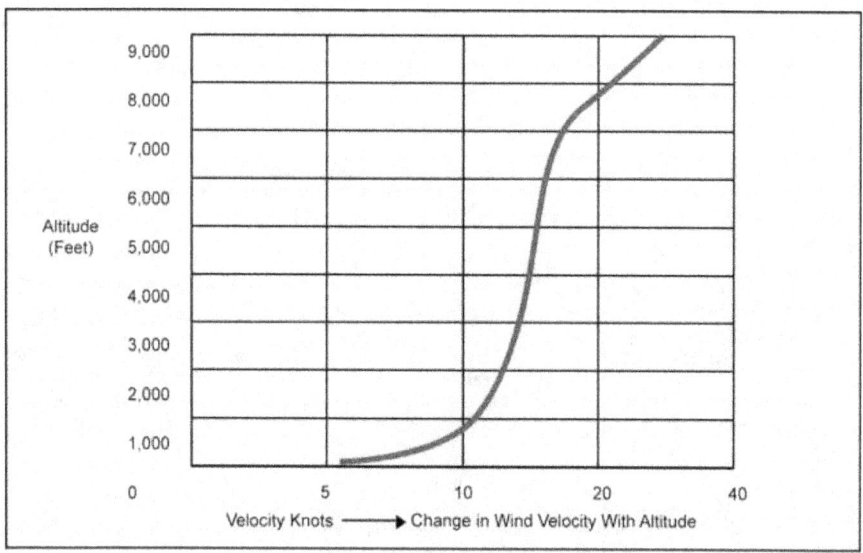

Figure 14-14. Change in velocity with altitude

LAND AND SEA BREEZES

14-65. These related weather phenomena are caused by the difference in heating and cooling rates of water and land. Generally associated with coastal regions, these effects also occur along larger lakes. The sea breeze occurs in morning/daytime and the land breeze occurs at dusk/early evening (Figure 14-15). Since land warms up faster than water, the air rising over the land creates a lower pressure area than that over the water. The air over the water moves toward the land to fill the low-pressure zone. The effect creates the onshore or so called sea breeze. The same sequence of events occurs to create the land breeze. At the end of the day water gives off its heat more slowly than land. After the land has cooled, heated air is still rising from the water. A low-pressure area is created offshore by the rising warm air. Air moving from the land to fill in the low-pressure area offshore causes the land breeze.

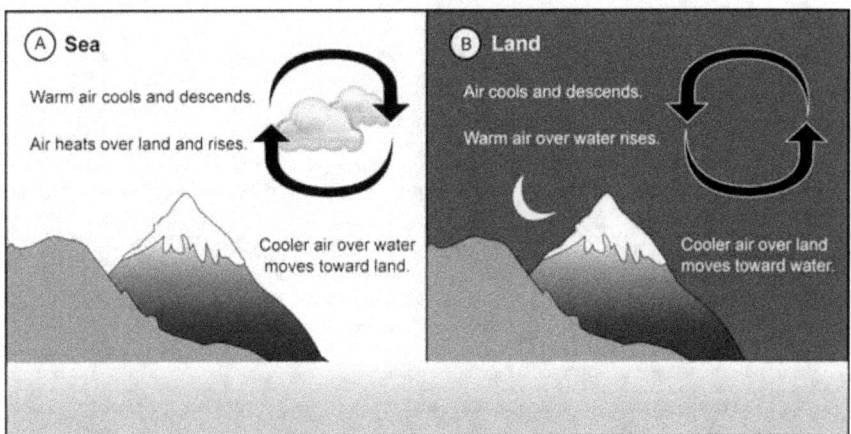

Figure 14-15. Land and sea breezes

14-66. An example of potential problems for MFF parachutists during these conditions would be a dusk infiltration onto a coastal DZ after a warm day. Without calculating the canopy OP to accommodate the land breeze's increased offshore wind velocity, an MFF jumpmaster might not factor enough canopy performance (K factor) to reach the beach. With degraded canopy performance over cold water, the parachutists could be seriously short of their intended beach DZ. In this case, the parachutists would be conducting an unintentional water landing just offshore from their beach desired impact point.

14-67. In mountains there is a parallel phenomenon that is more radical and therefore more potentially dangerous to the MFF parachutist under canopy. These are the valley and mountain breezes. For land and sea breezes, heat differential drove the mechanism. These two new events are based on the principle of warm air masses rising and cool air masses sinking. Valley breezes occur during the day after the sun has heated an exposed air mass against the side of a terrain feature. It rises causing an upslope breeze. At dusk into early evening, the air cools and flows downslope. A parachutist entering one of these moving air masses can be floated or otherwise involuntarily pushed under canopy as the air mass moves with this thermal event. The closer the parachutist is to the terrain feature, the more graphic the effect and the higher the potential for turbulence. When eddy winds and localized convective turbulence are factored into the combined effect, the probability of radical canopy performance is quite high.

14-68. Parachutists must be suspicious of altimeter readings due to the rapidly changing pressure differentials as air masses are pushed into the sides of larger terrain features. For HAHO parachutists in the clouds while topping mountains or ridgelines, it is possible to have altimeter readings up to 1,000 feet higher than the actual altitude AGL. The air mass is compressed against the terrain feature reflecting the air density of a lower altitude. This means that the parachutist can be up to 1,000 feet lower than the altitude reflected on the altimeter. As this also equates to roughly 30 millibars of pressure differential, there is also potential for barometrically activated EAADs/ARRs to fire at an unplanned, inopportune time.

14-69. The MFF jumpmaster must conduct a map survey for the highest terrain feature near the OP affecting opening altitude determination and assess canopy glide slope to accommodate terrain along the anticipated canopy flight route. Essentially, the MFF jumpmaster must consider and evaluate flight plans when conducting MFF operations in mountainous areas. The MFF jumpmaster accounts for terrain between the OP and the DZ, thermal air mass/flow, air pressure variations, air density altitudes, and environmental factors. The MFF jumpmaster identifies critical altitudes AGL and MSL and determines canopy performance parameters affecting the ability of the infiltrating team to arrive at their intended DZ. By assessing canopy K factors, the MFF jumpmaster must determine if the canopy performance under its loaded weight will, by a given point in the canopy flight path, provide enough altitude to clear terrain features.

14-70. All mountain phenomena are associated with radically unpredictable wind shears and buffeting. They are very serious hazards to the parachutist and a planning factor that must be addressed by the MFF jumpmaster. In general, HAHO or any canopy flight in and around mountains or hills can be hazardous. Cloud decks, which top mountains or ridgeline terrain features, also contain especially strong downdrafts and turbulence, especially on the leeward (downwind) side of the feature. Parachutists must also be aware that clouds do not have to be present for these turbulence conditions to occur. Turbulence is normally present around any cloud formation and can be found at any altitude—even in clear air a significant distance downwind of terrain.

EDDY WINDS

14-71. Commonly known as surface or ground turbulence, eddy winds are created when an air mass flows over physical obstructions on the earth's surface. Irregular terrain, large individual trees, tree lines along DZ edges, built-up areas, and individual buildings are common features which create turbulence that parachutists must negotiate when setting up for a landing.

14-72. Turbulence effects are a function of the angle of contact with and intensity (speed) of the airflow over and around the parachute's airfoil shape. In constant aerodynamic (laboratory) conditions, the parachutist is the only factor which changes the airfoil shape, and that only occurs when the parachutist pulls down on the steering lines. Since constant conditions do not exist in free air, the continuously changing aggregate effects of the airflow disrupt the equally pressurized and smooth flying of the canopy. Anything other than the hypothetical aerodynamically perfect condition is deflation or distortion of the canopy, no matter how slight or for how short a period of time. It is this turbulence disturbed airflow that can unpredictably cause the canopy to stop descending (float), involuntarily turn (side slide), dive, go back up, momentarily start to deflate (breathe), partially collapse (fold end cells under), radically deform the airfoil (deflate multiple cells of the canopy or fold the nose of the canopy under), or all of the above at what often appears to the parachutist to be simultaneously.

14-73. Other key factors are size (square footage), number of cells of the canopy construction, and type of nylon porosity in relation to the total weight suspended under it. Most MFF jumpmasters are concerned with combat-equipped parachutists overloading the maximum suspended weight recommendations of a canopy or parachute system. These are valid concerns to assist in avoiding structural damage to the canopy, especially at higher opening altitudes MSL. Thus, HAHO free-fall delays are used to minimize opening shock forces, especially for tandem bundle delivery systems; after the delivery of a cargo payload, the minimum weight for adequate pressurization of the tandem master's main canopy might not be met. Canopies under these conditions can be extremely unpredictable, especially the larger square footage tandem main parachutes.

14-74. For example, presentation of an underpressurized tandem canopy's stabilizers to the wind line when executing a landing approach, in conjunction with normal manipulation of the brake lines, can depressurize the canopy enough to collapse it. The reinflation time hazard is compounded in that for some tandem bundle delivery systems, payloads are dropped between 300 to 500 feet AGL. The tandem master is already close to the ground when turning crosswind onto the base leg of the landing approach. When a canopy without adequate pressurization is additionally subjected to environmental turbulence, its ability to maintain an aerodynamic airfoil shape is further degraded. It can abruptly collapse even when the turbulence effects appear to be insignificant. When eddy turbulence (gusting) is coupled with thermal uplifting, the danger of canopy collapse becomes serious, regardless of the adequacy of the canopy loading or the canopy's dimensions.

14-75. Turbulence eddies are also known as rotors and dust devils. They are naturally created, ranging in size from several inches to several kilometers in diameter. The most common that affect MFF parachute operations are those one meter to several hundred meters in diameter. The eddy's intensity is affected by wind velocity and the shape of the obstruction over which the air mass flowed to create it. Dust devils are dangerous in that the diameter of the locally rotating wind column is usually much larger than the visible portion. Where the eddy might be only marginally wider than the visible diameter at ground level, dust devils can also extend thousands of feet higher than the visible top and increase in diameter severalfold with elevation. The visible portion of the eddy is the uplifted soil and debris from the earth's surface.

14-76. For an MFF parachutist under canopy, flying into or in the vicinity of a dust devil almost certainly means a collapsed canopy. This is disastrously close to the earth when the canopy might not have time to reinflate or the rotating air column has twisted the parachute's line groups. Even a parachutist on the ground still in the harness must be prepared to release the RSL and cut away the main canopy if it were to start to be reinflated by eddy turbulence. Parachutists who have been picked up by a dust devil have been lifted between 10 and 40 feet AGL before being slammed back to the ground, resulting in death or serious injury.

14-77. Air density can also affect eddies. Higher air density correlates with a higher intensity in the manifested eddy. Cold air is more dense than warm air. Therefore, in winter conditions, one can expect slightly more pronounced turbulence effects (larger-diameter rotors, higher-intensity rotation, higher altitude effects) than those phenomena that are warm air/thermal conditions. Humid air is less dense than dry air, so eddy turbulence would be more pronounced in low humidity than in high humidity. DZs at higher field elevations have lower air density than at sea level and in general produce turbulence of lesser intensity. All of these parameters are theoretical. Each potential DZ's set of landing conditions must be evaluated by the MFF jumpmaster to determine the probable aggregate effect on the parachutists under supervision for the conditions anticipated for that operation.

14-78. Obstruction eddy characteristics vary depending on many factors. With wind speed less than 10 knots, small eddies (10 to 50 feet in depth) are created on both the downwind and upwind side of an obstacle. With wind speeds between 10 and 20 knots, significant turbulence eddies are created on both sides of an obstacle. Because of the wind's velocity, they are also carried several hundred feet downwind of the obstacle before they dissipate. Severity depends on the height and shape of the obstruction, as well as the wind speed. Obstructions with sharp definition/edges create more pronounced eddy currents. When winds exceed 20 knots, eddy currents are mostly formed on the leeward side of the obstruction and the turbulence zone is carried a considerable distance downwind of the obstacle. A rule of thumb for intermediate strength winds (10 to 20 knots) is that turbulence eddies are carried at least five times the distance downwind as the height of the obstacle that created them (Figures 14-16 and 14-17, page 14-26).

14-79. The rule of thumb for winds stronger than 20 knots is that the eddy turbulence is carried 10 times the distance downwind as the obstacle height. On large, open DZs with no obstructions (usually the case with DZs downwind of tree lines), eddies may be carried as much as one-half mile downwind of the obstruction. This phenomenon is also strengthened if the terrain slopes downhill away from the tree line causing the turbulence (Figure 14-18, page 14-27).

Chapter 14

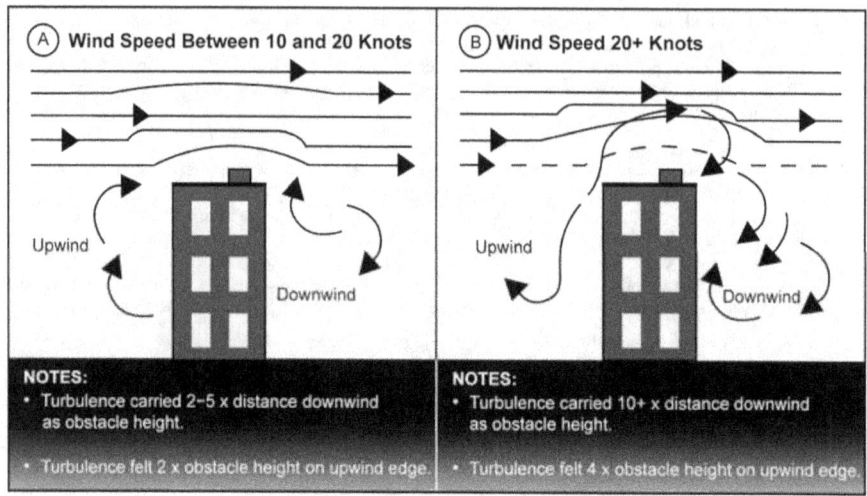

Figure 14-16. Single-obstacle eddy current

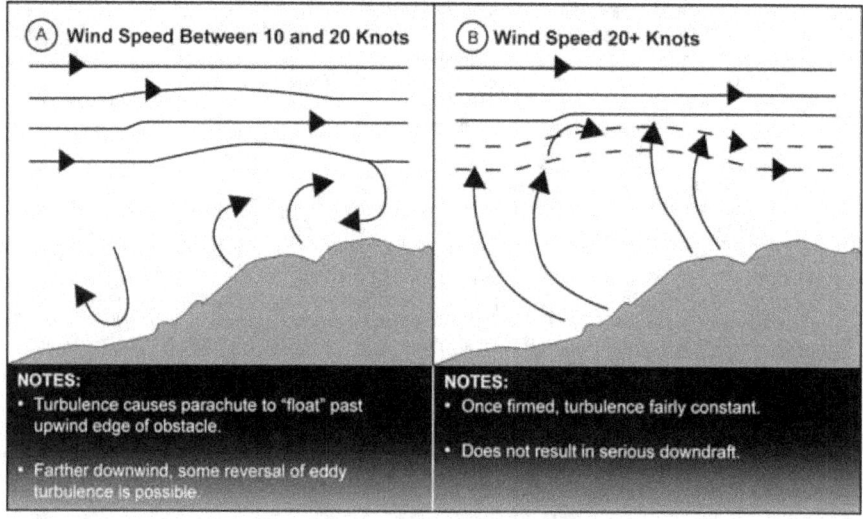

Figure 14-17. Terrain-induced eddy currents

Weather Factors for the Military Free-Fall Jumpmaster

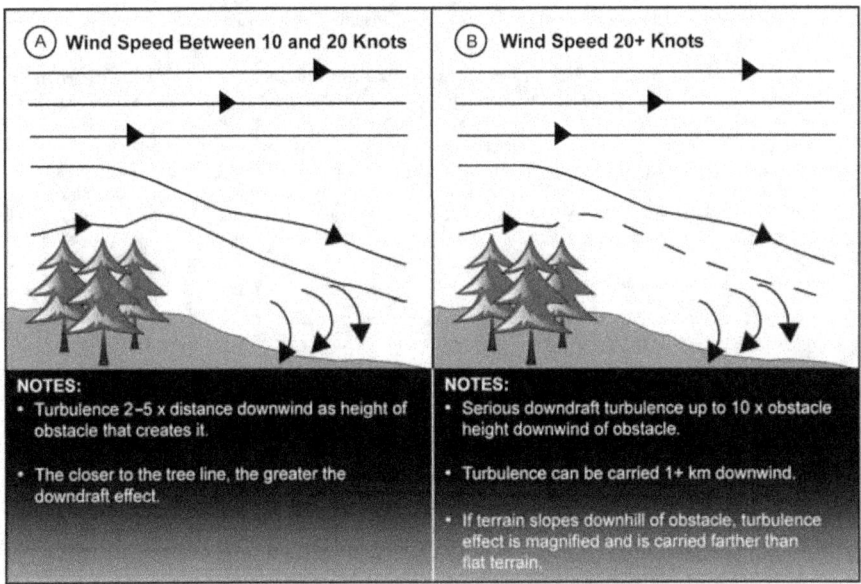

Figure 14-18. Tree-line-induced eddy currents

14-80. For winds stronger than 20 knots, the MFF parachutist may be backing in to the target area. As the parachutist descends into the turbulence area, eddy effects are unexpected because the object creating the eddy is a significant distance upwind and probably not in the visual field. Turbulence is also indicated by the behavior of the wind. Strong winds usually contain gusts, which are wind speed changes of 5 to 10 mph within a few seconds. Gusts can add to the random deflation and poor performance of an MFF ram-air canopy. All parachutists must visualize the upwind quadrant (wind cone) during their landing approach to assess the potential effects of wind speed and obstructions on the parachute's performance. Horizontal and vertical separation between parachutists should be increased during a strong wind condition landing approach to accommodate unexpected canopy movement at low altitudes.

14-81. Only minor eddy currents form over water surfaces. These are created by air movement friction with the wave tops. Regardless, parachutists must still be cautious in setting up their approach to land on or in the vicinity of a vessel at sea. Because a vessel is an obstruction, it or its superstructure can create unexpected downwind eddy currents. The larger the vessel, the greater the potential for eddy currents.

14-82. In HAHO operations, when air is unstable (conditions not consistent as altitude increases) and once eddy currents or turbulence forms, they tend to continue to grow. This is especially true when thermal conditions exist. For example, when fueled by thermal convection, ground-induced eddy currents can extend up to 6,000 feet AGL. Essentially, this is how dust devils can form and evolve to be several hundred meters in diameter at their base.

14-83. The lighter the parachutist (or total suspended weight), the more susceptible the canopy will be to turbulence-related problems. The forces acting on a canopy are proportional to the rate of change of wind velocity on the canopy. Therefore, it is dangerous to fly fast (canopy toggles in the full flight/up position) through turbulence. Flying through eddy turbulence should be done at 25- to 50-percent brakes. Too slow (more than 50-percent brakes) may also be dangerous because the canopy may stall without warning due to a gust. Increasing the canopy rate of descent through braking keeps the canopy's internal pressurization higher. This decreases its susceptibility to turbulence-induced instability, decreases the overall potential for collapse, and increases overall canopy control. It is an effective technique at any altitude.

14-84. The effects of eddy turbulence normally subside close to the ground, unless the parachutist is directly in contact with an eddy. Under roughly 50 feet AGL, turbulence should lessen, permitting a near-normal approach. Regardless, the approach should still be made in the 25- to 50-percent brake range as a precaution. Turbulence at the landing flare may side slide the canopy or enter it into a turn. The rate of turn may increase when lowered combat equipment is attached to one side of the parachute harness in lieu of being suspended from the center of mass (saddle). With significant weight pulling one side of the harness down, the canopy risers will be uneven. Because the canopy is already canted to one side, a gust is likely to send the canopy into a turn, usually to the side on which the equipment is suspended. When braking, the canopy can be controlled by manipulating the steering lines to offset the difference in riser lengths. Under these conditions, the MFF parachutist should be prepared to perform a PLF.

14-85. Other turbulence is also generated and experienced by MFF parachutists using ram-air canopies. Airfoil-generated vortices trail off the lateral corners of the canopy, much like those created by airplane wingtips. Canopy vortices come off the trailing edge at an upward 45-degree angle when the parachute is in full glide. In full brakes, the vortex angle is almost straight up due to the canopy's increased rate of descent. When in a partially braked configuration (such as in a stack setting up to land), the vortex angle produced is about 60 degrees from the canopy angle of attack at the ground. The vortices descend and dissipate at a rate slower than the parachutist's canopy. Closer distances between parachutists in the air result in a stronger vortex. The vortex intensity is greater with parachutists who have a higher total suspended weight because their canopies have a greater rate and steeper angle of descent. Upon passing into another canopy's vortex (wake), the turbulent air will hit the parachutist's body before it hits the canopy. The canopy will rock and then drop up to 20 feet, causing danger during landing. Near the ground or aloft, vortices drift with the wind, so it is possible to hit vortex turbulence from a canopy that has landed or turned upwind. If this occurs, the MFF parachutist must be consistent with the steering toggle manipulation and still flare at the normal time. If in doubt, the parachutist must be prepared to land using a PLF.

CLOUDS

14-86. There are four categories of clouds: high-altitude clouds, middle-altitude clouds, low-altitude clouds, and clouds with extensive vertical development (Figure 14-19, page 14-29). Each category has some typical weather characteristics that affect MFF parachute operations.

14-87. High clouds have a base that starts at 18,000 feet AGL and are valuable as indicators of approaching fronts and associated changes in weather. Middle clouds, found between 6,500 feet and 18,000 feet AGL, are mostly composed of ice crystals or supercooled water vapor, which produces high clouds and icing conditions. Middle and low clouds are most likely to impede parachuting. Low clouds extend from near surface to 6,500 feet AGL. Their cloud base changes rapidly. Clouds that have a base below 50 feet AGL or are in contact with the earth's surface are defined as fog.

14-88. Clouds with vertical development usually have cloud bases below 6,500 feet AGL, the tops of which can extend to over 60,000 feet AGL. This lifting action is caused by convective updrafts or frontal lift when two air masses meet. These conditions are extremely hazardous for parachutists. The turbulence inside cloud formations such as these is strong enough to cause structural damage to aircraft airframes (and therefore applies to parachutists under canopy). One example of these formations is the common thunderstorm cloud with its typical anvil head shape on top as it develops. A cumulonimbus cloud is essentially a thunderstorm without the thunder, but it has the same turbulence characteristics. Whether a cloud is exhibiting electrical or thunderstorm activity or not tells nothing about the turbulence in or around it. All clouds have some degree of turbulence associated with them. The faster a cloud is developing or changing, the more turbulence it possesses.

14-89. Cloud-induced or associated factors which affect MFF parachute operations are the cloud type and thickness (depth) or number of layers, amount of sky covered/obscuration, resultant vertical (MFF jumpmaster) and lateral/slant (pilot/HAHO parachutist) visibility, moisture type/precipitation contained in them, freezing/icing factors, and turbulence. On a macro scale, the three types of turbulence are thermal, mechanical, and frontal. Thermal turbulence is local vertical convective currents caused by surface heating or cold air masses moving over warmer ground or water. Mechanical turbulence results from wind passing over irregular terrain or obstructions. Frontal turbulence occurs when air masses are locally lifted by another air mass of a different temperature and density. This creates abrupt and usually radical wind shear, normally associated with approaching cold fronts. In general, frontal turbulence adversely affects HAHO parachute operations.

Base Altitude	International Cloud Classification, Abbreviations, and Weather Map Symbols				
	Cloud Type	Abbreviation	Symbol	Type: Moisture	
Bases of high clouds usually above 18,000 feet	Cirrus	Ci	—	Entirely ice particles	
	Cirrocumulus	Cc	2ᴗ	Mackerel sky—ice particles	
18,000 ft	Cirrostratus	Cs	⊔	Ice particles form halos around sun and moon	
Bases of middle clouds range from surface to 18,000 feet	Altocumulus	Ac	⊔	Puffy mass—water droplets	
	Altostratus	As	∠—	High gray sheet clouds—windy, often followed by rain or snow	
6,500 ft					
Bases of low clouds range from surface to 6,500 feet	*Cumulus	Cu	⌒	No precipitation normally	
	*Cumulonimbus	Cb	⛆	Violent vertical development, bottom portion of anvil head	
	Nimbostratus	Ns	⚌	Storm clouds with rain or snow	
	Stratocumulus	Sc	‿	Transitional weather conditions	
	Stratus	St	—	High fog/"low ceiling," rain, snow, sleet, fog	
Surface	*Cumulus and cumulonimbus are clouds with vertical development. Their base is usually below 6,500 feet but may be slightly higher. The tops of the cumulonimbus sometimes exceed 60,000 feet.				

Figure 14-19. Cloud classification

THUNDERSTORMS

14-90. Flight through thunderstorms and mission accomplishment are not compatible. This premise applies to rotary-wing aircraft. If it applies to powered flight from inside a protected cockpit, its impact is as pertinent, if not more so, for MFF parachutists under canopy.

14-91. Thunderstorms are mostly caused by thermal or convective uplifting. Their rate of updraft can range from a few feet per second to 6,000 feet per minute. It is not only turbulence, which is associated with a thunderstorm, but other weather variables that can cause adverse MFF conditions. For example, hail mostly melts before it reaches the ground. It can be marble through grapefruit size in diameter at HAHO altitudes. An example is a five-inch-diameter hailstone that came through an aircraft cockpit canopy at 29,000 feet AGL. Hailstones have been thrown five miles from their parent cloud formation. Generally, the atmospheric conditions that can float a two-pound ice ball are not conducive to stable ram-air canopy flight.

14-92. Icing conditions that can accumulate on a canopy and HAHO parachutist can occur as low as 14,000 feet MSL. Buildup degrades the canopy's glide slope (changes the K/drift factor) and overall performance. Ice forms when air bearing supercooled moisture comes into contact with a solid object. When a canopy and its parachutist move through this atmospheric condition, ice forms on all cell leading edges, suspension lines, slider, as well as surface obstructions, such as canopy seams, pilot chute, and deployment bag. Snow conditions are mostly found below 20,000 feet MSL and accumulation is not a concern. Environmental effects on the parachutist for any MFF operation are all that must be accommodated. Icing increases drag, which significantly decreases lift by deforming the canopy's airfoil shape. The most severe icing occurs between 0 and -10°C. Icing is rare below -20°C, but is possible in any cloud below 0°C. Freezing rain is more probable and occurs when supercooled water droplets (liquid water whose temperature is below freezing) hit any object and immediately freeze on contact. This condition is more common and dangerous than normal icing as dangerous amounts can build up in minutes. There are two documented instances of ice buildup resulting in a rate of descent and controllability problem severe enough to cause the concerned jumpers to jettison the main canopy.

14-93. Lightning is an obvious hazard. The energy to create a lightning stroke 10,000 feet long is 2 to 3 million electron volts and 500,000 to 600,000 amperes. The movement of a canopy through charged air can accumulate the static charge necessary to prompt a lightning stroke. Additionally, that amount of static electricity is enough to affect HAHO/magnetic compasses. This same accumulation or close lightning strokes can have enough effect to activate EAADs/ARRs, which employ electrically fired explosive cutters or pin pullers (for example, Sentinel, MARS, or Military CYPRES 2). EAADs/ARRs which employ mechanical and barometric mechanisms (for example, FF-2 or KAP-3) would not be electrically affected. Additionally, parachutists who find themselves in high electrical conditions should ensure that metal main rip cords are not left exposed, as these metallic objects could very easily act as a lightning rod and prompt a lightning strike. Figure 14-20 depicts air movement beneath a thunderstorm cell.

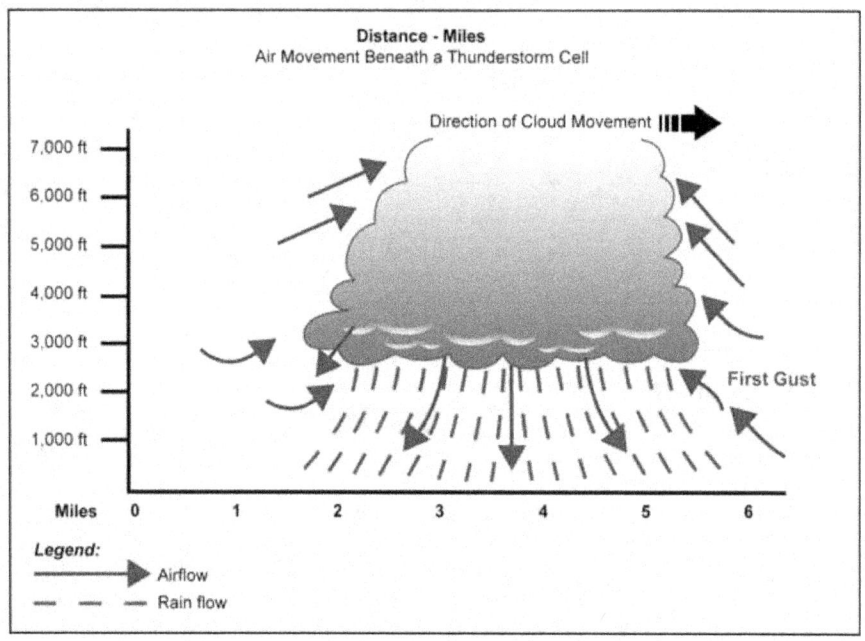

Figure 14-20. Air movement beneath a thunderstorm cell

14-94. Another adverse parachuting phenomena associated with thunderstorms is the first gust. This term applies to the rapid change in wind speed and direction close to the earth's surface immediately prior to the passage of a storm cell. The first gust results from the spreading of a thunderstorm's downdraft air current as it hits the surface of the earth. This downdraft is the reason canopy flight underneath a seemingly innocuous storm cell or squall cloud can be disastrous. The initial wind surge, as observed at the earth's surface, is the first gust. It can generate turbulence as far as 8 to 10 miles ahead of the storm core, depending on the extent of the cloud's development. First gust wind speeds may exceed 50 knots and can vary up to 180 degrees from previously prevailing wind direction. However, first gusts generally average 15 knots over previous velocities and average a 40-degree change in wind direction. The speed with which the first gust arrives is extremely hazardous for parachutists under canopy. First gusts are not limited to the area ahead of the storm cell's direction of movement. Additionally, they have associated with them an abrupt fall in barometric pressure as the storm cell approaches and a just as rapid rise in pressure as it passes. This atmospheric pressure change can easily be read on the parachutist's altimeter. Figure 14-21 depicts the first gust wind flow.

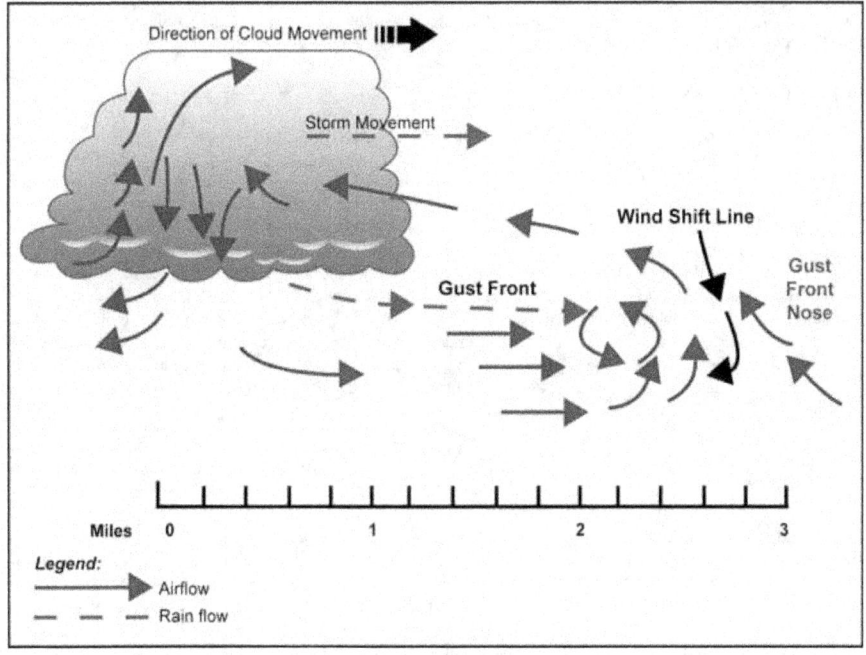

Figure 14-21. First gust wind flow

14-95. Fog occurs when the cloud base is within 50 feet of the earth's surface. Light wind is generally favorable to the formation of fog. It gives the moisture-bearing air mass a gentle mixing circulation which spreads the surface cooling action through a deeper layer of air, thickening the blanket of fog. Fog dissipates with heating (usually when the sun rises) or is blown away with stronger wind. When the relative humidity of the air layer that is directly in contact with the earth's surface drops, the amount of fog will decrease. The most significant impact of fog on MFF operations is the masking of the DZ surface on final landing approach. This is especially hazardous on DZs with lower air density altitudes where rates of descent are increased, on cool calm nights with no wind to land into, or when obstacles on the DZ are obscured. The MFF jumpmaster should be able to visualize and anticipate potential condition scenarios such as these. Based on isolation planning, the MFF jumpmaster can minimize their effect by choosing the time and place where weather factors best offset masking of the DZ surface and maximize the tactical realities of the infiltration DZ.

14-96. Parachutist depth perception is reduced over unbroken surfaces. With no relief features to provide visual clues, altitude estimation by the human eye is difficult. This occurs during the day and is worse during the night or other reduced-visibility conditions. Canopy flight over water, unbroken level ground (desert sand, grass fields, crops, and so on), and snow fields are examples. For instance, looking down on newly fallen snow, shadows are usually not visible or are not distinct enough to provide visual clues of surface elevation. Visual clues aid the human eye in determining angular perspective. Canopy flight planning for landing approach and flare altitude determination is made more difficult, especially at night over the same type of surfaces. With reduced depth perception, the tendency is to flare the canopy too late (plow in). This condition is exacerbated during night's reduced visibility in that there are also generally lighter wind conditions, hence a greater rate of descent and less wind against which to create a dynamic stall. During daylight conditions, amber goggles assist in giving the depth perception necessary to determine landing surface conditions. At night, staying on 100-percent oxygen all the way to the ground maximizes night vision's visual acuity and depth perception, and increases the probability of avoiding injury on landing.

APPLICATION OF GENERAL WEATHER PRINCIPLES

14-97. Knowledge of general weather rules and principles is helpful to the MFF jumpmaster. Opportunities to apply rules of thumb occur during the conduct of peacetime duties as the DZSO, during mission planning when the operation involves infiltrating blind, or when formulating mission abort criteria and contingency plans. For example, weather unfavorable to MFF parachuting operations is generally indicated by falling barometric pressure with winds prevailing from the west. Clearing weather and good parachuting conditions are usually characterized by winds which shift to western quadrants with a rising barometer. As the MFF jumpmaster studies weather data in preparation for an infiltration airborne operation, knowledge of these principles can assist in visualizing the weather conditions that will be in place over the anticipated AO or planned DZ. This gives the MFF jumpmaster a basis of fact with which to determine exit and opening points, probable upwind quadrants, and to establish realistic GPS navigation waypoints. Knowledge of general weather principles reduces the reliance on chance as the primary factor affecting the lives of the parachutists who are entrusted to the MFF jumpmaster for this critical infiltration portion of the mission.

14-98. Most of the following principles can be visualized by remembering that low-pressure systems (bad or less than ideal parachuting conditions) have counterclockwise airflow. High-pressure systems (relatively good MFF conditions) rotate clockwise frontally. The last principle is that these fronts rotate around their axes as the air mass moves from west to east. From the perspective of a ground observer, as the track of the frontal system crosses near that position, the direction of the wind changes depending on whether it is the leading edge, center, or trailing edge of the system that is passing. Knowing this and the speed of the front, one can fairly accurately determine from which direction the winds will be blowing at a specified time in the near future. At a minimum, the upwind quadrant can be determined and a wind cone established. In this manner, given no ground-sourced information, the MFF jumpmaster can determine on a blind drop where to exit the aircraft to be at least initially upwind of an anticipated DZ while under canopy.

14-99. When there is no ground source of wind or weather information, the drop is termed blind. The preceding techniques are the only tools the MFF jumpmaster has to pre-identify exit (release points) and opening points which will maximize the infiltrating team's odds of hitting a desired impact point on an unfamiliar DZ. By sound interpretation and application of basic weather principles, the MFF jumpmaster can enhance the team's odds of opening, at a minimum, upwind of the intended DZ and inside the wind cone. Despite the availability of meteorological information and the studied application of data, weather is still an inexact science. For this reason, operational plans that do not integrate alternate DZs, abort criteria, and contingency decision matrices are incomplete. The MFF jumpmaster is responsible for generating these plans.

14-100. MFF jumpmasters and especially their commanders must understand that landing off of an intended DZ on a blind drop or not landing at the desired impact point does not mean that the infiltration was unsuccessful. For directing commanders of MFF infiltrations and peacetime airborne operations, the definition of a successful MFF parachute infiltration is the team landing with the group intact, undiscovered, and knowing where they are, regardless of where they land. Peacetime training is too often conducted with detailed, current weather data and on the same familiar DZs. Those are not the conditions that teams will realistically experience on a tactical infiltration onto a blind DZ. Teams and their supporting staffs must exercise the same weather interpretation skills for training that they will be expected to employ on a contingency blind drop mission preparation. Directing commanders must support and understand this concept and the realities of the techniques being employed. All too often, landing out is construed as an

Weather Factors for the Military Free-Fall Jumpmaster

error on the part of the MFF jumpmaster instead of as an integral part of essential training. It is a tactical reality for which the directing commander must prepare during realistic peacetime training, as well as the cold facts of a blind tactical infiltration. Rarely does an infiltrating MFF element, dropping blind, land at a preselected desired impact point. A commander who expects otherwise does not understand the realities of MFF operations. A team can minimize the effects of weather-induced factors only through in-depth planning. It can never fully control or negate them. Regardless, the MFF jumpmaster's planning attempts to exploit them to tactical advantage.

14-101. As examples, two National Weather Service interpretation guidelines concerning prevailing conditions are listed which can be useful rules of thumb in blind drop planning. One states, "when the wind blows from a south to southeast direction and the barometer falls steadily, a storm is approaching from the west or northwest. Its center will pass near or north of the observer within 12 to 24 hours. After it passes, the wind will shift to the northwest by way of south and southwest." Visualizing the movement of a high-pressure air mass moving west to east will help explain this phenomenon. Storm centers (low-pressure air masses) rotate counterclockwise. Their eastward movement toward, while passing over, and to the east of the ground position explains the predictable shift of the wind direction (Figure 14-22).

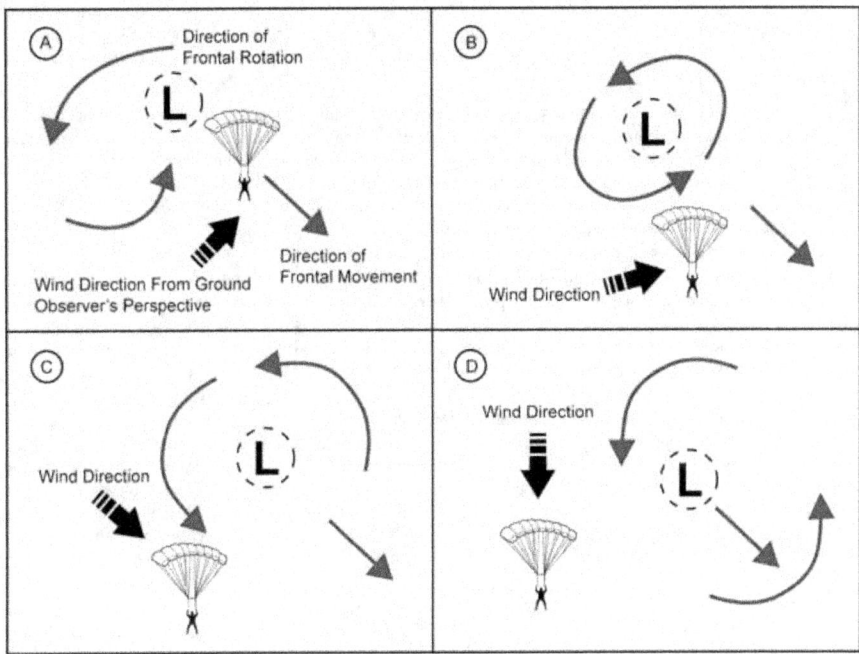

Figure 14-22. Wind shift as a front passes

14-102. A second adage is that "when the wind blows from the east or northeast and the barometer falls steadily, a storm is approaching from the south or southwest. Its center will pass near or to the south of the observer, after which the wind will shift to the northwest by way of north." Some other useful general rules concerning the barometer follow. The speed of a storm's approach will be indicated by the rate of fall of the barometric pressure. The storm's intensity is indicated by the amount of fall. A falling barometer and a rising thermometer often forecast rain. A barometer and thermometer rising together usually indicates good weather. A slowly rising barometer forecasts settled, stable weather. A steady, slow fall in barometric pressure indicates forthcoming unsettled or wet weather.

14-103. There are several linkages between barometric changes and wind velocity. It is generally true that a rapidly falling barometer forecasts the advent of strong winds. It is reflecting the development or approach of a low-pressure area or front. In a low-pressure center, the pressure gradient is usually steep, hence the higher wind velocities. Conversely, a rising barometer is associated with the prospect of lighter winds. This is true because it is indicating the arrival of a high-pressure center, where the pressure gradient is characteristically smaller, conditions more stable, and winds therefore lighter. Regardless, the barometer does not necessarily fall before or during a strong breeze. In fact, the wind can often blow hard without any appreciable accompanying change in pressure. This is the case where a steep pressure gradient exists (isobars close together), but the well-developed high- or low-pressure area is practically stationary. In this case, the strong winds will continue for some time. Any slackening or change in wind velocity will take place slowly.

14-104. Temperature can give some indications of approaching weather, but only as it applies to local conditions. For example, cold air carried down from a thunderstorm cloud as part of the first gust may be felt more than three miles in advance of the storm itself. This is a tangible warning confirming the approach of hazardous parachuting conditions. Stronger storm systems' downdrafts may be felt significantly further from the storm cell and are usually quite graphic in the amount and speed of temperature change.

14-105. Despite all of forecasting's scientific instruments, satellite imagery, and trained professionals, there is also a certain amount of guidance inherent in traditional weather proverbs. Where the origins have been lost over time and their originators did not know why they were true, their survival over the centuries attests to their general validity. They can be used, with caution, when adapted to local weather conditions. They illustrate some key weather principles. As such, the following is a distillation of the salient weather points derived from some weather proverbs and are known to have some value in making weather decisions and determining prevailing wind directions (the following apply generally to the midlatitudes):

- Fair weather and good MFF parachuting conditions are indicated when there are high-flying cirrus clouds that resemble wisps in a mare's tail. Only when the sky becomes heavy with cirrus clouds can one expect stormy weather. Thick cirrus clouds (mackerel skies) are the cirrocumulus clouds that resemble rippled sand on a beach. Less-than-ideal parachuting conditions (advancing low-pressure systems) can be expected within 8 to12 hours of the advent of heavy cirrus clouds.
- The proverb "Red sky in morning, jumper take warning; red sky at night, jumper's delight" draws on the following atmospheric facts. A red sunrise (obviously in the east) is caused by a dry and perhaps hazy or pollutant-laden air mass which scatters the shorter wavelengths of the sun's rays, allowing only the red wavelengths to pass through, causing a red tint to the sky and any clouds. With the dry air mass to the east, it is reasonable that the air mass to the west will be moist and perhaps stormy. With a generally west-to-east flow, it can be expected that the rainy/stormy weather would approach next. The opposite scenario occurs when the air mass to the west is dry with a red sunset; the approaching weather can be expected to be benign.
- A morning rainbow, when viewed in the west, is illuminated by the sun shining on it from the east. The moisture in the air creating the rainbow (and the storm behind it) will be moving to the east; therefore, one can expect rain. An evening rainbow, viewed in the east, would signify that the storm has passed. Another set of rainbow indicators are the wind conditions. If the rainbow is upwind, then one can expect its moisture to reach the observer. If the rainbow appears downwind, then the moisture has already passed and is moving away from the observer.
- Wind direction that shifts from east to west, or in the same arc as the sun moves, almost always results in clear skies. Wind that changes against the sun's movement, west to east, usually brings bad weather or less-than-ideal parachuting conditions.
- The shape and color of the moon can be indicators of coming weather changes. Whereas the moon has no appreciable control over the weather beyond a small tidal effect, it is another visible and reliable sign of weather change. It is not the moon's influence, but the atmospheric conditions that influence the moon's appearance. A moon halo is an excellent atmospheric sign

of rain. A moon's halo, especially after that day's pale sun, confirms the advent of rain. The moon's halo is created by the illuminated ice crystals of the high approaching cirriform clouds on the leading edge of a weather front. When the whole sky is covered with these cloud forms, a warm front is approaching, normally bringing rain within 8 to 12 hours.
- A halo visible around the sun indicates the approach of a storm within the next 36 hours. The storm will approach from the side of the sun that has the brightest (or the open) part of the halo. As cirrus and cirrostratus fronts push across the sky, they are backlit by the moon or sun. The halo first appears and subsequently becomes brightest in that part of the halo's arc from which the low-pressure system is approaching. Later, the halo becomes complete and the light is uniform throughout. As the storm advances, altostratus clouds arrive and obliterate the brightest part of the halo, which is the part closest to the direction from which the storm approaches. Hence, there are the two halo condition indicators.

14-106. Application of the principles described in this chapter has more value when referred against a weather map, however general or simplified. Understanding the atmospheric conditions which accompany depicted weather phenomena provides a fairly detailed and useful tool by which the MFF jumpmaster can plan parachute operations. Therefore, the following weather signs and rules of thumb (not in any specific order) are provided that can amplify some of the principles discussed and condense longer explanations of the physical events behind the weather. However, MFF jumpmasters must use great caution when considering these rules of thumb as most apply to midlatitude, interior-of-continents, Northern Hemisphere locations:

- Fair weather will generally continue when—
  - Summer fog clears off before noon.
  - Cloud bases along mountains increase in height.
  - Clouds tend to decrease in number.
  - The wind blows gently from west to northwest.
  - The barometer is steady or rising slowly.
  - There is a red sky at sunset when the sky overhead is clear.
  - The moon shines brightly and the wind is light.
  - There is a heavy dew or frost at night.
- Weather will generally change for the worse when—
  - Cirrus clouds change to cirrostratus, lower, and thicken.
  - Rapidly moving clouds increase in number and lower in altitude.
  - Clouds move in different directions at different altitudes.
  - Altocumulus or altostratus clouds darken the western horizon and the barometer begins to fall rapidly.
  - The wind shifts to the south or the east. The greatest change occurs when the wind shifts from the north to south via the east.
  - The wind blows strongly in the early morning.
  - The temperature rises abnormally in the winter.
  - The barometer falls steadily.
  - There is a downpour at night.
  - A cold, warm, or occluded front approaches.
- Weather will generally clear when—
  - Cloud bases increase in altitude.
  - The wind shifts to a westerly direction. The greatest change occurs when the wind shifts to the west from east via the south.
  - The barometer rises rapidly, except when it rises rapidly ahead of an approaching thunderstorm. Pressure rising rapidly ahead of an approaching thunderstorm is a sign of potentially severe weather.
  - A cold front has passed 3 to 6 hours previously.

Chapter 14

- Rain or snow will generally occur—
    - When a cold, warm, or occluded front approaches.
    - 18 to 36 hours after the first cirrostratus clouds are noted to thicken.
    - 12 to 24 hours after cirrostratus clouds are noted and there is a halo around the sun or moon.
    - Within 6 to 8 hours when the morning temperature is unusually high, the air is humid, and cumulus clouds are building.
    - Within an hour in the afternoon when there is static on a car radio and cumulus clouds are observed.
    - When the sky is dark and threatening in the west.
    - When a southerly wind increases in speed and clouds are moving from the west.
    - When the wind, especially a north wind, shifts in a counterclockwise direction from north to west to south.
    - The barometer falls steadily.
- The temperature will generally fall when—
    - The wind shifts into the north or northwest.
    - The wind continues to blow from the north or northwest.
    - The night is clear and the wind is light.
    - The barometer rises steadily in the winter.
    - A cold front has passed.
- The temperature will generally rise when—
    - The sky is overcast and there is a moderately southerly wind at night.
    - The sky is clear during the day and there is a light, southerly wind.
    - The wind shifts from the west or northwest to the south.
    - A warm front has passed.
- Fog will generally form when—
    - The sky is clear at sunset, the wind is light, and the air is humid.
    - Warm rain is falling through cold air ahead of a warm front.
    - There is a large temperature difference between relatively warm water and colder air above it.

MOON PHASES

14-107. In planning night MFF parachute operations, an MFF jumpmaster must be aware of the various moon phases and the light levels pertinent to each. The moon revolves eastward around the earth. It appears to move east to west because its rotational speed is slower than the earth's. A complete revolution around the earth takes 29 days, 12 hours, 44 minutes, and 28 seconds. Because the time required for a revolution does not vary, the same side of the moon is always exposed to the earth. Since the orbital plane of the moon is tilted 5 degrees, 9 minutes toward the earth's orbital plane, its orbit is closer to the Northern Hemisphere during winter months. As a result, moonlight is brighter in the winter than in the summer.

14-108. As the moon revolves on a vertical arc, the distance from a stationary point on the earth's surface to the moon varies as the moon moves on its easterly orbit. This distance is referred to as the altitude. The altitude and the moon phase are the two most important factors influencing night illumination. The rotation of the moon never changes and follows an exact time frame. Therefore, time tables for each moon phase (new moon, first quarter, full moon, and last quarter) can be accurately computed for any year. These are normally provided by the USAF AWS. Geographical location is not a consideration in computing moon phases. It is, however, pertinent to moonrise, moonset, percent illumination, and the use of NVGs. Various computer programs exist which will incorporate user-defined illumination parameters for NVG applications and elevations AGL.

Note: An example of MFF lunar data follows in Table 14-3, page 14-38.

Weather Factors for the Military Free-Fall Jumpmaster

14-109. The constant change in the moon's phase angle causes varying levels of light received from the moon. At low altitude (moon elevation above the horizon), the vertical component of moonlight incident to the earth's horizontal surface is small compared to the incident when the moon is at a higher altitude. Also, at low angles of incidence, light is further reduced in intensity by the longer distance it must travel through the earth's atmosphere. As the moon ascends in the sky, the distance light travels through the atmosphere decreases as the vertical component increases, thus providing greater illumination. The greatest light level is achieved when the moon is directly overhead. A description of the moon phases follows.

NEW MOON

14-110. A new moon phase occurs when the moon rotates to a position between the sun and the earth. This phase always begins during the day and is not visible at night. Visual observation of the new moon at night is not possible until the moon is approximately two days into this phase. The time required to complete the phase is approximately 8 days. During the first portion (5 days) of the phase, a low-light level will exist. As the phase progresses, illumination increases and light conditions will reach mid-light levels. Moonrise occurs during the daylight hours and moonset before midnight. Approximately 40- to 50-percent of the moon will be illuminated at the end of the new moon phase. Night parachute operations conducted during this phase will be under low-light levels, which will prevail most of the time. Best light conditions will exist shortly after darkness when the moon is at its highest observable altitude of the night.

FIRST QUARTER

14-111. The phase angle of the moon at the first quarter begins at the 90-degree position in relation to the earth. During this quarter, more than one half of the moon face, but not all the apparent disk, is illuminated. Approximately 4 days are required to complete the first quarter phase. During the first days of the first quarter and the last days of the new moon (approximately 5 days), the light will be in the mid-light range with increasing intensity. Moonrise occurs during daylight near the end of the day. Moonset changes from midnight to the early morning hours of darkness. The best time to conduct parachute operations will normally be about midnight when the moon is at its highest altitude. Light intensity is becoming brighter during this moon phase. When the moon is low on the horizon, avoiding backlighting a landing canopy stack should be included as a planning factor.

FULL MOON

14-112. A full moon occurs when the sun, moon, and earth are aligned. At this time, the moon is radiating its greatest percent of illumination. The full moon phase spans approximately 3 days before and 3 days after the full moon. High-light conditions begin during the last days of the first quarter and extend to the first days of the last quarter (approximately 12 days). During the early part of this phase, moonrise occurs just before nautical twilight and progressively moves into the hours of darkness. Moonset will occur during the early morning daylight hours. The optimum time to conduct parachute operations will be the first few hours after midnight.

LAST QUARTER

14-113. This phase of the moon is similar to the first quarter, only in reverse sequence. It begins when less than the entire disk is visible and ends when only half of the moon is visible. The last quarter will normally last approximately 5 days. Light will decrease from a high-light level (approximately 3 days) down to a mid-light level that extends into the transition phase (approximately 5 days). Moonrise will occur shortly before to just after midnight. The optimum time to conduct parachute operations is just prior to BMNT. Moonset occurs during daylight hours. Table 14-3, page 14-38, provides an example of MFF lunar data for Washington National Airport (latitude: 38 degrees 51 minutes north; longitude: 77 degrees 2 minutes west; drop zone altitude: 0). All times are Greenwich Mean Time plus 4 hours.

Chapter 14

Table 14-3. Military free-fall lunar data example

Day of Month	Moon Rise	Start NVG*	Stop NVG	Moon Set	% Illumination	Moon Phase
1	0906	1149	1339	2345	38	Transition phase continues
2	0949	1234	1339	****	29	
3	1037	****	****	0037	21	
4	1130	****	****	0126	13	
5	1228	****	****	0212	7	
6	1330	****	****	0254	3	End transition
7	1434	****	****	0332	1	New moon starts
8	1540	****	****	0408	0	
9	1647	****	****	0442	3	
10	1755	****	****	0516	7	
11	1904	****	****	0550	14	
12	2013	****	****	0626	23	
13	2121	****	****	0705	34	
14	2228	****	****	0749	45	First quarter starts
15	2331	****	****	0838	57	
16	****	0432	0508	0933	68	
17	0029	0448	0614	1033	78	
18	0121	0513	0742	1136	86	Full moon starts
19	0205	0536	0912	1241	93	
20	0245	0555	1033	1345	98	
21	0320	0618	1150	1449	100	
22	0352	0641	1302	1550	100	
23	0422	0704	1401	1650	97	
24	0452	0731	1402	1749	93	Full moon ends
25	0522	0800	1403	1847	87	Last quarter starts
26	0553	0832	1404	1944	80	
27	0626	0907	1405	2040	72	
28	0703	0945	1406	2134	63	
29	0744	1028	1407	2227	54	Last quarter ends
30	0829	1115	1408	2317	45	Transition phase starts
31	0920	1205	1408	****	35	

NOTES: NVG – Night vision goggle user-defined parameters.
* For NVG use, lunar altitude > 30 degrees above horizon, solar altitude < -6 degrees below horizon, and lunar illumination > 23%.
**** NVG illumination levels: outside of user-defined parameters.

TRANSITION PHASE

14-114. Although there is no term that describes the period following the last quarter, there is a period of approximately 7 days after the end of the last quarter before the new moon phase begins. It is similar to the new moon phase, but in reverse order. Illumination of the moon decreases from half of the disk to no visible form. Moonrise occurs a few hours before BMNT and moonset will always be during the daylight hours. Light will vary from a mid-light level during the first few days (approximately 3 days) to a low-light condition (approximately 4 days). To achieve any benefit from the moon illumination, parachute operations must be conducted 2 to 3 hours before BMNT. The longest period of time of low-light conditions exists from the transition phase to the first quarter. During this period, there are approximately 16 days when the moon is less than half illuminated, but is visible less than 50 percent of the hours of darkness.

REFERENCE TABLES

14-115. *Wind Estimation—Natural Indicators.* DZ reception personnel or MFF jumpmasters can use the following natural method (Table 14-4) to determine approximate wind velocity. For safety, however, personnel should add an additional 2.5 mph (3.7 kilometers per hour [kph] or 2 knots) to the estimate to avoid underestimation. Values, when overestimated and used in the computation of canopy drift distances, add safety margins to wind drift calculations.

Table 14-4. Approximate wind velocity by natural indicators

MPH	KPH	Knots	Natural Indicators
0	0	0	No motion; smoke rises vertically.
2	3.2	1.7	Leaves rustle; wind felt on face.
5	8	4	Smoke drifts.
10	16	9	Leaves and twigs in constant motion.
15	24	13	Small branches move; dust raised.
20	32	17	Small trees sway; crests raised on waves of inland water bodies.
28	45	24	Large branches in motion.
35	56	30	Whole trees in motion; ability to walk is affected.

14-116. *Estimating Wind Velocity—Handkerchief Method.* A second method of estimating wind velocity involves holding a handkerchief or similar mass cloth/streamer at the center and letting it hang free. By estimating the angle of the handkerchief to an imaginary vertical line by extending a line from the hand to the ground, it is possible, using the following chart (Table 14-5), to estimate wind velocity with reasonable accuracy. For safety, an additional 2.5 mph (3.7 kph or 2 knots) should be added to avoid underestimation.

Table 14-5. Handkerchief angle wind velocity

Angle (Degrees)	Speed (mph)	Speed (kph)	Speed (Knots)
15	5	8	4
30	10	16	9
50	15	24	13
60	20	32	17
70	25	40	22
80	30	48	26

14-117. *Conversion Tables.* Use the following tables (Tables 14-6 through 14-21, pages 14-40 through 14-44) to convert between measurements.

Table 14-11. Temperature

Convert From	Convert To
Fahrenheit	Celsius
	Subtract 32, multiply by 5, and divide by 9
Celsius	Fahrenheit
	Multiply by 9, divide by 5, and add 32

Table 14-12. Approximate conversion factors

To Change	To	Multiply By	To Change	To	Multiply By
Inches	Centimeters	2.540	Ounce-inches	Newton-meters	0.007062
Feet	Meters	0.305	Centimeters	Inches	3.94
Yards	Meters	0.914	Meters	Feet	3.280
Miles	Kilometers	1.609	Meters	Yards	1.094
Square inches	Square centimeters	6.451	Kilometers	Miles	0.621
Square feet	Square meters	0.093	Square centimeters	Square inches	0.155
Square yards	Square meters	0.836	Square meters	Square feet	10.76
Square miles	Square kilometers	2.590	Square meters	Square yards	1.196
Acres	Square hectometers	0.405	Square kilometers	Square miles	0.386
Cubic feet	Cubic meters	0.028	Square hectometers	Acres	2.471
Cubic yards	Cubic meters	0.765	Cubic meters	Cubic feet	35.315
Fluid ounces	Millimeters	29.573	Cubic meters	Cubic yards	1.308
Pints	Liters	0.473	Millimeters	Fluid ounces	0.034
Quarts	Liters	0.946	Liters	Pints	2.113
Gallons	Liters	3.785	Liters	Quarts	1.057
Ounces	Grams	28.349	Liters	Gallons	0.264
Pounds	Kilograms	0.454	Grams	Ounces	0.035
Short tons	Metric tons	0.907	Kilograms	Pounds	2.205
Pounds-feet	Newton-meters	1.356	Metric tons	Short tons	1.102
Pounds-inches	Newton-meters	0.11296	Nautical miles	Kilometers	1.852

Table 14-13. Area

To Change	To	Multiply By	To Change	To	Multiply By
Square millimeters	Square inches	0.00155	Square inches	Square millimeters	645.16
Square centimeters	Square inches	9.155	Square inches	Square centimeters	6.452
Square meters	Square inches	1,550	Square inches	Square meters	0.00065
Square meters	Square feet	10.764	Square feet	Square meters	0.093
Square meters	Square yards	1.196	Square yards	Square meters	0.836

Chapter 14

Table 14-6. Linear measure

Unit	Other Metric Equivalent	U.S. Equivalent
1 centimeter	10 millimeters	0.39 inch
1 decimeter	10 centimeters	3.94 inches
1 meter	10 decimeters	39.37 inches
1 decameter	10 meters	32.8 feet
1 hectometer	10 decameters	328.08 feet
1 kilometer	10 hectometers	3,280.8 feet

Table 14-7. Liquid measure

Unit	Other Metric Equivalent	U.S. Equivalent
1 centiliter	10 milliliters	0.34 fluid ounce
1 deciliter	10 centiliters	3.38 fluid ounces
1 liter	10 deciliters	33.81 fluid ounces
1 decaliter	10 liters	2.64 gallons
1 hectoliter	10 deciliters	26.42 gallons
1 kiloliter	10 hectoliters	264.18 gallons

Table 14-8. Weight

Unit	Other Metric Equivalent	U.S. Equivalent
1 centigram	10 milligrams	0.15 grain
1 decigram	10 centigrams	1.54 grains
1 gram	10 decigrams	0.035 ounce
1 decagram	10 grams	0.35 ounce
1 hectogram	10 decigrams	3.52 ounces
1 kilogram	10 hectograms	2.2 pounds
1 quintal	100 kilograms	220.46 pounds
1 metric ton	10 quintals	1.1 short tons

Table 14-9. Square measure

Unit	Other Metric Equivalent	U.S. Equivalent
1 square centimeter	100 square millimeters	0.155 square inch
1 square decimeter	100 square centimeters	15.5 square inches
1 square meter (centaur)	100 square decimeters	10.76 square feet
1 square decameter (are)	100 square meters	1,076.4 square feet
1 square hectometer (hectare)	100 square decameters	2.47 acres
1 square kilometer	100 square hectometers	0.386 square mile

Table 14-10. Cubic measure

Unit	Other Metric Equivalent	U.S. Equivalent
1 cubic centimeter	1,000 cubic millimeters	0.06 cubic inch
1 cubic decimeter	1,000 cubic centimeters	61.02 cubic inches
1 cubic meter	1,000 cubic decimeters	35.31 cubic feet

Table 14-13. Area (continued)

To Change	To	Multiply By	To Change	To	Multiply By
Square kilometers	Square miles	0.386	Square miles	Square kilometers	2.59

Table 14-14. Volume

To Change	To	Multiply By	To Change	To	Multiply By
Cubic centimeters	Cubic inches	0.061	Cubic inches	Cubic centimeters	16.39
Cubic meters	Cubic feet	35.31	Cubic feet	Cubic meters	0.028
Cubic meters	Cubic yards	1.308	Cubic yards	Cubic meters	0.765
Liters	Cubic inches	61.02	Cubic inches	Liters	0.016
Liters	Cubic feet	0.035	Cubic feet	Liters	28.32

Table 14-15. Capacity

To Change	To	Multiply By	To Change	To	Multiply By
Milliliters	Fluid drams	0.271	Fluid drams	Milliliters	3.697
Milliliters	Fluid ounces	0.034	Fluid ounces	Milliliters	29.57
Liters	Fluid ounces	33.81	Fluid ounces	Liters	0.030
Liters	Pints	2.113	Pints	Liters	0.473
Liters	Quarts	1.057	Quarts	Liters	0.946
Liters	Gallons	0.264	Gallons	Liters	3.785

Table 14-16. Statute miles to kilometers and nautical miles

Statute Miles	Kilometers	Nautical Miles	Statute Miles	Kilometers	Nautical Miles
1	1.61	0.86	60	96.60	52.14
2	3.22	1.74	70	112.70	60.83
3	4.83	2.61	80	128.80	69.52
4	6.44	3.48	90	144.90	78.21
5	8.05	4.35	100	161.00	86.92
6	9.66	5.21	200	322.00	173.80
7	11.27	6.08	300	483.00	260.70
8	12.88	6.95	400	644.00	347.60
9	14.49	7.82	500	805.00	434.50
10	16.10	8.69	600	966.00	521.40
20	32.20	17.38	700	1,127.00	608.30
30	48.30	26.07	800	1,288.00	695.20
40	64.40	34.76	900	1,449.00	782.10
50	80.50	43.45	1,000	1,610.00	869.00

Table 14-17. Nautical miles to kilometers and statute miles

Nautical Miles	Kilometers	Statute Miles	Nautical Miles	Kilometers	Statute Miles
1	1.85	1.15	60	111.00	69.00
2	3.70	2.30	70	129.50	80.50
3	5.55	3.45	80	148.00	92.00
4	7.40	4.60	90	166.50	103.50
5	9.25	5.75	100	185.00	115.00
6	11.10	6.90	200	370.00	230.00
7	12.95	8.05	300	555.00	345.00
8	14.80	9.20	400	740.00	460.00
9	16.65	10.35	500	925.00	575.00
10	18.50	11.50	600	1,110.00	690.00
20	37.00	23.00	700	1,295.00	805.00
30	55.50	34.50	800	1,480.00	920.00
40	74.00	46.00	900	1,665.00	1,033.00
50	92.50	57.50	1,000	1,850.00	1,150.00

Table 14-18. Kilometers to statute and nautical miles

Kilometers	Statute Miles	Nautical Miles	Kilometers	Statute Miles	Nautical Miles
1	0.62	0.54	60	37.28	32.38
2	1.24	1.08	70	43.50	37.77
3	1.86	1.62	80	49.71	43.17
4	2.49	2.16	90	55.93	48.56
5	3.11	2.70	100	62.14	53.96
6	3.73	3.24	200	124.28	107.92
7	4.35	3.78	300	186.42	161.88
8	4.97	4.32	400	248.56	215.84
9	5.59	4.86	500	310.70	269.80
10	6.21	5.40	600	372.84	323.76
20	12.43	10.79	700	434.98	377.72
30	18.64	16.19	800	497.12	431.68
40	24.86	21.58	900	559.26	485.64
50	31.07	26.98	1,000	621.40	539.60

Chapter 14

Table 14-19. Yards to meters

Yards	Meters	Yards	Meters	Yards	Meters
100	91	1,000	914	1,900	1,737
200	183	1,100	1,006	2,000	1,828
300	274	1,200	1,097	3,000	2,742
400	366	1,300	1,189	4,000	3,656
500	457	1,400	1,280	5,000	4,570
600	549	1,500	1,372	6,000	5,484
700	640	1,600	1,463	7,000	6,398
800	732	1,700	1,554	8,000	7,212
900	823	1,800	1,646	9,000	8,226

Table 14-20. Meters to yards

Meters	Yards	Meters	Yards	Meters	Yards
100	109	1,000	1,094	1,900	2,078
200	219	1,100	1,203	2,000	2,188
300	328	1,200	1,312	3,000	3,282
400	437	1,300	1,422	4,000	4,376
500	547	1,400	1,531	5,000	5,470
600	656	1,500	1,640	6,000	6,564
700	766	1,600	1,750	7,000	7,658
800	875	1,700	1,860	8,000	8,752
900	984	1,800	1,969	9,000	9,846

Table 14-21. Determination of altitude by barometric pressure (in inches of mercury)

Inches	.0	.1	.2	.3	.4	.5	.6	.7	.8	.9
					FEET					
13	22,638	22,430	22,223	22,018	21,815	21,612	21,412	21,213	21,015	20,819
14	20,624	20,431	20,238	20,048	19,858	19,670	19,483	19,298	19,114	18,931
15	18,749	18,568	18,389	18,211	18,033	17,858	17,683	17,509	17,337	17,156
16	16,995	16,825	16,657	16,490	16,324	16,158	15,994	15,831	15,669	15,507
17	15,347	15,187	15,029	14,871	14,715	14,559	14,404	14,250	14,097	13,945
18	13,793	13,643	13,493	13,344	13,196	13,049	12,902	12,756	12,611	12,467
19	12,324	12,181	12,039	11,898	11,758	11,618	11,479	11,340	11,203	11,066
20	10,930	10,794	10,659	10,525	10,391	10,259	10,126	9,995	9,864	9,733
21	9,604	9,474	9,346	9,218	9,091	8,964	8,838	8,712	8,587	8,463
22	8,339	8,216	8,093	7,971	7,849	7,728	7,608	7,488	7,368	7,249
23	7,131	7,013	6,896	6,779	6,662	6,546	6,431	6,316	6,212	6,088
24	5,947	5,861	5,749	5,637	5,525	5,414	5,303	5,193	5,083	4,947
25	4,865	4,756	4,648	4,540	4,433	4,326	4,220	4,114	4,009	3,903
26	3,799	3,694	3,590	3,487	3,384	3,281	3,179	3,077	2,975	2,874
27	2,773	2,672	2,572	2,473	2,373	2,274	2,176	2,077	1,979	1,882
28	1,794	1,688	1,591	1,495	1,399	1,303	1,208	1,113	1,019	925
29	831	737	644	551	458	366	247	182	91	0
30	-91	-181	-271	-361	-451	-540	-629	-718	-806	-894

Appendix A
Military Free-Fall Critical Task Lists

This appendix includes the critical task lists for the MFF Basic Course, the MFF Jumpmaster Course, the MFF Advanced Tactical Infiltration Course, and the MFF Instructor Course.

MFF Basic Course Module A: Ground Training (9 Tasks)	
Task Number	**Task Title**
331-MFF-1000	Perform Body Stabilization Techniques
331-MFF-1005	Pack the MC-4 Ram-Air Canopy
331-MFF-1010	React to Emergencies in the Aircraft During a Military Free-Fall Operation
331-MFF-1015	React to Emergencies While in Free Fall
331-MFF-1020	React to Emergencies During Canopy Descent Using a Ram-Air Personnel Parachute System (RAPPS)
331-MFF-1025	Don the MC-4 Ram-Air Personnel Parachute System
331-MFF-1030	Activate the Military CYPRES
331-MFF-1035	Rig Weapons and Combat Equipment for a Military Free-Fall Operation
331-MFF-1040	Respond to Aircraft Procedure Signals and Jump Commands During MFF Airborne Operation
Module B: Airborne Operations (7 Tasks)	
Task Number	**Task Title**
331-MFF-1045	Perform MFF Operation Wearing Combat Equipment
331-MFF-1050	Perform MFF Operation Wearing Combat Equipment and Portable Bailout Oxygen System
331-MFF-1055	Perform MFF Operation Wearing Combat Equipment and Portable Bailout Oxygen System as a Member of a Group
331-MFF-1060	Perform an MFF High-Altitude High-Opening (HAHO) Parachute Jump With Navigation Aids
331-MFF-1065	Perform MFF Operation Wearing Combat Equipment and Portable Bailout Oxygen System as a Member of a Group at Night
331-MFF-1070	Maneuver the MC-4 Ram-Air Canopy to a Designated Point as a Member of a Group
331-MFF-1075	Perform MFF Operations With Night Vision Goggles (NVGs)

Appendix A

MFF Jumpmaster Course Module A: Ground Training (12 Tasks)	
Task Number	**Task Title**
331-MFF-2000	Perform Jumpmaster Personnel Inspection (JMPI) on an MFF Parachutist
331-MFF-2005	Compute the High-Altitude Release Point (HARP) for a HALO Operation
331-MFF-2006	Compute the High-Altitude Release Point (HARP) for a HAHO Operation
331-MFF-2010	Determine the Altimeter Setting for a Military Free-Fall Operation
331-MFF-2015	Compute the Military CYPRES Electronic Automatic Activation Device (EAAD) Setting
331-MFF-2020	Rig Specialized Weapons and Combat Equipment for MFF Operations
331-MFF-2025	Perform the Duties of an MFF Jumpmaster
331-MFF-2030	Operate the Six-Man Prebreather Portable Oxygen System
331-MFF-2045	Conduct Military Free-Fall Parachute Refresher Training
331-MFF-2050	Conduct Military Free-Fall Jumpmaster Refresher Training
331-MFF-2055	Perform Sustained Airborne Training for MFF Operations
331-MFF-2065	Supervise the Donning of the MC-4 Ram-Air Personnel Parachute System (RAPPS)
Module B: Airborne Operations (4 Tasks)	
Task Number	**Task Title**
331-MFF-2027	Determine the Exit Order (Solo Jumpers With MTTB or Autonomous Bundle Systems)
331-MFF-2035	Issue Jump Commands Used in MFF Parachute Operation
331-MFF-2040	Direct an Aircraft to the Release Point
331-MFF-2060	React to Emergencies in the Aircraft During an MFF Operation as a Jumpmaster
MFF Advanced Tactical Infiltration Course Module A: Ground Training (6 Tasks)	
Task Number	**Task Title**
331-MFF-4005	Perform Body Stabilization Techniques With Night Vision Goggles (NVGs), Special Weapons, and Equipment
331-MFF-4007	Rig the Parachutist Helmet With Night Vision Goggles for MFF Operations
331-MFF-4008	Program an Electronic Navigation Board System for MFF Operations
331-MFF-4009	Rig an Electronic Navigation Board System for MFF Operations
331-MFF-4012	Construct an Autonomous Precision Airdrop Bundle System
331-MFF-4013	Program the Autonomous Precision Airdrop Bundle System

Military Free-Fall Critical Task Lists

MFF Advanced Tactical Infiltration Course
Module B: Airborne Operations (14 Tasks)

Task Number	Task Title
331-MFF-1005	Pack the MC-4 Ram-Air Canopy
331-MFF-2000	Perform Jumpmaster Personnel Inspection (JMPI) on an MFF Parachutist
331-MFF-2027	Determine the Exit Order (Solo Jumpers With the Autonomous Bundle Systems)
331-MFF-4001	Perform a Military Free-Fall Operation With Special Weapons and Equipment
331-MFF-4003	Perform MFF Operation With Night Vision Goggles, Special Weapons and Equipment, and Portable Oxygen System
331-MFF-4004	Perform an MFF High-Altitude High-Opening (HAHO) Parachute Jump Utilizing Electronic Navigation and Combat Equipment
331-MFF-4010	Operate the MFF Electronic Navigation System While Under Canopy
331-MFF-4011	Maneuver the MC-4 Ram-Air Canopy to a Designated Point on the Drop Zone Utilizing Electronic Navigation Equipment
331-MFF-4014	Employ the Autonomous Precision Airdrop Bundle System
331-MFF-4015	Perform MFF Operations With Night Vision Goggles, Special Weapons, Combat Equipment (CE), and a Portable Bailout Oxygen System as a Member of a Group
331-MFF-4016	Compute the High-Altitude Release Point (HARP) for a HAHO Operation Utilizing Electronic Navigation Equipment
331-MFF-4017	Perform a Blind Drop for an MFF Operation
331-MFF-4018	Conduct Military Free-Fall Parachute Special Weapon and Equipment Refresher Training
331-MFF-4019	Conduct Military Free-Fall Jumpmaster Special Weapon and Equipment Refresher Training

MFF Instructor Course (19 Tasks)

Task Number	Task Title
331-MFF-1005	Pack the MC-4 Ram-Air Canopy
331-MFF-1010	React to Emergencies in the Aircraft During a Military Free-Fall Operation
331-MFF-1020	React to Emergencies During Canopy Descent Using a Ram-Air Personnel Parachute System (RAPPS)
331-MFF-1035	Rig Weapons and Combat Equipment for a Military Free-Fall Operation
331-MFF-1065	Perform an MFF Parachute Jump with CE and Portable Bailout Oxygen System as a Member of a Group at Night
331-MFF-1070	Maneuver the MC-4 Ram-Air Canopy to a Designated Point as a Member of a Group
331-MFF-2035	Issue Jump Commands Used in MFF Parachute Operations
331-MFF-3005	Conduct Vertical Wind Tunnel Training for Military Free-Fall Parachutist Course (MFFPC) Students
331-MFF-3010	Pack the Instructor-Certified Ram-Air Parachute System (ICRAPS)
331-MFF-3015	Don the Instructor-Certified Ram-Air Parachute System (ICRAPS)

Appendix A

MFF Instructor Course (19 Tasks) (continued)	
Task Number	Task Title
331-MFF-3020	Inspect the Instructor-Certified Ram-Air Parachute System (ICRAPS)
331-MFF-3025	Perform an MFF Jump With the Instructor-Certified Ram-Air Parachute System (ICRAPS)
331-MFF-3030	Conduct the Jump Briefing to Military Free-Fall Parachutists
331-MFF-3035	Conduct Jump With Military Free-Fall Parachutist Course (MFFPC) Student
331-MFF-3040	Perform Corrective Action to an MFFPC Student During Free Fall
331-MFF-3045	Evaluate a Military Free-Fall Student During an MFF Parachute Jump
331-MFF-3050	Lead MFFPC Student Under Canopy to the Drop Zone
331-MFF-3055	Conduct Post Jump After Action Report (AAR) With MFFPC Student
331-MFF-3060	Conduct Drop Zone Safety Officer (DZSO) Duties During Advanced Military Free-Fall Parachutist Course (AMFFPC)

Appendix B
Military Free-Fall Parachutist Qualification and Refresher Training Requirements

MFF parachuting skills are highly perishable. MFF personnel maintain these skills through regularly scheduled training periods to develop the necessary degree of proficiency. Otherwise, mission capability and parachutist safety will be reduced.

MEDICAL AND PHYSIOLOGICAL TRAINING REQUIREMENTS

B-1. Each MFF parachutist must have met the following minimum requirements to participate in MFF operations:
- Must have a current HALO physical examination IAW Service regulations. Students attending the MFF course must have a HALO physical IAW the USAJFKSWCS standard.
- Must have a current physiological training card (AF Form 1274 [Physiological Training]) dated within the last 5 years. A physiological training card is maintained by undergoing physiological training every 5 years.
- Must be a graduate of a USAJFKSWCS-recognized MFF parachutist course.
- Must be a current MFF parachutist.

CURRENCY REQUIREMENTS

B-2. Currency does not equate to proficiency. Parachutists cannot consider MFF airborne operations to meet pay requirements as proficiency jumps unless the mission profile follows a tactical insertion profile. MFF jumpmaster currency standards are outlined in Chapter 13.

B-3. To meet the minimum MFF currency standards, the parachutist must have—
- A current HALO physical (per Service requirements).
- A current USAF physiological training card (AF Form 702 [Individual Physiological Training Record] or AF Form 1274).
- Conducted an MFF jump within the last 180 days.

MILITARY FREE-FALL PARACHUTE REQUALIFICATION AND REFRESHER TRAINING

B-4. Previously qualified MFF parachutists who, after meeting medical and USAF chamber currency requirements, do not meet the proficiency and currency requirements listed above, will undergo the following training to become requalified:
- Watch an emergency procedures video (if available).
- Review arm-and-hand signals, aircraft procedures, and jump commands.
- Review DZ markings.
- Attend wind tunnel training (if available).
- Attend a packing class.
- Attend Military CYPRES 2 class.
- Attend an oxygen class.
- Review exit procedures and body stabilization.
- Attend emergency procedures class and suspended harness drills.

Appendix B

- Attend combat equipment rigging (combat pack and weapon) class.
- Attend canopy control and grouping under canopy class.
- Perform one daylight jump without combat equipment, stressing a stable exit, maintaining heading, and pulling the rip cord at the prescribed pull altitude while maintaining heading (±500 feet).
- Perform one daylight jump with rifle and combat equipment, executing a stable exit, making a left and right turn, stopping on heading, and pulling the rip cord at the prescribed pull altitude (±500 feet) while maintaining heading, and landing within 50 meters of the group leader.
- Perform one night jump with rifle, combat pack (rucksack), and complete oxygen system, executing a manual parachute activation at the prescribed pull altitude (±500 feet), and landing within 50 meters of the group leader.

Note: At any time the jumpmaster may stop a parachutist from going to the next level if he determines that the parachutist has not satisfactorily performed the task. Jumpmaster will conduct after action review (AAR) with all MFF refresher jumpers.

MILITARY FREE-FALL HIGH-ALTITUDE HIGH-OPENING PARACHUTIST REQUALIFICATION AND REFRESHER TRAINING

B-5. Previously qualified MFF parachutists who do not meet proficiency and currency requirements will, after becoming current as an MFF parachutist, undergo the training outlined below. The intent of the following recommendations is to build upon the training progression listed in the previous paragraphs. In addition, the intent is to provide safe training and to increase parachutist skills, ability, and confidence, culminating in a HAHO night combat equipment oxygen jump. Recommendations include the parachutists make—

- One daylight MFF ram-air parachute jump with combat equipment from not higher than 13,000 feet AGL with opening not lower than 10,000 feet AGL. They must land within 100 meters of the group leader.
- One daylight MFF ram-air parachute jump with combat equipment and complete oxygen system with opening not higher than 18,000 feet AGL nor lower than 16,000 feet AGL. They must land within 100 meters of the group leader.
- A daylight combat equipment jump at altitudes above 20,000 feet MSL, depending upon the availability of USAF physiology technicians. For familiarization purposes, prebreathing can still take place below 20,000 feet MSL.

Note: Parachutists should perform same day jump sequence as above for night MFF refreshers.

Note: Altitude requirements are a recommendation; not all installations have the ability to get to these altitudes for training jumps during requalification and refresher training.

Appendix C
Recommended Military Free-Fall Training Programs

Commanders conduct oxygen-training jumps below 20,000 feet MSL to eliminate the need for prebreathing. They conduct proficiency jumps as a part of other training operations, such as field training exercises or Army training and evaluation programs, to take advantage of available training assets.

MINIMUM QUARTERLY TRAINING

C-1. Commanders follow a minimum program consisting of nine parachute jumps per quarter (Table C-1). They do not plan more than four proficiency jumps for any one day.

Table C-1. Minimum quarterly training guide

Jump Number	Type of Jump	Type of Jump Definition
1	HA	HALO/administrative-nontactical
2	HEO	HALO/combat equipment/oxygen
3	HEN	HALO/combat equipment/night
4	HEON (Minimum one per month)	HALO/combat equipment/oxygen/night
5	HEON/NVG	HALO/combat equipment/oxygen/night/NVG
6	SA	HAHO/stand-off/administrative-nontactical
7	SEON (Minimum one per quarter)	HAHO/stand-off/combat equipment/oxygen/night
8	SEN/NVG	HAHO/stand-off/combat equipment/night/NVG
9	SEON/NVG	HAHO/stand-off/combat equipment/oxygen/night/NVG

Note: Commanders must remember that for safety and parachutist confidence, parachutists require a jump refresher before executing night combat equipment jumps after prolonged periods of nonjumping. Commanders may not be able to include the nine jumps depicted in Table C-1 in the quarterly training plan; however, they follow the intent of the progression where possible. For example, after a 3-month layoff, an element should make a daylight jump before a night combat equipment jump.

Note: Units can fulfill oxygen-training requirements at altitudes below 20,000 feet MSL. A mission profile that is consistent with prebreathing requirements can be flown without requiring the coordination with or the presence of USAF physiological technicians. Training missions using full oxygen equipment can be flown at altitudes below 13,000 feet MSL. Flights at these altitudes would be consistent with any altitude's oxygen use requirements. These training mission profiles might occur in areas where airspace restrictions are in force or when there are not enough aircrew personnel.

RECOMMENDED TACTICAL MILITARY FREE-FALL HAHO-SPECIFIC TRAINING PROGRAM

C-2. HALO proficiency does not equate to HAHO proficiency. Parachutists should only consider MFF jumps with tactical application as proficiency jumps. Nontactical jumps are for currency and not

Appendix C

necessarily for proficiency. While HALO is an integral part of MFF operations, HAHO/stand-off operations provide the tactical commander a unique method for infiltration. The tactical commander may infiltrate these elements by parachute without requiring the aircraft to fly over the intended target area. These elements can be released at an offset release point and navigate long distances under canopy.

C-3. The desired end state for a combat-ready MFF team is to have the ability to land at the designated landing point as a detachment, with all required organic weapons systems, individual load-carrying equipment and issued personal protective equipment (PPE/body armor), mission-appropriate rucksack or assault pack, and tactical communications, by both HALO and HAHO means. The detachments will use available navigational aids and supplemental oxygen systems and organic parachute assemblies. The units should use the maximum altitudes available for training with a culminating jump conducted at (or close to) 24,999 MSL using on-board oxygen consoles, during hours of darkness, onto an austere and poorly lit landing area while using NVGs.

C-4. The MFF-coded detachments may fulfill these requirements at lower altitudes, but units will use the mission profile with the on-board consoles to ensure jumpers are maintaining proficiency of the full spectrum of oxygen equipment. Examples of justification to utilize lower altitudes for proficiency jumps include the following:

- Lack of USAF physiological technicians.
- Training in areas where airspace restrictions are in force.
- Aircrew limitations or restrictions.
- Limiting factors of winds and weather.

Note: Simply using bailout bottles and masking prior to exit would not be considered meeting the oxygen system requirement for a "culminating proficiency jump." Even if jumpers exit at 7,500 feet AGL, jumpers will conduct a culminating proficiency jump using on-board consoles and procedures.

C-5. All MFF detachments will conduct, at a minimum, one (1) HAEON per month and one (1) SAEON per quarter based on operations tempo. This is done to maintain a combat-ready status (also known as Level I).

C-6. The goal of the first three days of training should give MFF detachments the opportunity to identify differences in canopy descent rates, weaknesses in canopy control skills, and to give jumpers the chance to make familiarization jumps with new equipment during daylight before progressing to night jumps.

C-7. For example, a current jumper who has not jumped with a navigational aid will use these 3 days to become familiar with navigational aids. A jumper who has never used NVGs will use these first 3 days to become familiar with flying the canopy while wearing NVGs in daylight hours before attempting to use them at night.

C-8. Identifying different descent rates between jumpers will allow the detachment leadership to better plan for cross-loading of equipment and chalk order during exit so the detachment can minimize the amount of time it takes to group under canopy, build the stack, and navigate to the desired landing area. Table C-2, page C-3 provides a template for a two-week MFF training plan that focuses on stand-off parachute infiltration.

Note: The suggested 10-day combat-ready training program (Table C-2) is not meant to be a basic train-up plan done as a requalification event. It is meant as an advanced proficiency train up for fully trained elements. Anyone executing this recommended schedule should already be a *current SOF common standard* free-fall jumper.

Recommended Military Free-Fall Training Programs

Table C-2. Suggested 10-day combat-ready training program

Jump Number	Type Jump	Maximum Recommended Exit Altitude (AGL)	Minimum Recommended Pull Altitude (AGL)	Notes
1	SA	8,000	6,000	Jumps Focus on Fundamental Canopy Skills
2	SAE	8,000	6,000	Proper Equipment Lowering Procedures and New Equipment
3	SAEO	8,000	6,000	SFODA Can Adjust Cross Load of Equipment to Even the Canopy Descent Rates
4	SAEO	8,000	6,000	
5	SAN	12,500	6,000	A Seasoned SFODA May Exit at 12,500 Feet and Pull Soon After Exit Where a Lower-Experienced Level May Dictate Pulling as Low as 6,000 Feet
6	SAEN	12,500	6,000	
7	SAON	12,500	6,000	
8	SAEON	12,500	6,000	
9	SAN	12,500	10,000	SFODA Focuses on Use of Navigational Aids, Night Vision Aids, and Maintaining a Tight Canopy Stack
10	SAEN	12,500	10,000	
11	SAON	12,500	10,000	
12	SAO	17,500	15,000	SFODA Focuses on Use of Navigational Aids and Maintaining a Tight Canopy Stack
13	SAEO	17,500	15,000	
14	SAEO	17,500	15,000	
15	SAEO	24,999 (MSL)	23,000 (MSL)	Pre-Dawn Exit
16				Commander's Time
17				
18	SAEON	24,999 (MSL)	23,000 (MSL)	Culmination Exercise (CULEX)/ Full Mission Profile (FMP)
19				Commander's Time
20				

NOTE: The following codes will be used on USASOC Form 1099-E (Military Free Fall Jump Record) to indicate the type jump performed. One or more code symbols may be used. (For example: T-S-O-N-J indicates a Tactical, Stand-off jump using Oxygen performed at Night as Jumpmaster.)

A	Administrative/Nontactical	NVG	Night Vision Goggle
C	Combat	O	Oxygen
E	Combat Equipment	S	HAHO/Stand-off
H	HALO	T	Tactical
J	Jumpmaster	TB	Tandem Bundle
N	Night	TP	Tandem Personnel

RECOMMENDED MILITARY FREE-FALL JUMP PROGRESSION FOR NIGHT VISION GOGGLE/MILITARY FREE-FALL TRANSITION PROCESS

C-9. Vertical wind tunnel (VWT) training is not directed but highly recommended. Advanced Tactical Infiltration Course (ATIC) graduates and jumpmasters should progress their jumpers through the VWT in the same manner as for live jumps (dummy parachute with NVGs, dummy parachute and equipment with NVGs, and emergency procedures with NVGs). Jumpers should concentrate on the modified pull sequence and emergency procedures during all phases of VWT training with NVGs in both the UP and DOWN

Appendix C

positions. If VWT training is not feasible, then, at a minimum, table drills will be conducted in an unlit room with NVGs, with a parachute and lit altimeter simulating the modified pull sequence and emergency procedures.

LIVE JUMP NIGHT VISION GOGGLE TRAINING

C-10. ARSOF will conduct the following minimum jump progression to complete the NVG/MFF transition process:
- Perform one daylight administrative jump as a member of a group with NVGs mounted on the jumper's helmet. Jumper will execute a stable exit, maintain heading, and execute a minimum of three practice rip cord touches while utilizing the new standard pull sequence. Performance will focus on the new standard pull sequence, pulling the rip cord at the prescribed pull altitude (±500 feet) while maintaining heading. Jumper will execute proper canopy control procedures and land within 50 meters of the designated group leader.
- Perform one daylight jump as a member of a group with rifle, combat equipment, and NVGs mounted on the jumper's helmet. Jumper will execute a stable exit, maintain heading, and execute a minimum of three practice rip cord touches while utilizing the modified pull sequence. Performance will focus on the modified pull sequence, pulling the rip cord at the prescribed pull altitude (±500 feet) while maintaining heading. Jumper will execute proper canopy control procedures and land within 50 meters of the designated group leader.
- Perform one daylight jump as a member of a group with complete oxygen system and NVGs mounted on the jumper's helmet. Jumper will execute a stable exit, maintain heading, and execute a minimum of three practice rip cord touches while utilizing the modified pull sequence. Performance will focus on the modified pull sequence, pulling the rip cord at the prescribed pull altitude (±500 feet) while maintaining heading. Jumper will execute proper canopy control procedures and land within 50 meters of the designated group leader.
- Perform one daylight jump as a member of a group with rifle, combat pack, complete oxygen system, and NVGs mounted on the jumper's helmet. Jumper will execute a stable exit, maintain heading, and execute a minimum of three practice rip cord touches while utilizing the modified pull sequence. Performance will focus on the modified pull sequence, pulling the rip cord at the prescribed pull altitude (±500 feet) while maintaining heading. Jumper will execute proper canopy control procedures and land within 50 meters of the designated group leader.
- Perform one night administrative jump as a member of a group with NVGs mounted on the jumper's helmet. Jumper will execute a stable exit, maintain heading, and execute a minimum of three practice rip cord touches while utilizing the modified pull sequence. Performance will focus on the modified pull sequence, pulling the rip cord at the prescribed pull altitude (±500 feet) while maintaining heading. Jumper will execute proper canopy control procedures and land within 50 meters of the designated group leader.
- Perform one night jump as a member of a group with rifle, combat equipment, and NVGs mounted on the jumper's helmet. Jumper will execute a stable exit, maintain heading, and execute a minimum of three practice rip cord touches while utilizing the modified pull sequence. Performance will focus on the modified pull sequence, pulling the rip cord at the prescribed pull altitude (±500 feet) while maintaining heading. Jumper will execute proper canopy control procedures and land within 50 meters of the designated group leader.
- Perform one night jump as a member of a group with complete oxygen system and NVGs mounted on the jumper's helmet. Jumper will execute a stable exit, maintain heading, and execute a minimum of three practice rip cord touches while utilizing the modified pull sequence. Performance will focus on the modified pull sequence, pulling the rip cord at the prescribed pull altitude (±500 feet) while maintaining heading. Jumper will execute proper canopy control procedures and land within 50 meters of the designated group leader.
- Perform one night jump as a member of a group with rifle, combat pack, complete oxygen system, and NVGs mounted on the jumper's helmet. Jumper will execute a stable exit, maintain heading, and execute a minimum of three practice rip cord touches while utilizing the modified

pull sequence. Performance will focus on the modified pull sequence, pulling the rip cord at the prescribed pull altitude (±500 feet) while maintaining heading. Jumper will execute proper canopy control procedures and land within 50 meters of the designated group leader.

SUSTAINMENT TRAINING

C-11. MFF-qualified parachutists using NVGs during MFF operations will follow the standard currency guidelines set forth in Chapter 8-4 of USASOC Regulation 350-2, *Airborne Operations*.

AUTHORIZED TRAINERS

C-12. The initial train-the-trainer qualifications will be done by Soldiers who have completed the Tactical Infiltration Course at Yuma Proving Ground (completion dates inclusive from January 2010) or through members of the MFF School Tactical Application Detachment. Once initial training throughout the force has commenced, any current MFF jumpmaster who has completed the approved NVG training support package may administer MFF/NVG training.

This page intentionally left blank.

Appendix D
Suggested Military Free-Fall Sustained Airborne Training

Sustained airborne training must be conducted within the 24-hour period before station time of any MFF parachute operation. At a minimum, MFF sustained airborne training must consist of the jumpmaster troop briefing, a mock aircraft rehearsal, action procedures in free fall and canopy flight, emergency procedures, canopy entanglement procedures, and landing procedures. Figures D-1 through D-4, pages D-1 through D-4, provide outlines of the material to be covered during sustained training.

- In-Flight Rigging Procedures.
- Actions at the Time Warnings.
- Oxygen Procedures.
- Aircraft Procedure Signals and Jump Commands.
- Bundle Ejection Control.
- Aircraft Exit Procedure.
- Automatic Ripcord Release Arming and Disarming.
- In-Flight Emergency Procedures.

NOTE: The jumpmaster uses field-expedient mock aircraft to conduct the rehearsal. The rehearsal is performance-oriented and conducted exactly as the actual mission will occur.

Figure D-1. Mock aircraft rehearsal

GROUP PROCEDURES
- In Free Fall.
- Under Canopy.

COMMUNICATIONS
(Air-to-Air, Air-to-Ground, Ground-to-Air)
- Call Signs.
- Frequencies.
- Time Windows.
- Transponder Codes.
- Drop Zone Ground Marking Patterns.
- Visual Authentication Codes.
- Abort Signals.

Figure D-2. Actions in free fall and canopy flight

MANIFEST CALL
- Identification Cards.
- Identification Tags.
- Uniform Rigged Equipment and Bundle Inspection.

> **WARNING**
> Parachutists must not conduct MFF operations for a period of 24 hours or longer after scuba diving.

INTRODUCE ASSISTANTS AND OXYGEN SAFETY PERSONNEL
- Spare Parachute Systems.
- Spare Altimeters.

BRIEF OVERVIEW OF THE TACTICAL PLAN

CRITICAL TIMES
- Weather Decision.
- Load Time.
- Station Time.
- Prebreathing Time.
- Takeoff Time.
- Time Over Target.

MARSHALING PLAN
- Location of Sustained Airborne Training.
- Movement to the Departure Airfield.
- Aircraft Parking Location.
- Parachute Issue Location and Time.
- Jumpmaster Personnel Inspection Location and Time.
- Joint Mission Briefing Location and Time.
- Rigging of Oxygen Consoles and Equipment.

OPERATIONAL INFORMATION
- Type Aircraft.
- Type Airdrop (HALO or HAHO).
- Type Release (Jumpmaster-Directed Release).
- Type Exit (Door or Ramp).
- Number of Parachutists and Exit Sequence.
- Automatic Ripcord Release Millibar Setting.
- Equipment Bundles.
- In-Flight Rigging.
- Altimeter Setting.
- Oxygen Procedures.

Aircraft Flight Information
- Flight Route and Checkpoints.
- Duration of Flight.
- Drop Heading, Exit Altitude, and Airspeed.
- High Altitude Release Point.

Figure D-3. Sample jumpmaster troop briefing

Canopy Flight Information
- Wind Speed at Opening Altitude.
- Forecasted Altitude Winds (Direction and Speed).
- Cloud Layers and Temperatures Aloft.
- Opening Altitude and HAHO Delay.
- Headings Under Canopy and Checkpoints.
- Other Navigational Aids.
- Radio Frequencies.

Drop Zone Information
- Name and Location (Primary and Alternates).
- Drop Zone Dimensions.
- Drop Zone Markings (If Used).
- Obstacles on or Near the Drop Zone.
- Forecasted Ground Winds (Direction and Speed).
- Cloud Ceiling or Other Obscurants.

Assembly Plan
- Assembly Area Location.
- Assembly Aids (If Used).
- Disposition of Air Items.
- Medical Evacuation Procedures.

Special Instructions
- Life Preservers.
- Off-the-Drop-Zone Procedures.

Figure D-3. Sample jumpmaster troop briefing (continued)

Appendix D

PROBLEMS AND MALFUNCTIONS IN FREE FALL AND UNDER CANOPY
- Collision on Exit.
- Instability in Free Fall.
- Rucksack Shifts.
- Accidental Opening.
- Altimeter Failure or Loss.
- Lost Goggles.
- Clouds.
- Floating Ripcord.
- Hard Rip Cord Pull.
- Pack Closure.
- Pilot Chute Hesitation.
- Horseshoe.
- Bag Lock.
- Hung Slider.
- Snivels.
- Closed-End Cells.
- Premature Brake Release.
- Broken Control Lines.
- Broken Suspension Lines.
- Line Twists.
- Holes or Tears.
- Tension Knots.
- Pilot Chute Over the Nose of Canopy.
- Dual Main and Reserve Deployments.
- Premature Activation of Parachute in Free Fall.

CUTAWAY PROCEDURES
- Total Malfunction.
- Partial Malfunction.

POSTOPENING PROCEDURES
- Controllability Check.
- Penetration and Rate of Descent.

Figure D-4. Emergency procedures

D-1. Sustained airborne training must be conducted within the 24-hour period before station time of any MFF parachute operation. The jumpmaster must read and ensure all MFF parachutists understand all aspects of the sustained airborne training and MFF operation. The jumpmaster should also observe all parachutists to ensure they understand and are fully involved in practicing all emergency procedures they could encounter.

> *Note*: Parachutists must ensure that they have their identification (ID) cards and ID tags with them during airborne operations. They must not wear the ID tags around their necks. At no time will parachutists have tobacco, chewing gum, or anything else in the mouth during airborne operations.

AIRCRAFT PROCEDURES AND JUMP COMMANDS

D-2. The jumpmaster will explain the aircraft procedure signals and jump commands that follow.

Suggested Military Free-Fall Sustained Airborne Training

NONOXYGEN MILITARY FREE-FALL OPERATIONS

D-3. Actions for aircraft procedure signals and jump commands are as follows:
- DON HELMETS: Put on helmets, fasten chinstraps, and ensure seatbelts are securely fastened.
- UNFASTEN SEATBELTS: At approximately 1,000 feet AGL.
- 20 MINUTES: Stay alert and don equipment as directed.

Note: The 20 MINUTES command may be given prior to takeoff.

- 10 MINUTES: Keep eyes trained on the jumpmaster.
- WINDS: Surface wind speed on the DZ is expressed in knots. All jumpers conduct a pin check of their fellow jumpers and ensure the CYPRES has the correct millibar setting.
- STAND UP: 2 minutes from the release point. Jumpers stand up, face jumpmaster, check the CYPRES setting and pins on jumper to their front, and give that individual a "thumbs up." Jumpers then conduct a check of their own handles and equipment.
- MOVE TO THE REAR: 1 minute from the release point. Jumpers move to the ramp hinge (or within 1 meter of the door). All jumpers ensure goggles are covering their eyes at this time.
- STAND BY: 15 seconds from the release point. The jumpmaster gives a "thumbs-up" signal. Jumpers return the signal and move to within approximately 1 foot of the exit point.
- GO: At the release point. The jumpmaster points out the door or ramp. Jumpers exit as prescribed by the jumpmaster.
- ABORT: If the ABORT command is given, observe the jumpmaster for additional commands. Keep eyes on the jumpmaster for instructions. If the ABORT command is given before the command to STAND UP, jumpers will don their helmets, fasten their seatbelts, and prepare for landing.

OXYGEN MILITARY FREE-FALL OPERATIONS

D-4. After the JMPI, jumpers may not remove their helmets. Actions for aircraft procedure signals and jump commands are as follows:
- DON HELMETS: Put on helmets, fasten chinstraps, and ensure seatbelts are securely fastened.
- UNFASTEN SEATBELTS: At approximately 1,000 feet AGL.
- 20 MINUTES: Stay alert and don equipment as directed.

Note: The 20 MINUTES command may be given prior to takeoff.

- 10 MINUTES: Keep eyes trained on the jumpmaster.

Note: For oxygen jumps requiring oxygen console or aircraft-supplied oxygen, the oxygen safety and/or the USAF Physiological Technician must make periodic checks of all oxygen equipment used during the flight.

- WINDS: Surface wind speed on the DZ is expressed in knots. All jumpers conduct a pin check of their fellow jumpers and ensure the CYPRES has the correct millibar setting.
- MASK: If not using a console, jumpers turn on their bailout system first. Jumpers secure their masks to their helmets, fit their masks to their faces, observe the jumpmaster, and give a "thumbs-up" signal if they are receiving good oxygen flow. Hold the "thumbs-up" signal until the signal is returned by the jumpmaster.

Appendix D

> **WARNING**
>
> Jumpers must not sleep while breathing on console or after the mask signal is given. Jumpers must stay alert. If at any time the jumper experiences difficulty with the oxygen system, the individual must extend an arm straight out to the front, palm facing down, and wait for assistance. Jumpers should not break the mask seal if receiving oxygen.

- OXYGEN CHECK: Keep eyes on the jumpmaster, return the "thumbs-up" signal, and hold it until the jumpmaster returns the signal.
- STAND UP: 2 minutes from the release point. Jumpers stand, face the jumpmaster, and check the CYPRES setting, pins, and oxygen bottles of the jumper to their front, and gives the individual a "thumbs up." Once this check is complete, jumpers check their own handles and equipment. If using console oxygen, jumpers place their right hand on the bailout system ON/OFF lever, and place their left hand on the console hose at the AIROX VIII connection.
- MOVE TO THE REAR: 1 minute from the release point. If using console oxygen, jumpers turn on the oxygen bailout bottle system, disconnect from the console, place the hose next to the console, and move to the ramp hinge or within 1 meter of the door. All jumpers ensure goggles are covering their eyes at this time.
- STAND BY: 15 seconds from the release point. The jumpmaster will give a "thumbs-up" signal. Jumpers return the signal and move to within approximately 1 foot of the exit point.
- GO: At the release point. The jumpmaster points out the door or ramp. Jumpers exit as prescribed by the jumpmaster. Jumper—
  - Upon exiting the aircraft, conducts a practice pull. During the pull sequence, if the oxygen hose interferes with the ability to locate and pull the main rip cord, the jumper makes a second attempt by utilizing the head-tilt method, looking at the rip cord stow pocket.
  - Locates the rip cord housing with the right hand.
  - Traces the rip cord cable housing down and locates the rip cord cable, which may be protruding from the housing.
  - Pulls the cable. If this attempt is unsuccessful, the jumper performs cutaway procedures.

Note: The jumper must keep the mask on while under canopy. Jumpers must not disconnect one side, allowing the mask to dangle.

  - Upon landing, turns off his oxygen.
  - Places elastic cover on mask and puts the mask in the plastic bag.
  - Covers the AIROX VIII with the plastic bag.
  - Conducts normal recovery procedures.
- ABORT: If the ABORT command is given, observe the jumpmaster for additional commands. Keep eyes on the jumpmaster for instructions. If the ABORT command is given before the command to STAND UP, jumpers will don their helmets, fasten their seatbelts, and prepare for landing.

AIRCRAFT EMERGENCIES

D-5. The jumpmaster will explain the aircraft emergencies that follow.

Suggested Military Free-Fall Sustained Airborne Training

AIRCRAFT EMERGENCIES ON THE GROUND AND PRIOR TO TAKEOFF

D-6. Jumpers take all commands from the primary jumpmaster.

D-7. Jumpers exit the aircraft and assemble 100 meters in a safe direction, as directed by the primary jumpmaster. Once assembled, jumpers report to the primary jumpmaster.

AIRCRAFT EMERGENCIES IN FLIGHT: EMERGENCY LANDINGS

D-8. For incidents occurring below 1,000 feet AGL, all jumpers will land with the aircraft.

D-9. Six short rings or a verbal warning from the aircrew will alert the jumpers to prepare for a crash landing. One long continuous bell or a verbal warning from the aircrew or the jumpmaster will indicate that a crash is imminent.

D-10. Jumpers assume the emergency crash landing position by interlocking fingers behind their head and placing their head between their legs. Jumpers must not unfasten seatbelts until the aircraft comes to a stop. Once the aircraft comes to a complete stop, jumpers offload the aircraft and assemble 100 meters in a safe direction, as directed by the primary jumpmaster. Once assembled, jumpers report to the primary jumpmaster.

AIRCRAFT EMERGENCIES IN FLIGHT: EMERGENCY BAILOUT

D-11. Three short bell rings or a verbal warning from the aircrew will alert the jumpers to prepare to bail out. The signal to bail out is one long, continuous bell or verbal warning from the aircrew. The following actions are to be taken for an emergency exit during flight.

D-12. The jumpmaster gives the emergency bailout signal by extending his arm straight up with the index finger extended and moving his arm in a circular motion. Jump commands may be given at this time, if time permits. If there is no time for the full jump commands, abbreviated signals will be given immediately after the bailout signal. Commands and signals to be given and action to be taken are as follows:

D-13. Aircraft emergencies in-flight include the following:
- Between 1,000 feet AGL and 3,000 feet AGL, a clenched fist placed by the reserve rip cord and thrust out to the side means "CLEAR AND PULL THE RESERVE RIP CORD." The jumper performs the following actions:
  - Perform a dive exit at the jumpmaster's command.
  - Deploy the reserve parachute immediately.
  - Attempt to land with other jumpers.
  - Report to the jumpmaster.
- Above 3,000 feet AGL, a clenched fist placed by the main rip cord and thrust out to the side means "CLEAR AND PULL THE MAIN RIP CORD." Jumpers perform the following actions:
  - Perform dive exit at the jumpmaster's command.
  - Immediately clear airspace and deploy the main parachute.
  - Attempt to land with other jumpers.
  - Report to the primary jumpmaster.

ACTIONS IN FREE FALL

D-14. The jumpmaster will explain the procedures for actions in free fall that follow.

COLLISION ON EXIT

D-15. Jumper maintains his arch, gently pushes off of the other jumper, regains his stability, checks altitude, checks the rip cords, and continues the jump as planned.

Appendix D

SPINNING

D-16. Jumpers counter, relax, arch, check hands and feet, and maintain altitude awareness. If unable to maintain altitude awareness and hopelessly spinning, jumpers wave off and pull.

TUMBLING

D-17. Jumpers relax, arch, keep head up, check hands and feet, and maintain altitude awareness. If unable to maintain altitude awareness and still tumbling, jumpers wave off and pull.

ENTERING A CLOUD OR LOSS OF VISIBILITY

D-18. Jumpers stop all movement and return to a stable relaxed arch. Jumpers maintain altitude awareness and pull at the prescribed altitude, even if still in the cloud.

RUCKSACK SHIFTS

D-19. Jumpers counter any turns by turning in the opposite direction. If the rucksack strap moves below the knee, jumper makes one attempt to replace it while maintaining stability. If unsuccessful, jumper relaxes and attempts to fly. If unable to maintain altitude awareness and gain control, jumper should wave off and pull.

COLLISION AVOIDANCE

D-20. The lower jumper has the right-of-way. Jumpers must never get over the top of another jumper. The jumper should use forward movement or side slide to get off another jumper's back.

LOST OR LOOSE GOGGLES

D-21. Jumper makes one attempt to replace them. If unsuccessful, jumper continues the free fall as planned and squints eyes in order to maintain vision. Jumper maintains altitude awareness and pulls at the prescribed altitude. If unable to maintain altitude awareness, jumper waves off and pulls.

ALTIMETER FAILURE OR LOST ALTIMETER

D-22. If the altimeter fails or is lost prior to exit, jumper informs the primary jumpmaster/assistant jumpmaster and the defective altimeter will be exchanged with an onboard spare. If the onboard spare is in use or both altimeters fail prior to exit, the jumper will move forward, be seated, and air land.

D-23. If the altimeter fails or is lost during free fall, regardless of whether conducting day or night operations, jumper immediately clears airspace, waves off, and pulls.

PULL

D-24. At the prescribed pull altitude, the jumper maintains his arch, looks at the main rip cord, and, with the right hand, traces the main rip cord cable housing to the main rip cord handle while moving the left hand into the counter position. The jumper pulls the rip cord to full-arm extension and raises his right shoulder to ensure the pilot chute has launched and waits 2 seconds. Pull priorities are as follows:
- Priority one—pull.
- Priority two—never sacrifice altitude for stability.

ACTIONS UNDER CANOPY

D-25. The jumpmaster will explain the postopening procedures for actions under canopy that follow.

Suggested Military Free-Fall Sustained Airborne Training

CHECK CANOPY

D-26. Jumper checks canopy and grasps the rear risers. Jumper checks canopy for the three Ss:
- Square.
- Stable.
- Steerable.

LOCATE DROP ZONE

D-27. Jumper locates the DZ using the rear risers.

CLEAR AIRSPACE

D-28. Jumper clears his airspace using the rear risers to steer to avoid other canopies and turning right to avoid collisions unless left is closer. Jumper activates strobe light at night.

GAIN CONTROL OF TOGGLES

D-29. Jumper unstows the brake lines and gains control of the canopy by pulling down on the steering toggles to release the brakes.

CONDUCT CONTROLLABILITY CHECK

D-30. If canopy controllability is ever in question, jumper performs a canopy controllability check as follows:
- Release the brakes.
- Look left, clear airspace, and turn left 90 degrees.
- Look right, clear airspace, and turn right 90 degrees.
- If the canopy requires more than 50-percent opposite toggle to counter a turn, the canopy is uncontrollable.
- Determine the stall point. If the canopy stalls prior to the 50-percent brake setting, it is uncontrollable. The jumper should execute cutaway procedures and deploy the reserve parachute.

WARNING

The jumper must avoid turbulent air directly behind and above another ram-air parachute by flying offset to the parachute to his front. The jumper must maintain 25 meters of separation to the rear and above of other canopies. The jumper must not make sharp turns (greater than 45 degrees) on the final approach unless it is to avoid other jumpers or obstacles.

FLY THROUGH CLOUDS

D-31. Jumper stops all turns, stays alert, and flies straight through the cloud at half brakes.

CONDUCT LANDING PATTERN

D-32. Jumpers fly the downwind leg along the wind line, passing the target area approximately 1,000 feet AGL and offset about 45 degrees to the left or right of the target. The jumper continues the downwind leg until approximately 750 feet AGL, where he gently turns into the base leg.

D-33. Jumpers fly the base or crosswind leg across the wind line. This leg may be flown using the brakes, depending on wind conditions. Jumpers may shorten or extend the base leg as necessary. When the jumper reaches his turning altitude of approximately 500 feet AGL, he makes a braked turn toward the target.

D-34. Jumpers fly the final approach into the wind. At 200 feet, the jumper eases the toggles to full-flight position to allow airspeed to build. At 10 to 15 feet AGL, the jumper conducts a flared landing by pulling both toggles down to the full-brakes position. The jumper keeps his arms and elbows to his side and keeps his feet and knees together in the event he needs to conduct a PLF.

> **WARNING**
>
> On a misjudged flare attempt or if the parachute enters a stall, the jumper should hold the toggles in the flare position and be prepared to perform a PLF. The canopy may travel backward because of high winds.

PARACHUTE EMERGENCIES AND EMERGENCY PROCEDURES

D-35. The jumpmaster will explain the parachute emergencies and emergency procedures that follow.

PREMATURE ACTIVATION OF PARACHUTE INSIDE THE AIRCRAFT WITH THE RAMP OR DOOR CLOSED

D-36. If a premature activation of the parachute occurs within the aircraft with the ramp or doors closed, the procedures are as follows:
- Shout "PILOT CHUTE" and contain the pilot chute.
- Move the jumper to the front of the aircraft.
- Notify the jumpmaster to ensure the ramp or doors are not opened.

D-37. If the pilot chute activates or container comes open, the container will be reclosed and the jumper will be seated. The jumper WILL NOT jump unless ordered to bail out by the jumpmaster. If the D-bag and the suspension lines fall out, the jumper will disconnect the RSL, and then cut away the main and place it in the kit bag. The jumper will land with the aircraft.

D-38. If the reserve pilot chute is deployed, the procedures are as follows:
- Shout "PILOT CHUTE" and contain the pilot chute.
- Move the jumper to the front of the aircraft.

The jumper will remove his equipment and place the parachute system inside the kit bag. The jumper will then be seated and fasten his seatbelt. The jumper will air land.

PREMATURE ACTIVATION OF PARACHUTE INSIDE THE AIRCRAFT WITH THE RAMP OR DOOR OPEN

D-39. If a premature activation of the parachute occurs within the aircraft with the ramp or doors open, the jumper will perform the same actions as with the ramp or door closed.

D-40. If the pilot chute or parachute is pulled outside the aircraft, the jumper and jumpers in front of that jumper must exit immediately.

PREMATURE ACTIVATION OF PARACHUTE IN FREE FALL

D-41. In the event of a premature activation of a parachute in free fall, the jumper must first determine which parachute has activated.

Suggested Military Free-Fall Sustained Airborne Training

Main Parachute Only Deploys

D-42. If the main has activated, there will be a three-ring assembly on the risers, D-bag and pilot chute trailing.

Reserve Parachute Only Deploys

D-43. If the reserve has activated, there WILL NOT be a three-ring assembly on the risers nor a pilot chute trailing.

Both Main and Reserve Completely Deploy

D-44. In the event both the main and reserve canopies completely deploy, the jumper must identify the scenario and separate the risers. The jumper ensures that his main and reserve risers are not entangled. If they are entangled (or if the jumper is unsure), he should fly both canopies to the ground for landing using the rear risers. Canopies should be kept touching one another (to avoid a down plane) by using rear risers. If the risers are not entangled, the jumper may achieve canopy separation by reaching up and pulling down on the front-left riser of the front canopy. The jumper maintains this front-left riser hold and, at the same time, with his right hand, grabs the red cutaway pillow and executes cutaway procedures while simultaneously releasing the left riser with his left hand. The jumper performs a controllability check and continues to fly the reserve canopy for a landing on the intended DZ.

Main Parachute Fully Deploys and Reserve Parachute Partially Deploys

D-45. If the jumper has not released the brakes, he leaves them stowed. If brakes are released, the jumper should release the toggles all the way up. The jumper should attempt to pull in the reserve deployment bag and hold it between his legs. Should the reserve fully inflate, the jumper should wait for the reserve parachute to rise above shoulder height and execute cutaway procedures.

CUTAWAY PROCEDURES

> **WARNING**
>
> Jumpers must cut away the main parachute before pulling the reserve rip cord handle.

D-46. Once the jumper initiates the cutaway sequence, it must be continued through to completion. Cutaway procedures are as follows:
- Throw away the main rip cord.
- Look to identify the red cutaway pillow on the right main lift web, chest high, inboard.
- Grab the red cutaway pillow with the right hand.
- Look to identify the reserve rip cord handle on the left main lift web, chest high, inboard.
- Grab the reserve rip cord handle with the left hand.
- Arch, head up.
- Pull the red cutaway pillow to a full-arm extension.
- Throw away the cutaway pillow.
- Pull the silver reserve rip cord handle to a full-arm extension.
- Throw away the reserve rip cord handle.
- Raise right shoulder to ensure the reserve pilot chute has launched.
- Perform postopening procedures.
- Conduct a canopy controllability check.

Appendix D

TOTAL MALFUNCTIONS

D-47. A total malfunction occurs when the rip cord has been pulled and the D-bag and canopy remain in the container. Common total malfunctions are described below.

Hard Rip Cord Pull

D-48. If the initial pull is unsuccessful, jumper reaches across with the left hand in a punching motion and pushes the right hand and rip cord out. If this does not pull the rip cord, jumper executes cutaway procedures.

Note: The first attempt is made during normal pull procedures.

Floating Rip Cord or Unable to See Rip Cord

D-49. Jumper arches, looks at the rip cord stow pocket, and locates the rip cord housing with the right hand. If jumper cannot see the rip cord or it is floating, he locates the cable housing on the right shoulder with his right hand. Jumper traces the cable housing down to where the rip cord cable protrudes out. He makes a circle with his index finger and thumb, and pulls to full-arm extension. Jumper makes one attempt; if unsuccessful, he performs cutaway procedures.

Pack Closure

D-50. If the jumper pulls the rip cord and raises his right shoulder, noting that no pilot chute deploys, the jumper immediately raises the right shoulder again. If the problem is not corrected, the jumper must execute cutaway procedures.

PARTIAL MALFUNCTIONS

D-51. A partial malfunction occurs when the container assembly opens and the canopy does not fully or properly deploy. Common partial malfunctions are described below.

Note: Jumper makes only two attempts to clear partial malfunctions.

Horseshoe Malfunction

D-52. In the event of a horseshoe malfunction, jumper makes no attempt to clear the malfunction. He immediately executes cutaway procedures.

Bag Lock

D-53. In the event of bag lock, jumper makes no attempt to clear the malfunction. He immediately executes cutaway procedures.

Pilot Chute Hesitation

D-54. If, after pulling the rip cord, raising his right shoulder, and conducting a two-second count, the jumper notes that the pilot chute is caught in the partial vacuum behind his body, the jumper will raise right shoulder again in an attempt to disrupt the vacuum followed by another two-second count. If the main parachute does not deploy, jumper executes cutaway procedures.

Snivels

D-55. Jumper reaches up and releases the brakes. He pulls both toggles down to the full-brakes position for 3 to 4 seconds, letting up slowly to the 50-percent brake setting, clearing the streamer. If unsuccessful, jumper makes one more attempt with the same procedure. If the malfunction still has not cleared, jumper performs cutaway procedures.

Hung Slider

D-56. If, after pulling the main rip cord, the jumper notes that the main parachute elongates from the bag but the slider does not lower below the cascade point, he grasps the toggles and pulls down to a full-brakes position for 3 to 4 seconds, and then releases the toggles to 50-percent brakes. If the slider does not descend below the cascade point on the lines allowing the main parachute to fully inflate after two attempts, the jumper executes cutaway procedures. If the slider does lower below the cascades but not all the way down, jumper performs a canopy controllability check.

Broken Suspension Lines (A, B, C, D)

D-57. If jumper has one broken suspension line, he will perform a controllability check. If the jumper encounters two or more broken suspension lines, he will perform cutaway procedures.

> **WARNING**
>
> If ever canopy controllability is in question, jumpers must perform a canopy controllability check. If the canopy is uncontrollable, the decision to cut away must be made by 2,500 feet AGL. Jumpers must not initiate cutaway procedures below 1,000 feet AGL. If the malfunction cannot be resolved and cutaway procedures have not been initiated by 1,000 feet AGL, the jumper must immediately deploy his reserve parachute.

Broken Control Lines

D-58. If encountering broken control lines, jumper releases the brake and steers with the good toggle and the rear riser of the side with the broken control line and continues with postopening procedures. Overuse may fatigue the arms. The jumper determines the stall point at a safe altitude using the rear risers, and flares the parachute for landing with both rear risers.

Note: The canopy responds much quicker when using the rear risers.

Pilot Chute Over the Nose of the Canopy or Through the Suspension Lines

D-59. Jumper performs postopening procedures and performs a canopy controllability check. If uncontrollable, jumper executes cutaway procedures.

Closed End Cells

D-60. Jumper brings both toggles to the full-brakes position for 3 to 4 seconds and then lets the toggles up slowly. This process may be repeated a maximum of two times. If this maneuver is unsuccessful, jumper performs a canopy controllability check. If uncontrollable, he executes cutaway procedures.

Premature Brake Release

D-61. If one control line releases on opening, jumper immediately releases the other control line and performs postopening procedures.

Appendix D

Line Twists

D-62. Jumper reaches up, grabs the risers (thumbs facing down), pulls the hands apart separating the risers, and uses a kicking motion to untwist the suspension lines. Jumper does not unstow the brakes until line twists are cleared, and he maintains altitude awareness while clearing line twists. If still hopelessly twisted at 2,500 feet AGL, jumper must execute cutaway procedures.

Holes or Tears

D-63. When performing postopening procedures, if the jumper notices that there is a hole or tear in the lower skin of the canopy, he should perform a canopy controllability check. If uncontrollable, he executes cutaway procedures. If there is a hole in the top skin of the canopy, the jumper must immediately execute cutaway procedures.

Tension Knots

D-64. During postopening procedures, if the jumper notices a tension knot in his lines, he will reach up and grab the affected line group and pull it down to his chest, releasing the lines in a snapping motion in an attempt to clear the knot. Jumper will repeat only twice. If procedure fails to clear, he performs a controllability check. If uncontrollable, jumper executes cutaway procedures.

CANOPY COLLISION AND ENTANGLEMENT EMERGENCIES

D-65. The jumpmaster will explain the canopy collision and entanglement emergencies that follow.

ACTIONS TO BE TAKEN TO AVOID AN ENTANGLEMENT WITH ANOTHER JUMPER

D-66. Jumper maintains a minimum of 25 meters horizontal and vertical separation from other jumpers. The jumper attempts to steer clear by looking right, clearing right, and turning to the right unless the left is closer. The lower jumper has the right-of-way.

ACTIONS TO BE TAKEN IF COLLISION IS IMMINENT

D-67. Jumper assumes the modified spread-eagle position with his left arm across his torso, protecting the reserve rip cord handle and cutaway pillow. If the entanglement occurs, the jumper attempts to free himself.

CANOPY ENTANGLEMENT EMERGENCIES

D-68. If entanglement occurs, "positive" communication of altitude and intent is critical for a successful disengagement. The jumper should never tell the other jumper to NOT do something by using such a word as "don't." Jumpers should only use the words "cut away" if the intent is to have the other jumper execute cutaway procedures.

D-69. Actions to be taken if the lower jumper becomes entangled with the higher jumper and the higher jumper has a good canopy are described below.

Entanglements Above 2,000 Feet AGL

D-70. In such entanglements, the higher jumper attempts to clear the entanglement. If the lower canopy is cleared, it should reinflate within 150 to 200 feet. If the lower canopy cannot be cleared, jumpers should check their altitude. At 2,000 feet AGL or above, the lower jumper disconnects his RSL and executes cutaway procedures.

D-71. If the lower jumper does not want to cut away, the higher jumper must make every effort to maintain control of the lower jumper's canopy. The lower jumper should jettison his combat equipment. The higher jumper must fly the final approach at half brakes and land with half brakes. Both jumpers should be prepared to perform a PLF.

Suggested Military Free-Fall Sustained Airborne Training

WARNING

The lower jumper must disconnect the RSL prior to performing cutaway procedures to prevent his reserve from deploying into the entanglement.

Entanglements Between 1,000 and 2,000 Feet AGL

D-72. In such entanglements, the lower jumper has two options:
- The lower jumper can execute cutaway procedures.
- If the lower jumper does not want to cut away, the higher jumper must make every effort to maintain control of the lower jumper's canopy. The lower jumper should jettison his combat equipment. The higher jumper must fly the final approach at half brakes and land with half brakes. Both jumpers should be prepared to perform a PLF.

Entanglements Below 1,000 Feet AGL

D-73. In such entanglements, the higher jumper must make every effort to maintain control of the lower jumper's canopy. The lower jumper should jettison his combat equipment. The higher jumper should fly the final approach and land with half brakes. Both jumpers conduct a PLF.

Two Bad Canopies

D-74. Jumpers take the following actions at any altitude if both jumpers become entangled and neither has a good canopy:
- The higher jumper has cutaway priority. He should clear himself of entangled lines and cut away, altitude permitting.
- The lower jumper should cut away after the higher jumper, altitude permitting. The higher jumper may be fatally engulfed in the canopy if the lower jumper performs a cutaway first.
- If all else fails and impact with the ground is imminent, both jumpers should deploy their reserve parachutes by 1,000 feet AGL in an attempt to slow descent. If only one reserve parachute deploys, the jumper with the good reserve must bring the other jumper to the ground.
- If both reserves deploy, both jumpers must attempt to cut away the main parachutes.

Note: Communication between the jumpers and altitude awareness are critical for a successful disengagement.

WARNING

Jumpers should not cut away the main parachute below 1,000 feet AGL unless both reserves are fully deployed. There is insufficient altitude or airspeed for the reserve parachute to properly deploy.

EMERGENCY LANDINGS

D-75. The jumpmaster will explain the various types of emergency landings that follow.

Appendix D

ACTIONS FOR TREE LANDINGS

D-76. If a tree landing is expected, jumpers—
- Jettison the rucksack if it has already been lowered. If the rucksack has not been lowered, it should remain attached.
- Continue wearing goggles over the eyes.
- Keep the oxygen mask on, if worn.
- Turn the canopy into the wind and attempt to land between the trees.
- Control the parachute until landing. Vertical descent into the trees is desired and should be accomplished with rear risers.
- Assume a PLF position as the feet enter the branches, ensuring that the forearms are rotated in front of the face for protection, and the feet and knees are kept together.
- Prepare for a PLF in case there is a clear pass through the branches and contact with the ground is made.
- Signal for assistance if suspended in trees and wait for help.

ACTIONS FOR WIRE LANDINGS

D-77. Jumpers should avoid wires at all costs by using proper canopy-control techniques. Any other landing, to include a downwind landing, is preferred to a wire landing. Jumpers should execute the following actions:
- After postopening procedures and prior to descending below 1,000 feet AGL, locate alternate landing sites along their canopy route that will permit a safe into-the-wind landing.
- Avoid crossing power lines while under canopy at an altitude below 1,000 feet AGL.

D-78. If contact with wires is expected, the jumper should—
- Attempt to fly parallel with and pass through the wires.
- Be prepared to flare and/or perform a PLF should he miss or pass through the wires.
- If the canopy is entangled with the wires and contact with the ground is made, immediately cut away from the main parachute and quickly move away from the wires.
- If suspended in the wires, remain motionless and wait for help. The jumper should not let anyone touch him and he should not cut away.

ACTIONS FOR WATER LANDINGS

D-79. If a water landing is expected, the jumper should—
- Attempt to land as close to the shore as possible.
- Jettison his helmet, rucksack, weapon, and oxygen equipment.
- Unfasten waist straps, chest straps, and disconnect the RSL.
- Attempt to land facing into the wind using normal canopy-control procedures, flaring the parachute to land.
- Be prepared to perform a PLF if the water is shallow.
- Upon entering the water, release leg straps, remove harness, and swim upwind or upstream away from the canopy.
- If trapped under the canopy, follow a seam to the edge of the canopy.

Note: If the jumper lands with the harness attached and is being pulled through the water, he cuts away the main canopy, releases the leg straps, and swims free of the harness.

Actions for High-Wind Landings

D-80. The jumpers should always attempt to land into the wind with the canopy level with the ground. When landing in high winds, the jumper should perform the following actions:
- Once contact with the ground is made, release one toggle and pull down on the other.
- Pivot in the direction of the toggle (for example, pull the right toggle down all the way and pivot to the right).
- Continue pulling the control line hand-over-hand until the canopy collapses.

D-81. To recover from a drag, the jumper releases one toggle completely and pulls the other control line hand-over-hand until the canopy collapses. If unable to recover from a drag, the jumper disconnects the RSL and cuts away the main.

Actions for Dust Devils and Turbulent Air

D-82. Jumpers should stay alert under canopy for signs of swirling or erratic wind conditions. The DZSO may use red smoke or flares to warn of visible turbulence, such as dust devils. Jumpers avoid turbulence at all costs by maneuvering away under canopy. If the jumper is unable to avoid the turbulence, he should maintain full flight and remove all slack from the brake lines to prepare for a possible canopy collapse. If the canopy does begin to collapse, the jumper should quickly conduct a 12 to 24-inch strike on the toggles to prevent collapse. Depending on the altitude, the jumper should reattempt this procedure until the canopy reinflates or landing is imminent. As the jumper approaches the ground, he should be prepared to conduct a PLF.

D-83. If the jumper lands and is overtaken by a dust devil, he should—
- Try to gather up the canopy.
- Lay down on top of the canopy.
- If unable to control the canopy, disconnect the RSL and cut away.

Actions for Off-Drop-Zone Landings

D-84. If jumpers are unable to make it to the DZ, they identify a landing area with enough altitude to permit a safe into-the-wind landing. They land on high ground and avoid gullies, ravines, and landing uphill or downhill. Jumpers land along the side of the slope. Then they gather their equipment and move in the direction of the DZ or nearest road. If the road must be crossed, they cross on the high ground where traffic can be observed.

Actions for Other Obstacles

D-85. Jumpers should attempt to steer clear of all other obstacles, including trees, cacti, buildings, and vehicles (on or off the DZ). If unable to avoid the obstacle, jumpers attempt to make contact with both feet and perform a PLF. If a jumper lands on the road or field landing strip, he should gather the canopy and evacuate the road or field landing strip immediately.

ACTIONS WHEN COMBAT EQUIPMENT IS USED

D-86. The jumpmaster will explain the actions when combat equipment is used that follow.

Actions in Free Fall

D-87. When using combat equipment in free fall, the jumper should—
- Fly body as usual, ensuring he maintains positive legs.
- If the rucksack shifts and causes a turn, counter the turn by turning the opposite direction.
- If the rucksack strap moves below the knee, make one attempt to replace it while maintaining stability. If unsuccessful, relax and continue. Counter any turns by turning into the opposite direction.

Appendix D

ACTIONS UNDER CANOPY

D-88. When using combat equipment under canopy, the jumper should—
- Check canopy.
- Gain canopy control using rear riser.
- Clear airspace.
- Orient himself to the wind cone.
- Maintain air awareness and heading.

D-89. If jumping a PDB with HSPRs, the jumper——
- Maintains left rear riser control.
- Physically checks right equipment attachment snap by tracing the right main lift web, and physically checks the lowering line.
- Gains control of right-side rear riser.
- Physically checks the left equipment attachment snap.
- Gains control of toggles.
- If needed, performs a controllability check.
- Continues to fly normal pattern.
- At 1,500 feet AGL, brings toggles to 25-percent brakes, performs grip switch into left hand, and again physically checks lowering line with right hand.
- Continues flying the canopy to the DZ.
- Turns final approach at 500 feet AGL.
- Looks below for fellow jumpers.
- Pulls the yellow release handle so the PDB will fall. If the PDB does not fall, the jumper kicks his legs in an attempt to free the PDB.

Note: All manipulation of the PDB must stop by 200 feet AGL.

- Ensures he is at full flight and prepares to land.
- Flares as normal for landing.
- Performs PLF, if necessary.

Note: If the lowering line is observed to be disconnected, jumper reconnects it if it is accessible and above 500 feet AGL. If not, jumper follows normal landing procedures and lands with his PDB still connected. He flares as normal and conducts a PLF.

D-90. If jumping equipment using quick-release snaps, the jumper—
- Maintains left riser control.
- Physically checks the right quick-release snap (by tracing their right main lift web).
- Physically checks the lowering line.
- Loosens the right-side shoulder strap.
- Gains control of the right-side rear riser.
- Loosens the left-side shoulder strap.
- Disconnects the left quick-release snap.
- Gains control of toggles.
- Continues to fly normal pattern.
- At 1,500 feet AGL (prior to entering downwind leg), crosses his legs and brings toggles to 25-percent brakes (eye level).
- Performs grip switch into his left hand.

- Using his right hand, physically checks the lowering line.
- Disconnects the right-side quick-release snap.
- Catches the PDB between his legs.
- Flies normal approach pattern.
- At 500 feet, looks below for fellow jumpers, uncross his legs, and lowers the PDB off his feet.
- Flares as normal for landing.

ACTIONS IF COMBAT EQUIPMENT WILL NOT LOWER

D-91. If combat equipment will not lower, jumpers maintain altitude awareness under canopy. If the jumper cannot lower his equipment to his feet by 500 feet AGL, he ensures he is facing into the wind and makes one attempt to free the equipment by kicking his legs. If still unable to free the equipment, the jumper will land with his equipment. The jumper flies his canopy and flares as normal (during daylight) or at 50-percent brakes (during night hours) into the wind. The jumper should be prepared to perform a PLF.

Note: All manipulation of the PDB must stop at 200 feet AGL. The jumper must ensure he is at full flight and be prepared to land and conduct a PLF, if necessary. Any controllability issues or malfunctions take precedence over lowering procedures.

GROUPING PROCEDURES

D-92. In order to perform grouping procedures, jumpers—
- Exit as per primary jumpmaster instructions.
- Maintain 25-meter separation. Jumpers must never get over another jumper's back.
- Apply positive legs; jumpers must not backslide.
- At the designated altitude, turn 180 degrees away from the group and forward glide to increase separation.
- Conduct pull procedures at prescribed altitude.
- While the canopy is opening, grab the rear risers and check canopy.
- Clear the airspace in all directions. If a jumper is heading toward another jumper, he should use his rear risers to turn away.
- Perform postopening procedures and get into the canopy formation.
- Maintain 25-meter separation.
- If unable to see the group leader, follow the lower jumper to the front. If a jumper becomes the low man, he should assume group-lead position and follow landing procedures as prescribed in sustained airborne training.
- Follow the group to the designated landing area.

Note: Jumpers must not perform 360-degree turns.

HIGH-ALTITUDE HIGH-OPENING PROCEDURES

D-93. Jumpers perform HAHO procedures as follows:
- Exit single file at one-arm interval IAW the group leader's instructions.
- Pull at the designated pull altitude or time delay, as briefed.
- While the canopy is opening, grab the rear risers and check canopy, and then clear the airspace in all directions. If jumper is heading toward another jumper, he uses his rear risers to turn away.
- Perform postopening procedures and assume position in the canopy formation.
- Maintain 25-meter vertical and horizontal separation.

D-94. The group leader or low man leads the formation to the intended landing area.

Appendix D

NIGHT MILITARY FREE-FALL OPERATIONS

D-95. Jumpers conducting night MFF operations illuminate chemlights just prior to receiving a JMPI unless directed otherwise. At the 10-minute warning, all jumpers ensure their chemlights are visible. A minimum of three chemlights must be worn—one on the altimeter and two additional lights (placed IAW unit SOP).

D-96. If jumpers are unable to determine the proper flare altitude, they should go to 50-percent brakes, keep their feet and knees together, and be prepared to conduct PLFs. This technique is recommended for night landings.

D-97. During nontactical night jumps, if a jumper lands in the lights, he should quickly move away from the lights so other jumpers may land near them. If a jumper disturbs the lights, he should quickly return the lights to their original position.

D-98. If a jumper lands off of the DZ, he should—
- Attempt to land facing into the wind according to the direction of the wind arrow or briefed wind direction.
- Attempt to locate a landing area that is free of obstacles (that is, open areas away from roads). Jumpers must remember that many roads have power lines running parallel which are difficult to see at night.

D-99. In a nontactical environment, the jumper—
- Attempts to move toward the DZ.
- If unsure of the DZ location or if there is no means of radio contact, moves to the nearest trail, road, or high ground, and reestablishes contact according to unit SOP.
- If he does not have his equipment, marks the place on the trail where he entered it using his second chemlight.

D-100. In a tactical environment, the jumper conducts linkup procedures according to the premission brief.

RECOVERY PROCEDURES FOR ADMINISTRATIVE MILITARY FREE-FALL OPERATIONS

D-101. Jumpers must daisy-chain the lines while walking toward the parachute without pulling the parachute toward themselves. Jumpers—
- Gather up the canopy with the D-bag and pilot chute.
- Move to the DZSO while keeping a sharp lookout for landing parachutists.
- Report to the DZSO.
- If making multiple jumps from the same location, repack the parachute and reset the CYPRES.
- After the last jump of the day, turn off the CYPRES. Place the canopy in the kit bag first, daisy-chain the suspension lines, then place the container inside the kit bag and snap it closed. Failure to turn off the CYPRES could result in a reserve deployment.
- Account for all their equipment. Report any missing equipment to the jumpmaster immediately.
- Leave the altimeters on their wrists until returning to the packing/recovery area.

INJURIES

D-102. All injuries must be reported immediately to the DZSO/primary jumpmaster/medic. Jumpers must take all necessary precautions to prevent all injuries. Information on injuries must be reported on DA Form 285-AB (U.S. Army Abbreviated Ground Accident Report).

D-103. Jumpers must remember the basic rules of the air:
- Stay off other jumpers' backs.
- The lower jumper has the right-of-way.
- Pull at the prescribed altitude.
- Land into the wind.
- When in doubt, apply half brakes and perform a PLF.

Appendix E
Sample Accident Report

This appendix provides a sample accident report at Figure E-1, pages E-2 through E-4. It provides an example of the amount of information that elements should provide in such reports.

Appendix E

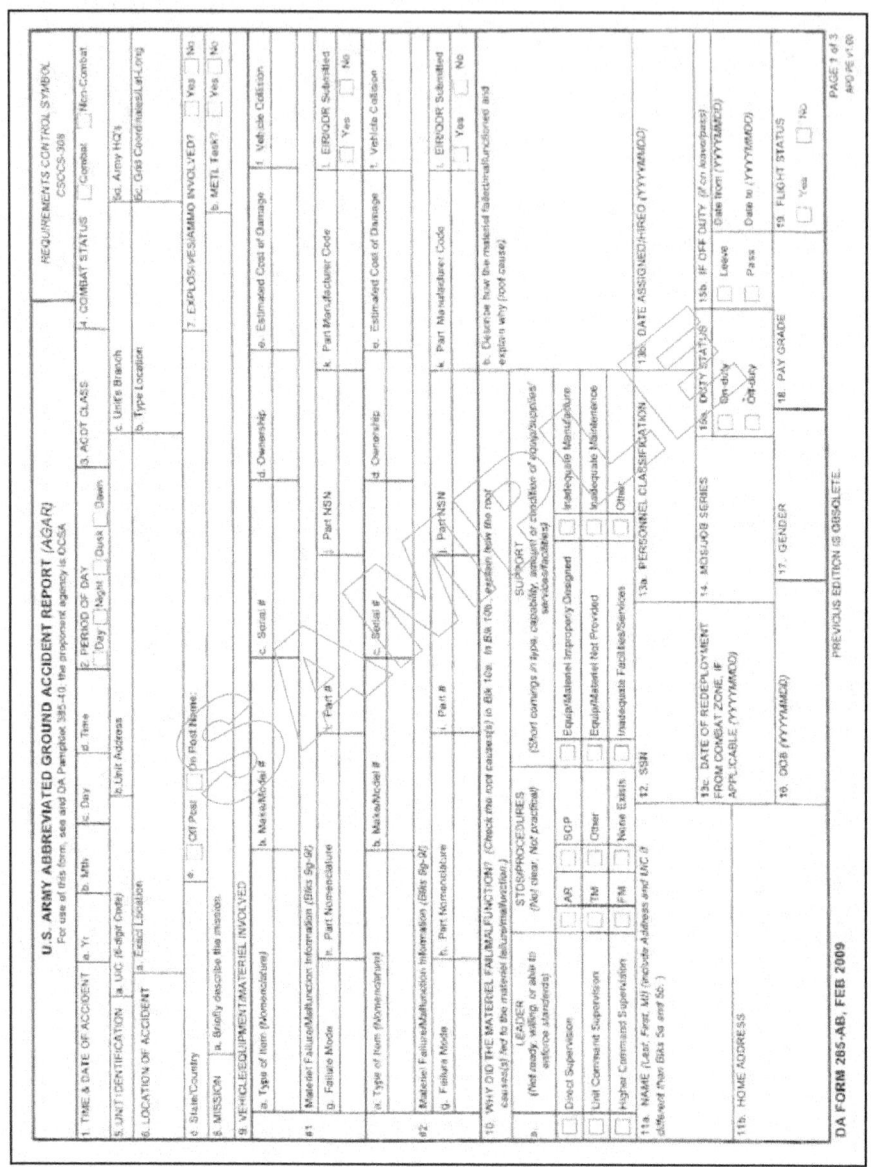

Figure E-1. Sample accident report

Sample Accident Report

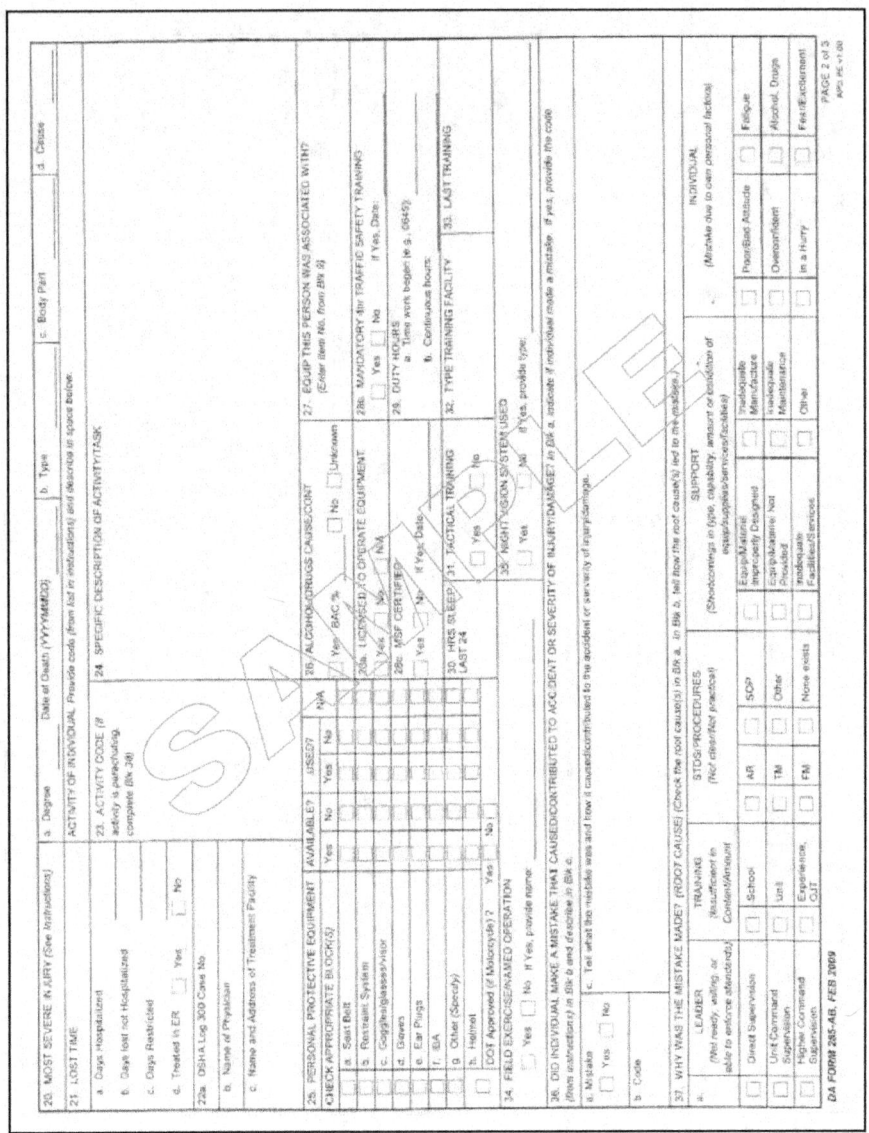

Figure E-1. Sample accident report (continued)

Appendix E

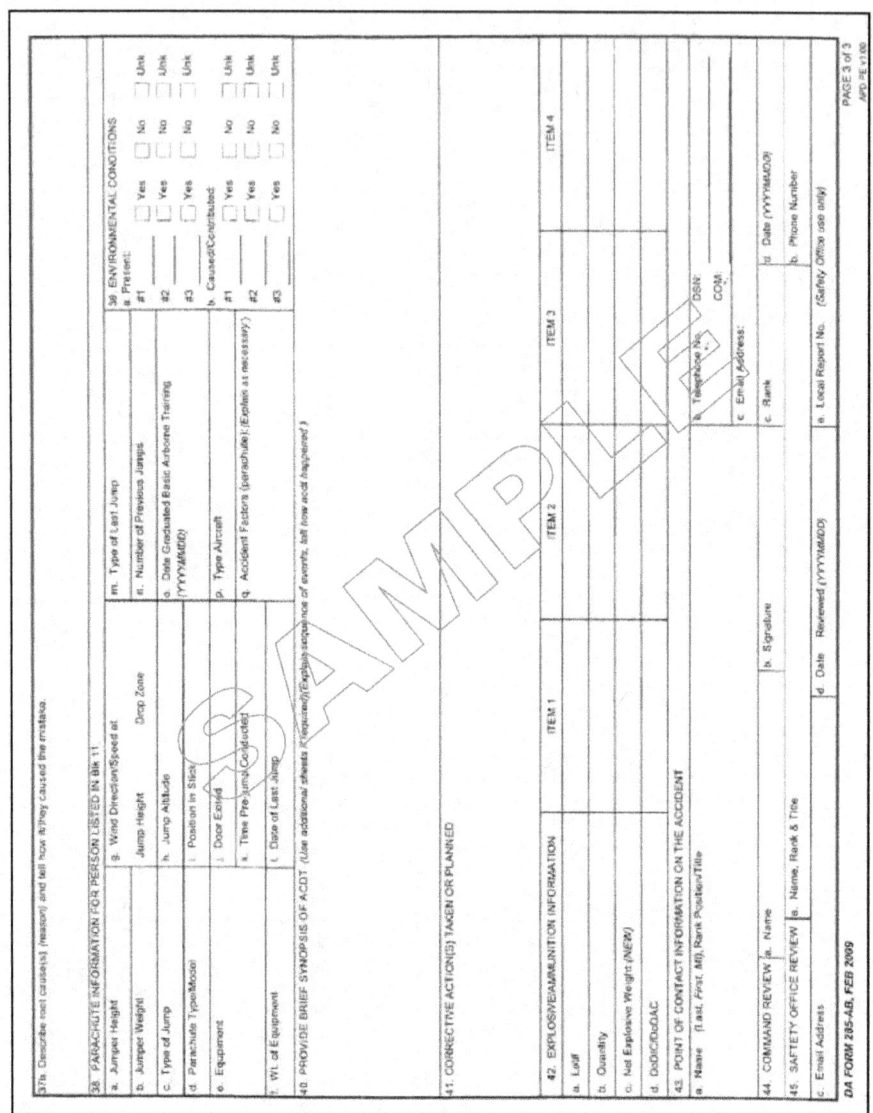

Figure E-1. Sample accident report (continued)

Appendix F
High-Altitude Release Point Calculation

The effects of variable wind directions and speed must be accounted for when determining the HARP for each MFF mission. Accurate wind data is essential to calculate the HARP precisely. Commanders are cautioned against planning pinpoint landings on targets when wind data is questionable due to the source, timeliness of reporting, or other dynamic meteorological conditions (for example, thunderstorms or changing fronts). Wind will affect the parachutist during free fall and canopy performance after deployment.

OBTAINING WIND DATA

F-1. Military airfields, civilian airports or weather services, artillery meteorological sections, or pilot teams in the operational areas can provide wind data. Aircrew personnel can also determine wind data during flight as the aircraft passes through different flight levels. (It is not advisable to use this technique for actual infiltrations, as the data obtained en route to the objective area may not reflect conditions at the objective area.)

RECORDING WIND DATA

F-2. The jumpmaster records the reported wind data according to altitude in feet, direction in degrees (True), and speed (velocity) in knots as follows:

- *HALO*: He records the wind data for canopy flight every 1,000 feet of altitude from surface to pull altitude and every 2,000 feet of altitude from pull altitude to exit altitude for free fall.

Note: If the pull altitude is greater than 6,000 feet AGL, the jumpmaster will record the winds for, and calculate for, a HAHO operation.

- *HAHO*: He records the wind data for canopy flight every 1,000 feet of altitude from surface to 10,000 feet and every 2,000 feet of altitude from 10,000 feet to exit altitude.

CALCULATING AND PLOTTING THE HIGH-ALTITUDE RELEASE POINT

F-3. The jumpmaster calculates and plots the HARP's location in reverse sequence (Figure F-1, page F-2). First, he calculates the distance and direction from the desired impact point to the parachute opening point. Second, he calculates the distance and direction from the parachute opening point to the preliminary release point (PRP). Third, he calculates the distance and direction from the PRP (to compensate for forward throw) to the HARP.

F-4. Calculation of the HARP during HAHO operations may or may not require calculation of free-fall drift, depending upon the length of free fall required. For HAHO missions requiring less than 2,000 feet of free fall, the jumpmaster disregards free-fall drift.

F-5. When plotting the HARP on a map, the jumpmaster converts the wind direction from True North to a grid azimuth using the declination diagram.

Appendix F

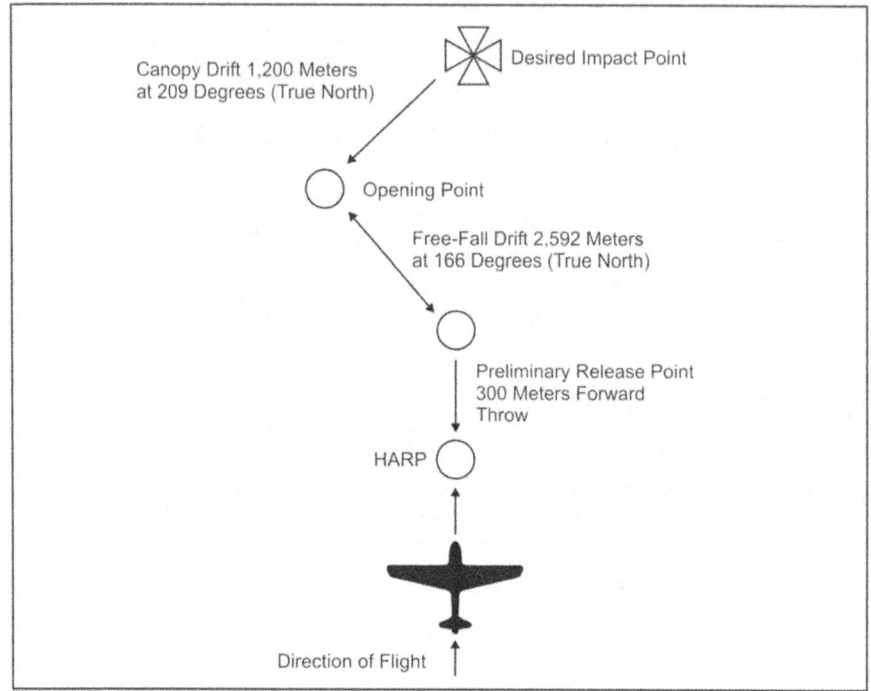

Figure F-1. Plotting the HARP, free-fall, and canopy drift for a 20,000-foot HALO mission profile

USING THE WIND DRIFT FORMULA AND CONSTANTS

F-6. The jumpmaster uses the wind drift formula $D = KAV$:
- D = distance in meters.
- K = constant (drift in meters per 1,000-foot loss of altitude in a 1-knot wind).
- A = altitude in thousands of feet.
- V = average wind speed (velocity).

The jumpmaster also uses the following wind drift constants (K factors):
- $K = 3$ (parachutist in free fall).
- $K = 25$ (MC-3 parachute system and RAPPS [HALO]).
- $K = 48$ (RAPPS [HAHO]).

F-7. Table F-1, page F-3, defines the HAHO K factors for Department of Defense RAPPSs.

Note: The jumpmaster calculating the HAHO wind drift uses the constant of the least performing canopy; for example, the U.S. Navy MT-1SS uses the S-type reserve that has a K factor of 66. Therefore, if a parachutist has to activate his reserve parachute, he will still be able to glide to the DZ.

High-Altitude Release Point Calculation

Table F-1. HAHO K factors for Department of Defense Ram-Air Personnel Parachute Systems

K-Factor	Parachute Systems	Remarks
48	MC-4, MC-5, MJ-1, MT-2XX/SL	Large High-Glide RAPPS
66	MT-1SS, PD230	Small Low-Glide RAPPS
46	TP-400	Tandem Offset Resupply Delivery System (TORDS)
Multimission Parachute Systems (MMPSs)		
K-Factor	Parachute Systems	Remarks
46	MP-360	Large High-Glide Zero-Porosity RAPPS
31	HG-380	Operations in High-Glide Mode
39	HG-380	Operations in Parachute Mode

NOTE: The jumpmaster always calculates for the lowest performance parachute (largest K factor) to be used on that MFF operation.

CALCULATING HALO FREE-FALL DRIFT AND DIRECTION

F-8. To determine the parachutist's drift in free fall, the jumpmaster calculates the average wind speed (velocity) and average wind direction from the exit to the opening altitude. Opening altitude (4,000 feet in this example) is not included since that is where the free fall stops. The wind data from 4,000 feet to 1,000 feet is calculated using the canopy drift constant.

EXAMPLE:

Altitude	Velocity	Direction
20,000	85	160
18,000	75	160
16,000	75	165
14,000	65	165
12,000	50	155
10,000	45	150
8,000	20	185
6,000	<u>20</u>	<u>190</u>
	435 knots	1330 degrees

The jumpmaster determines the averages by—

- Determining the total free-fall distance from the exit (20,000) to the opening (4,000).
  A = 20,000 − 4,000 = 16,000, or A = 16.
- Dividing the sum of the wind velocities (435) by the number of velocities (8).
  V = 435 ÷ 8 = 54.375, or V = 54 (rounded to nearest whole number) knots average wind speed (velocity).
- Dividing the sum of the wind directions (1330) by the number of directions (8).
  Direction = 1330 ÷ 8 = 166.25, or Direction = 166 degrees (rounded to nearest whole number) average wind direction.

Note: Jumpmasters use the following rounding guidelines:

- 0.0 to 0.4: Round down to the nearest whole number.
- 0.5 to 0.9: Round up to the nearest whole number.

F-9. The jumpmaster substitutes the numerical values for the letters of the D = KAV formula.
- D = (3) (16) (54).
- D = 2,592 meters at 166 degrees (True North).

Appendix F

Note: If the jumpmaster uses wind directions from 271 degrees to 089 degrees to calculate the average wind direction, incompatible averages may result. (All rules for erroneous winds and doglegs apply.) To compensate, the jumpmaster adds 360 degrees to directions of 001 to 089 degrees and averages the wind direction. If the resulting average is greater than 360 degrees, the jumpmaster subtracts 360 to obtain the correct average wind direction.

EXAMPLE:	Direction (incorrect)	Direction (correct)	Direction (average greater than 360 degrees)
	345	345	345
	350	350	355
	345	345	005 (+ 360) = 365
	010	010 (+ 360) = 370	020 (+ 360) = 380
	015	015 (+ 360) = 375	025 (+ 360) = 385
	350	350	035 (+ 360) = 395
	1415 degrees	2135 degrees	2225 degrees

Direction = 1415 ÷ 6 = 235.83 degrees or D = 236 degrees (incorrect).

Direction = 2135 ÷ 6 = 355.83 degrees or D = 356 degrees (correct).

Direction = 2225 ÷ 6 = 370.83 degrees or D = 371 (- 360) = 011 degrees.

CALCULATING CANOPY DRIFT

F-10. To determine the parachutist's drift under canopy, the jumpmaster calculates the average wind speed (velocity) and direction from 1,000 feet to the opening altitude.

EXAMPLE:	Altitude	Velocity	Direction
	4,000	15	190
	3,000	14	220
	2,000	11	205
	1,000	9	220
		49	835

Note: Disregard surface winds for calculation. Winds from 1,000 feet to surface are not used to allow the parachutist to maneuver in the landing pattern.

The jumpmaster determines the averages by—
- Dividing the sum of the velocities (49) by the number of velocities (4).
  V = 49 ÷ 4 = 12.25, or V = 12 (rounded to nearest whole number) average wind speed (velocity).
- Dividing the sum of the wind directions (835) by the number of directions (4).
  Direction = 835 ÷ 4 = 208.75 degrees, or 209 degrees (rounded to the nearest whole number) average wind direction.

The jumpmaster substitutes the numerical values for the letters of the D = KAV formula.
- D = (25) (4) (12).
- D = 1,200 meters at 209 degrees (True North).

CALCULATING FORWARD THROW

F-11. Compensation must be made for the distance a parachutist's body initially travels into the direction of flight due to forward speed (velocity). The forward throw distances used in HALO and HAHO are—
- *300 meters* for a high-performance aircraft with exit speeds above 120 knots.
- *150 meters* for a low-performance aircraft with exit speeds below 120 knots.

CALCULATING DOGLEGS

F-12. A dogleg is a situation in which the wind direction changes 90 degrees or more for two (or more) consecutive recorded altitudes. Doglegs require separate calculations from the altitude where the wind direction changes.

Note: A single 90-degree or greater change in wind direction is treated as an erroneous wind and will not be included in wind direction or velocity calculations; the altitude will still be included in the D = KAV formula.

CALCULATING THE HAHO HIGH-ALTITUDE RELEASE POINT

F-13. To calculate the HAHO HARP, the jumpmaster uses the modified D = KAV formula, as the intention is to maximize the linear distance traveled using the gliding capability of the RAPPS.

Note: For doglegs with less than 6,000 feet of vertical descent, the jumpmaster may use the standard D = KAV formula; however, it will be less accurate.

The jumpmaster uses the following HAHO gliding distance formula:

- $D = \frac{(A - SF)(V + 20.8)}{K}$
- D = gliding distance in nautical miles (nm).
- A = altitude in thousands of feet.
- SF = safety factor in thousands of feet.
- V = average wind speed (velocity) in knots.
- 20.8 = canopy speed constant.
- K = 48 (canopy drift constant).

Note: Jumpmasters use the following rounding guidelines:

- 0.0 to 0.4: Round down to the nearest whole number.
- 0.5 to 0.9: Round up to the nearest whole number.

F-14. The jumpmaster calculates the safety factor, which provides a buffer area after exit to permit the parachutists to assemble under canopy and to establish the landing pattern over the DZ. For example, the element commander desires 1,000 feet for canopy assembly after exit and 2,000 feet to establish the landing pattern. The safety factor is 3,000 feet. Therefore, SF = 3.

F-15. The jumpmaster calculates the total canopy gliding distance in nautical miles. He does not round up or down. Instead, he truncates the result to the tenth of a nautical mile; for example, 12.666 = 12.6 and 18.37486 = 18.3. To convert nautical miles to kilometers (km), the jumpmaster multiplies by 1.85 and again truncates the result.

F-16. When an element exits the aircraft in stick formation, the jumpmaster compensates for dispersion between the parachutists. He obtains this figure by dividing the total number of parachutists by 2 and then multiplying the result obtained by 50 meters. He plots the calculated distance back into the aircraft's line of flight. This procedure places the middle of the stick on the desired opening point.

F-17. The jumpmaster plots back into the aircraft's line of flight to compensate for forward throw (300 meters for high-performance aircraft and 150 meters for low-performance aircraft). The following are examples of HAHO HARP calculations.

Appendix F

EXAMPLE 1: HAHO HARP Calculation

Situation. The exit altitude is 14,000 feet. Twelve parachutists will exit the aircraft in stick formation. The element commander desires 1,000 feet for canopy assembly and a 1,000-foot arrival altitude over the DZ. Wind speed and direction at altitude are—

Altitude	Velocity	Direction
14,000	25	090
12,000	22	080
10,000	21	090
9,000	21	090
8,000	20	085
7,000	18	080
6,000	18	080
5,000	17	085
4,000	16	080
3,000	12	075
2,000	12	080
1,000	08	080
	210 knots	995 degrees

F-18. The jumpmaster—

- Determines the average wind speed: $V = 210 \div 12 = 17.50$, or $V = 18$ (rounded to nearest whole number) average wind speed.
- Determines the average wind direction: $D = 995 \div 12 = 82.91$, or $D = 83$ (rounded to nearest whole number) degrees (True North) average wind direction.
- Determines the safety factor is 2 (minimum).
- Substitutes the numerical values for the letters of the formula:
  $D = (12 - 2) (20.8 + 18) \div 48$.
  $D = (10) (38.8) \div 48$.
  $D = 388.0 \div 48$.
  $D = 8.0$ nm at 83 degrees (True North).
- Determines the gliding distance: 8.0 nm x 1.85 = 14.8 km.
- Determines dispersion: $(12 \div 2) \times 50 = 300$ meters.
- Determines forward throw: 300 meters.
- Converts the average wind direction from degrees (True North) to a grid azimuth and plots it on the map to determine the canopy opening point.
- Plots the dispersion and forward throw from the PRP to determine the HARP.
- Determines the grid azimuth from the opening point to the desired impact point. Converts the grid azimuth to a magnetic azimuth. The magnetic azimuth is the compass heading followed by the parachutists to the DZ.

Note: If there is no free fall prior to canopy deployment, the opening point is the PRP.

EXAMPLE 2: HAHO HARP Calculation With a Dogleg.

Situation. The exit altitude is 15,000 feet. Twelve parachutists exit the aircraft in stick formation. The element commander desires 1,000 feet for canopy assembly and a 2,000-foot arrival altitude over the DZ. A change of wind direction creates a dogleg at 9,000 feet AGL. Wind speed and direction at altitude are—

Altitude	Velocity	Direction
14,000	33	210
12,000	30	210
10,000	29	180
	92 knots	600 degrees
9,000	26	075
8,000	24	080
7,000	22	085
6,000	20	090
5,000	18	090
4,000	14	085
3,000	12	090
2,000	10	085
1,000	8	080
	154 knots	760 degrees

JUMPMASTER CALCULATIONS (BELOW THE DOGLEG FROM 9,000 TO 1,000 FEET)

F-19. The jumpmaster calculates the gliding distance and direction from the desired impact point to the dogleg at 9,000 feet. He—

- Determines that the average wind speed (velocity) from 1,000 feet to 9,000 feet is 17.11 or $V = 17$ (rounded to the nearest whole number) knots average wind speed.
- Determines that the average wind direction from 1,000 feet to 9,000 feet is 84.44 or 84 (rounded to the nearest whole number) degrees (True North).
- Determines that the safety factor is 3. He must remember that in a formula for a HAHO dogleg, the safety factor is 2 on the base leg and 1 on the dogleg to equal a total safety factor of 3.
- Establishes that altitude = 9,000 feet, or $A = 9$.
- Substitutes the numerical value for the letters of the formula:
  $D = (9 - 2) (20.8 + 17) \div 48$.
  $D = (7) (37.8) \div 48$.
  $D = 264.6 \div 48 = 5.5$ nm x $1.85 = 10.1$ km gliding distance at 84 degrees (True North).

JUMPMASTER CALCULATIONS (ABOVE THE DOGLEG FROM 10,000 TO 14,000 FEET)

F-20. The jumpmaster calculates the gliding distance and direction from 10,000 feet to the exit altitude. He—

- Determines that the average wind speed (velocity) from 10,000 feet to 15,000 feet is 30.66 or 31 (rounded to the nearest whole number) knots.
- Determines that the average wind direction from 10,000 feet to 15,000 feet is 200 degrees (True North).
- Determines that the safety factor is 1.

Appendix F

- Establishes that altitude = 5,000 feet, or A = 5.
- Substitutes the numerical value for the letters of the formula:
  $D = (5 - 1)(20.8 + 31) \div 48$.
  $D = (4)(51.8) \div 48$.
  $D = 207.2 \div 48 = 4.3$ nm x $1.85 = 7.9$ or 8 km (rounded to the nearest whole number) gliding distance at 200 degrees (True North).

F-21. The jumpmaster converts the True North azimuths to grid azimuths. He plots the glide path from the desired impact point to the dogleg, and plots the glide path from the dogleg to the opening point. He calculates the dispersion for 12 parachutists (300 meters) and plots the PRP from the opening point. The jumpmaster compensates for forward throw and plots the HARP.

F-22. The jumpmaster determines the grid azimuth from the opening point to the desired impact point. He converts the grid azimuth to a magnetic azimuth. The magnetic azimuth is the compass heading followed to the DZ. By holding a single compass heading, the parachutist will maintain direction and follow a curving path from the opening point to the DZ, rather than a path with distinct turns.

Note: The safety factor above the dogleg and below the dogleg, when combined, mathematically incorporates the desired effect over the complete group.

Appendix G
Jumpmaster Personnel Inspection

Before each MFF parachute operation, the jumpmaster conducts a systematic inspection of each parachutist's parachute and combat equipment for proper wear, fit, and attachment. All equipment being airdropped will receive a JMPI. The jumpmaster must never sacrifice safety for speed.

> **DANGER**
> Improper or incomplete jumpmaster personnel inspections may result in DEATH, serious injury, or equipment loss and damage.

JMPI OF THE MC-4 HARNESS AND CONTAINER SYSTEM

G-1. The jumpmaster uses the following sequence to detect and identify deficiencies. With hands and eyes working together, he starts at the front of the parachutist and moves to the rear, from top to bottom, right side to left side (Figure G-1, page G-2).

Note: If making an oxygen jump, the jumpmaster first performs the oxygen inspection sequence on page G-6. Then he continues with the following:

Note: If jumping in the vicinity of a water hazard, the jumpmaster follows the inspection sequence for flotation devices on page G-9. Then he continues with the following:

- Harness: Checks for proper fit before continuing the JMPI.
- Helmet and goggles:
    - Uses correct helmet: MC-3, Gentex HGU-55/P, Gentex lightweight parachutist helmet, Bell helmet, Protec with free-fall liner, or ACH helmet.
    - Makes sure helmet fits properly and is serviceable.
    - Uses approved goggles (Kroop; military-issue sun, wind, and dust goggles; or Gentex only).
    - Makes sure the lenses are clear and not cracked or scratched.
    - Makes sure the goggle strap is secured if worn outside of helmet.
    - Checks that bayonet receivers are present and securely attached.
    - Makes sure the two adjustment screws are present on the receiver covers.
    - Checks chin strap for proper attachment and serviceability, with excess stowed.
    - Right riser: Makes sure no twists are present in front or rear riser from riser cover to 3-ring release assembly.

Appendix G

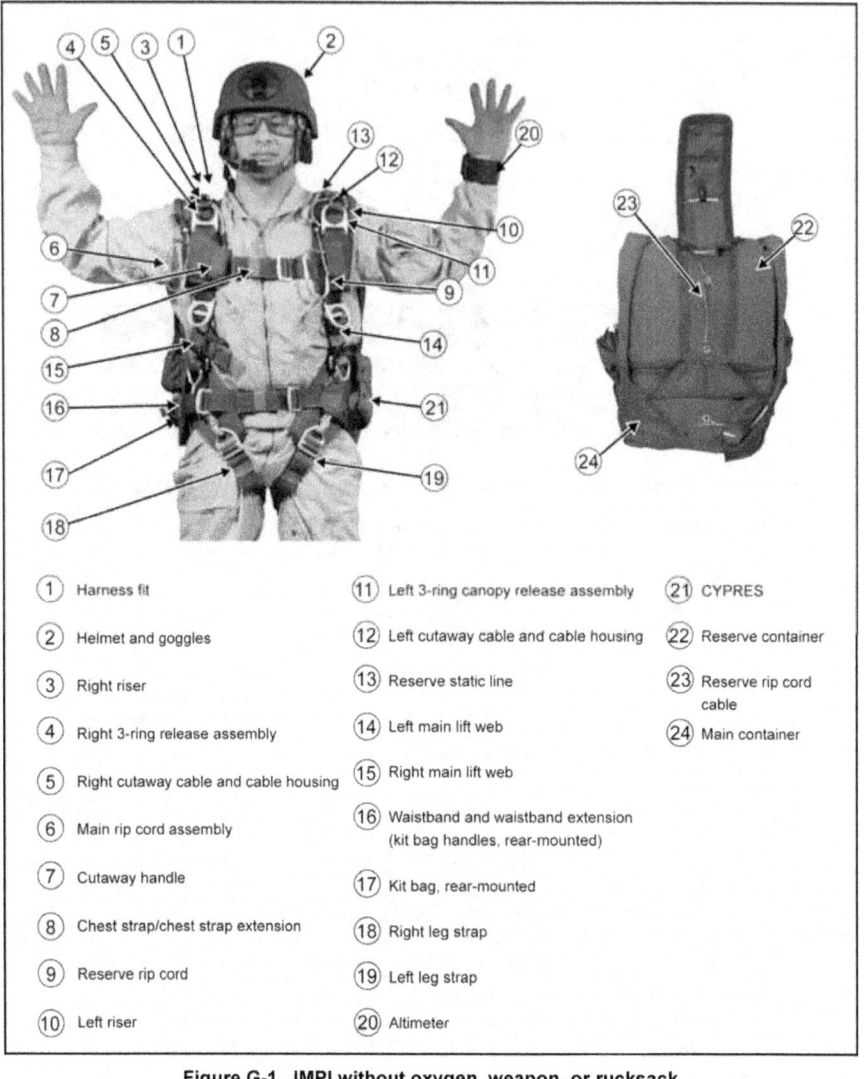

Figure G-1. JMPI without oxygen, weapon, or rucksack

- Right 3-ring release assembly:
    - Checks for correct assembly—small ring, medium ring, and base ring (elongated snowman effect).
    - Gives small and medium ring a one-quarter turn to check for free movement.

- Right main canopy release cable and cable housing:
  - Inspects for tacking and proper routing.
  - Makes sure the 3-ring locking loop is through the small ring and the grommet on the riser and the eye on the cable housing (without any twists or frays).
  - Rotates riser toward the parachutist's neck, ensuring the release cable is routed through the locking loop and the running end is stowed in the stowage flute.
- Main rip cord assembly:
  - Makes sure the housing is tacked properly.
  - Makes sure there are no broken strands on main rip cord cable.
  - Makes sure the two swage balls are present on the end of the rip cord cable.
  - Checks that the main rip cord handle is properly seated in the elastic pocket.

Note: The jumpmaster resumes JMPI sequence of the CMWH after inspection of the main canopy release handle (page G-10, JMPI for the MC-4 RAPPS using the center-mounted weapons harness).

- Cutaway handle (main canopy release rip cord):
  - Makes sure that the cutaway cables are not twisted more than 180 degrees.
  - Checks that the handle is seated in its pocket and the Velcro is properly mated.
- Chest strap:
  - Makes sure there are no twists and it is properly routed (to include the chest strap extension).
  - Makes sure the excess is rolled under and stowed in the slack retainer.
  - Makes sure it is properly routed through the friction adapter.

Note: If jumping with a weapon, jumpmaster follows the inspection sequence on page G-10. Then he continues with the following:

- Reserve rip cord:
  - Makes sure it is properly seated in the elastic pocket.
  - Checks that the two swage balls are present on end of the reserve rip cord cable.
  - Makes sure there are no broken strands.
  - Makes sure the cable is properly routed to the cable housing.
  - Makes sure the cable housing is tacked.
  - Makes sure RSL is free and clear of reserve rip cord cable (first free and clear).
- Left riser: Makes sure there are no twists in the front or rear riser from the riser cover to the 3-ring release assembly.
- Left 3-ring release assembly:
  - Checks for correct assembly—small ring, medium ring, and base ring (elongated snowman effect).
  - Gives the small and medium rings a one-quarter turn to check for free movement.
- Left main canopy release cable and cable housing:
  - Inspects for tacking and proper routing.
  - Makes sure the 3-ring locking loop is through the small ring and the grommet on the riser and the eye on the cable housing (without any twists or frays).
  - Rotates riser toward parachutist's neck, making sure the release cable is routed through the locking loop and the running end is stowed in the stowage flute.

Appendix G

- Reserve static line:
    - Makes sure the RSL quick-release lanyard is attached and snapped.
    - Makes sure the RSL loop is attached to the release shackle and routed correctly.
    - Makes sure RSL is free and clear of canopy release cable housing (second free and clear).
- Left main lift web:
    - Checks that the large equipment attachment ring and V-ring are present.
    - Makes sure the running end of the adjustment strap is rolled and stowed in the slack retainer.
    - Makes sure there are no twists.

Note: If jumping with a rucksack, jumpmaster follows the inspection sequence on page G-11. Then he continues with the following:

- Right main lift web:
    - Checks that the large equipment attachment ring and V-ring are present.
    - Makes sure the running end of the adjustment strap is rolled and stowed in the slack retainer.
    - Makes sure there are no twists.
    - Checks free-floating strap and oxygen fitting block for proper attachment, and makes sure the four screws are present on the back of the fitting block.
- Waistband, waistband extension, and kit bag handles (rear-mounted):
    - Makes sure the right wing flap is secured to the waistband.
    - Checks that there are no twists from its attachment point on the right side of container to the left wing flap.
    - Makes sure the excess is rolled under and stowed in the slack retainer.
    - Checks for proper routing through the waistband extension friction adapter.
    - Checks the waistband extension is routed through the kit bag handles (rear-mounted).
    - Checks the kit bag is positioned between the jumper's back and the main pack tray.
    - Makes sure waistband is routed over all equipment.
- Right leg strap, kit bag handle (front-mounted):
    - Makes sure the snap hook gate closes and has proper spring tension.
    - Makes sure the excess is rolled under and stowed in the slack retainer.
    - Checks for correct routing, with no twist in leg strap or saddle.
    - Ensures the leg strap is routed through one kit bag carrying handle (front-mounted).
- Left leg strap, kit bag handle (front-mounted):
    - Makes sure the snap hook gate closes and has proper spring tension.
    - Makes sure the excess is rolled under and stowed in the slack retainer.
    - Checks for correct routing, with no twist in leg strap or saddle.
    - Ensures the leg strap is routed through one kit bag carrying handle (front-mounted).
- Altimeter, MA2-30:
    - Makes sure it is located on parachutist's left wrist, that it fits snugly, and it is properly attached (with 0 to the top).
    - Checks for proper free-fall altimeter setting.
    - Tells the parachutist to turn and continues the JMPI.
- Altimeter, MA-10:
    - Makes sure it is located on parachutist's left wrist, that it fits snugly, and it is properly attached (with 0 to the top).
    - Looks at the altimeter and ensures the altimeter is on (steady yellow light).

Jumpmaster Personnel Inspection

- Checks for proper free-fall altimeter setting.
- Tells the parachutist to turn and continues the JMPI.
• Reserve container:
    - Peels open the reserve rip cord protective flap.
    - Makes sure the reserve rip cord cable housing is tacked down.
    - Checks that the RSL is routed correctly, and that the reserve rip cord cable runs through the RSL guide and assist ring.
• Reserve rip cord cable:
    - Checks that the reserve rip cord cable has no broken strands.
    - Makes sure it is routed on the left side of the grommets.
    - Makes sure the top pin is inserted at a 45-degree angle.
    - Makes sure the closing loops are not frayed.
    - Makes sure both pins are not seated past their shoulders.
    - Tells the parachutist to bend.
• Main container:
    - Opens both protective flaps.
    - Makes sure the closing flaps are closed in the proper sequence (bottom, left, right, top).
    - Makes sure the main rip cord cable housing is tacked.
    - Checks that main rip cord cable and EAAD/ARR power cable are not twisted around each other.
    - Makes sure the 2-inch cable extension with swage ball is at the 12 o'clock position (top).
    - Makes sure the closing loop is not frayed.
    - Makes sure the main pin is not seated past its shoulder.
• The Military CYPRES 2: Jumpmaster does the normal JMPI sequence through the inspection of the parachutist's altimeter. After the altimeter inspection, the jumpmaster continues with the following:
    - Opens the reserve protector flap and pins it up and out of the way.
    - Traces the control cable and inspects for any damage and proper routing.
    - Makes sure the control cable is properly routed through the binding tape guide.
    - Makes sure the binding tape is properly tacked to the reserve top-closing flap.
    - Makes sure the control unit is set at the proper default for the current free-fall operation.
    - Makes sure the control unit LED indicator light is not lit.
    - Checks that the control unit digital readout screen shows the proper millibar setting or 0▼.
    - Inspects the reserve rip cord cable for proper routing and no broken strands.
    - Makes sure the reserve rip cord cable runs through the assist ring (little ring) of the RSL and then through the guide ring (big ring).
    - At the top reserve locking pin, ensures the reserve rip cord cable is to the right of the grommet.
    - Inspects the pin to make sure it is not shouldered inside the grommet and the pin is not bent.
    - Makes sure the continuous Military CYPRES 2 closing loop of the reserve is not frayed.
    - Continues to inspect the reserve rip cord cable to the bottom locking pin, making sure it is properly routed and there are no broken strands.
    - At the bottom reserve locking pin, inspects the pin to make sure it is not shouldered inside the grommet and the pin is not bent.
    - Makes sure the continuous Military CYPRES 2 closing loop of the reserve is not frayed.
    - Makes sure the closing flaps are closed in the proper sequence (bottom, left, right, top).
    - Makes sure the main rip cord cable housing is tacked.
    - Inspects the cable for proper routing and no broken strands.

Appendix G

- Continues past the locking pin and inspects the 2-inch extension for proper routing and no broken strands.
- Ensures the 2-inch extension terminates with a single steel swage ball.
- Pinches the swage ball to ensure the main locking pin does not come loose.
- Inspects the pin to make sure it is not shouldered inside the grommet and the pin is not bent.
- Makes sure the main closing loop is not frayed.
- Slaps the bottom of the container to indicate completion of the JMPI.

JMPI FOR THE MC-4 RAPPS WITH THE 106-CUBIC-INCH PORTABLE BAILOUT OXYGEN SYSTEM

G-2. The jumpmaster inspects the entire oxygen system before inspecting the harness/container system. The recommended inspection sequence for the MC-4 RAPPS with the oxygen system follows (Figure G-2, page G-7):

- Inspects the inside of the mask, making sure there is no debris, the four self-sealing screws are present, the combination valve retainer is present, and the portion that matches with the parachutist's face is not torn or damaged in any way that would cause the parachutist to have an improper seal or fit.
- Attaches the mask to the left-side bayonet receiver.
- Checks for proper fit and seal. Makes sure there is no damage to the hard-shell or soft-shell portion of the mask. Makes sure the four capped tee nuts secure the four attaching straps to the hard-shell portion of the mask, and the excess is either taped or tacked. Checks that the combination valve is of the correct type (green exhalation port flaps only), the delivery tube clamp is present and attached properly, and there is no damage to the delivery tube at its attachment point to the combination valve.
- Detaches the oxygen mask from the left-side bayonet receiver. Inspects the oxygen mask delivery tube to make sure there is no damage (checks for holes, discoloration, or deterioration). Makes sure the delivery tube retainer is present and attached correctly. Checks that the elastic slack retainer is around the chest strap and that the Velcro is mated around the delivery tube.
- Moves to the quick-disconnect assembly. Inspects the delivery tube clamp to make sure that it is present and attached properly. Makes sure there is no damage to the delivery tube at its attachment point to the quick disconnect. Disconnects the delivery tube from the AIROX VIII. Inspects the quick disconnect to make sure that there is no debris inside the quick disconnect and that the antisuffocation valve moves freely, has correct spring tension, and returns to the closed position. Inspects the gasket (O-ring) to make sure it is present and the beveled lip portion is up (not reversed).
- Inspects the AIROX VIII. Disconnects the oxygen mask delivery tube. Makes sure the dust cover is present and serviceable. Checks that the debris screen is present and is not damaged or corroded. Checks that there is no debris inside the AIROX VIII. Reconnects the oxygen mask delivery tube, making sure the quick-disconnect assembly is fully seated.
- Grasps the AIROX VIII and moves the entire assembly gently up and down to check that the dovetail mounting plate is correctly mated with the oxygen fitting block. Inspects the oxygen fitting block to make sure it is assembled correctly and the four attachment screws are present and secure on the back of the oxygen fitting block.
- Checks the ambient air port of the AIROX VIII. Visually inspects the inside of the ambient air port to make sure the debris screen is present and is not damaged or corroded. Checks that there is no debris inside the ambient air port. Inspects the antisuffocation valve to make sure it has correct spring tension and returns to the closed position. Inspects the gasket (O-ring) to make sure it is present and that the beveled lip portion is up (not reversed).
- Inspects the blue antitamper seal (blue dot of paint). Makes sure it is present and aligned.
- Checks to ensure both flow indicator windows are not damaged.

- Grasps the "B" nut, giving it a slight turn to make sure it is tight. (The "B" nut attaches the delivery hose [medium pressure] to the AIROX VIII.)
- Follows the delivery hose (medium pressure) from its point of connection on the AIROX VIII and checks for proper routing. Makes sure the delivery hose is routed from the AIROX VIII over the outside of the waistband. Checks that the delivery hose then makes a 180-degree bend and runs under the waistband and between the parachutist's body and his right main lift web. Checks that it then runs to the union elbow.

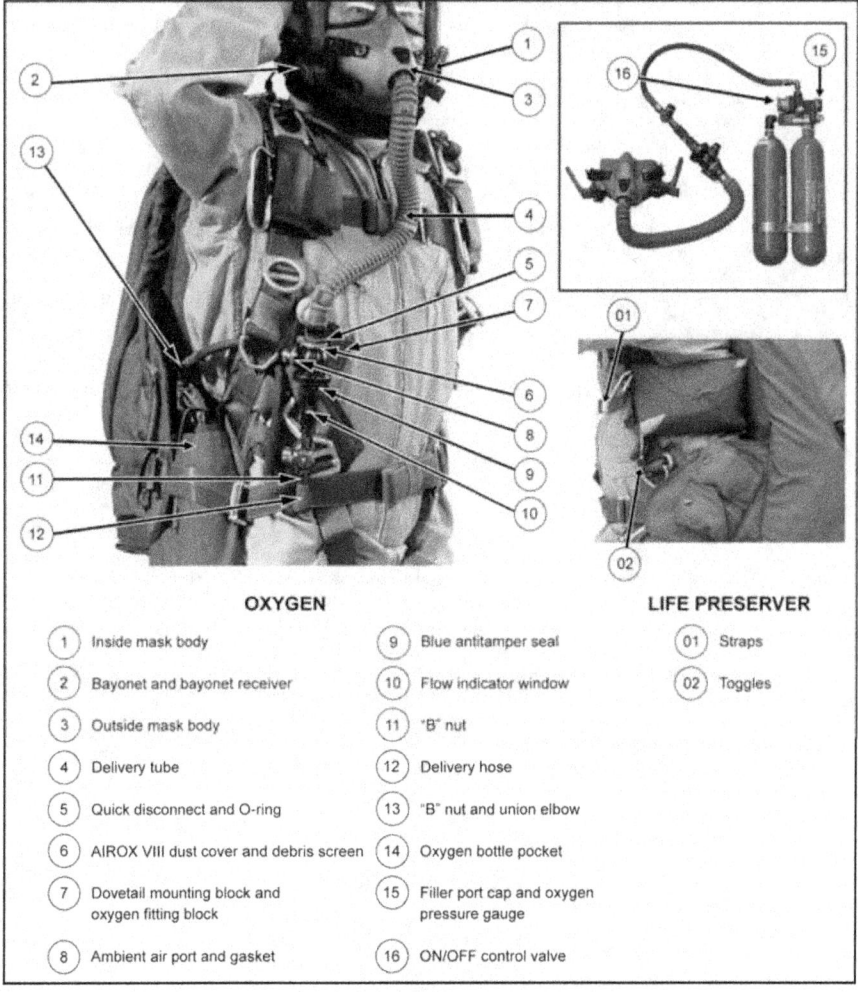

OXYGEN

1. Inside mask body
2. Bayonet and bayonet receiver
3. Outside mask body
4. Delivery tube
5. Quick disconnect and O-ring
6. AIROX VIII dust cover and debris screen
7. Dovetail mounting block and oxygen fitting block
8. Ambient air port and gasket
9. Blue antitamper seal
10. Flow indicator window
11. "B" nut
12. Delivery hose
13. "B" nut and union elbow
14. Oxygen bottle pocket
15. Filler port cap and oxygen pressure gauge
16. ON/OFF control valve

LIFE PRESERVER

01. Straps
02. Toggles

Figure G-2. JMPI with oxygen and life preserver

Appendix G

- Checks the "B" nut at its point of attachment to the union elbow for tightness by giving it a slight twist. Then gives the union elbow a slight twist, checking for proper tightness to the reducer manifold.
- Pushes up on the bottom of the oxygen bottle pocket with the left hand while the right hand is on the manifold and pulls the bottles away from the parachutist's body. Moves to the overpressure relief valve making sure it is seated by pushing in on the cap. While in this position with the oxygen bottles away from the parachutist's body, inspects the waistband from its point of attachment on the container to the right wing flap friction adapter. Makes sure the waistband is not twisted and the waistband is routed through both of the center loops on the oxygen bottle pocket. Checks that the oxygen system is between the waistband and right wing flap.
- Tells the parachutist to bend. Inspects the filler port cap making sure it is present and finger-tight. Checks that the oxygen pressure gauge indicates adequate pressure. The needle on the oxygen pressure gauge must be on the number 1 of 1800 psi or higher to be correct.
- Tells the parachutist to stand erect. Turns the ON/OFF control valve on and listens for a flow of oxygen out of the oxygen mask. Makes sure the ON/OFF control valve can be locked in the ON position. Turns the ON/OFF control valve off, making sure it can be locked in the OFF position.

G-3. This sequence completes the JMPI of the 106-cubic-inch PBOS. The jumpmaster returns to the normal JMPI sequence for the MC-4 RAPPS.

JMPI FOR THE POM MFF MASK AND JUMP BOTTLE SYSTEM

G-4. The jumpmaster begins the inspection by reaching, with his right hand, the pressure reducer on the bottle and turning the ON/OFF toggle switch to the ON position. Then he continues with the following:

- Moves to the front of the jumper and visually checks that the mask is attached to the left side of the jumper's helmet.
- With his right hand, grasps the mask on the outside portion on the hard shell and rotates the mask making the inside visible. Visually inspects the inside to ensure cleanliness. Looks for the presence of the pressure-demand relief valve (brass ring), microphone element, exhalation valve, and antisuffocation valve.
- Using his left index finger as a guide, places it inside the mask at the top and rotates it in a clockwise direction while peeling back the lip and exposing the inside of the mask. He should be inspecting for tears, dirt, or damage to the inner soft-shell portion. He continues this process until he comes back up to the 12 o'clock position. Then he does another sweep on outer portion of the inner soft shell making sure there are no damages to the mask that would hinder a good seal to the jumper's face.
- With his right hand, gently pulls out and rotates the mask on the jumper's face. With the left hand, attaches the bayonet fitting into the receiver on the right side of the jumper's helmet and seats it with 2 clicks.
- From the nose of the jumper, looks in a clockwise direction around the mask and inspects for any problems with fit and the mask edges are not pinched or rolled over the jumper's face. Gives the mask a shake ensuring a good seal and sounds off with key word "PROPER FIT."
- Brings his right index finger to the friction buckle at the 2 o'clock position on the mask. Makes sure it is present and excess webbing of the attaching strap is stowed properly by either tape or tacking. In a clockwise motion, inspects the remaining three buckles and attaching straps and sounds off with key words "BUCKLE TAPE" (four times).
- Inspects the hard shell for cracks and ensures exhalation valve cover is secure. Visually looks under the cover for the spring and the overall cleanliness. Ensures intercom block on the top of the mask has two screws present and the intercom cord is attached if necessary. Checks regulator for damage and cleanliness, and gently shakes to ensure it is secure.
- Places his right hand to the right of the quick-disconnect fitting on the regulator. Grasps and pushes the hose toward the mask to ensure it is properly seated. Pulls on the hose to make sure that it is securely attached and sounds off with key words "PUSH IN - PULL OUT." Instructs

the jumper to breathe in and out, listens for oxygen flow (it should stop when the jumper exhales). Next, he moves to the union elbow and with his left hand, grasps it; with his right hand, he grasps the blue "B" nut and attempts to turn it to the right ensuring tightness, sounding off with key word "TIGHT."

- From the mask end, traces the medium-pressure delivery hose as it routes behind the jumper's neck. The hose should be routed through the heavyweight retainer band that is half-hitched to the carrying handle on the top center of the jumper's container. Moves to the right side of the jumper. The hose should be seen running from the carrying handle to and through a half-hitched retainer band attached to the equipment tie-down loop. Traces the hose from the retainer to the top of the swivel "T."
- Grasps the nut that connects the medium-pressure delivery hose to the swivel "T" and ensures it is connected and that the unit rotates freely.
- Places his left hand on the bottom of the bottles. At the same time, places his right hand on the manifold at the top of the bottles and lifts up and out.
- While pulling out on the bottles, looks behind them to visually inspect the origin. Places his right hand behind the bottles and sweeps the waistband to ensure there are no twists and it routes through the middle loops on the oxygen pouch. Sounds off with key word "ORIGIN."
- Rotates the bottles forward and with his left hand checks for the presence of the filler cap and ensures it is tight.
- Looks at the bottle gauge and states in reference to the 1800 on the gauge face, key words "ONE," or "ABOVE" or "BELOW ONE."
- Rotates the switch to the OFF position and sounds off with key word "LOCK OFF."
- Removes the mask from the jumper's face and starts his inspection of the jumper's parachute as normal.

JMPI FOR THE MC-4 RAPPS WITH FLOTATION DEVICES

G-5. The recommended JMPI sequence for the MC-4 with flotation devices follows (Figure G-2, page G-7):
- B-7 and LPU-10/P life preservers:
    - Checks that the life preserver straps are over the uniform and under the parachute harness (B-7 chest strap fastened with a quick release).
    - Ensures flotation packets fit under the armpits, with the flaps to the outside, and the toggles down and to the front. Makes sure no part of the flotation packet is under the parachute harness.
- UDT life vest:
    - Makes sure the life vest is worn around the neck with all straps under the parachute harness, including the parachute harness chest strap. The vest is secured with a rubber band to prevent interference with the cutaway handle and the reserve rip cord.
    - Makes sure the inflatable portion of the vest does not go under the chest strap.
    - Unscrews the CO_2 cartridge to make sure it has not been fired. Reinserts the cartridge into its fitting and ensures it is finger-tight. Makes sure the protective flap does not cover the toggle.

G-6. Jumpmaster returns to the normal JMPI sequence for the MC-4 RAPPS.

JMPI FOR THE MC-4 RAPPS WITH WEAPON (M16A1/A2 AND M4A1)

G-7. The jumpmaster follows the normal JMPI sequence until he encounters the weapon sling over the chest strap extension. The recommended inspection sequence for the MC-4 RAPPS with weapon follows (Figure G-3, page G-10):
- Makes sure the sling is routed over the chest strap extension and under the left main lift web.
- Makes sure the sling is routed over the parachutist's shoulder.
- Checks that the weapon tie-down is secured around the weapon sling about 6 inches from the swivel on the stock of the weapon.

Appendix G

- Makes sure the sling is routed to the outside of the weapon butt stock and that the weapon magazine is to the parachutist's rear.
- Checks that the weapon is placed between the left wing flap and the parachutist with the waistband extension routed through the weapon-carrying handle.

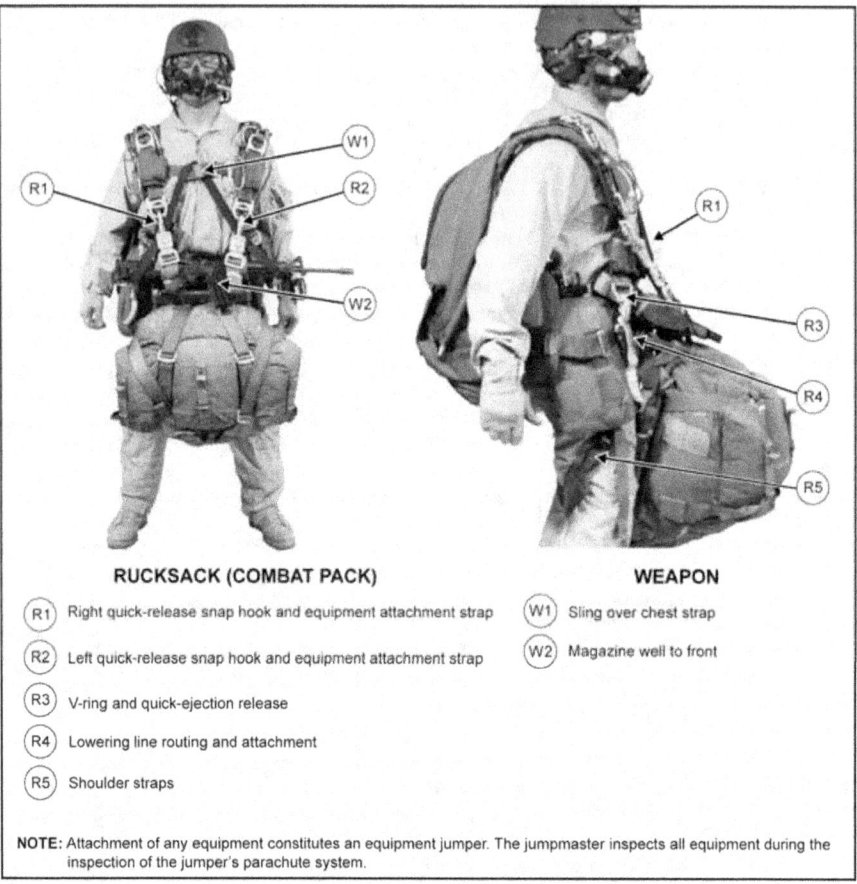

RUCKSACK (COMBAT PACK)

- R1 Right quick-release snap hook and equipment attachment strap
- R2 Left quick-release snap hook and equipment attachment strap
- R3 V-ring and quick-ejection release
- R4 Lowering line routing and attachment
- R5 Shoulder straps

WEAPON

- W1 Sling over chest strap
- W2 Magazine well to front

NOTE: Attachment of any equipment constitutes an equipment jumper. The jumpmaster inspects all equipment during the inspection of the jumper's parachute system.

Figure G-3. JMPI for weapon, front-mounted rucksack

JMPI FOR THE MC-4 RAPPS USING THE CENTER-MOUNTED WEAPONS HARNESS

G-8. The sequence of the JMPI procedures for the center-mounted weapons harness remains the same for the helmet, oxygen, and right three-ring assembly, main rip cord grip, and cutaway handles. The sequence remains the same regardless of the weapons system checked; for example, M-4, M24, M249, M240B, and AT-4. The jumpmaster continues with the following:

- Ensures proper routing of the chest strap through the friction adapter and picks up the inspection at the weapon's sling. Jumpmaster states the key word, "Weapon."

Note: In all Category B weapons—M24, M249, M240B, and AT-4—the sling must be routed through the chest strap.

- Ensures the weapon's sling is routed under the chest strap on the main rip cord side and over the chest strap on the reserve rip cord side. Jumpmaster states the key words, "Under, Over."
- Traces the weapon's sling from the chest strap down to the sling tie-down point and ensures that it is secured to the weapon and all excess webbing has been taped or secured using a heavyweight retainer band.
- At the point of attachment, ensures the weapon's muzzle is oriented to the jumper's left, the sling is routed in front of the weapon, and the main lift web attachment buckle is outboard of the left main lift web. Jumpmaster states the key words, "Muzzle Left Outboard."
- Traces the weapons system ensuring that the forward assist and charging handle are located away from the jumper, the waistband is behind the weapon, and the harness is securely attached to the weapon.

Note: When jumping weapons that have bipods and/or feed tray covers, the weapon's attachment straps must cover and secure these components to mitigate an inadvertent release.

- Upon reaching the right main lift web, ensures the main lift web attaching point is rotated outboard of the right main lift web. Jumpmaster states the key word, "Outboard."
- Upon reaching the right sling attachment point, traces the sling ensuring proper routing and all excess material is secured using tape or a heavyweight retainer band, and the sling is routed behind the chest strap.

G-9. The jumpmaster goes to the reserve rip cord handle and continues the normal JMPI.

JMPI FOR THE MC-4 RAPPS WITH REAR- OR FRONT-MOUNTED COMBAT PACK (RUCKSACK)

G-10. The recommended inspection sequence for the MC-4 parachute assembly with the combat pack (rucksack) follows (Figures G-3 [front-mounted], page G-10, and G-4 [rear-mounted], page G-12). The jumpmaster follows the normal JMPI sequence until he arrives at the equipment attachment ring on the left main lift web. Then the jumpmaster—

- Makes sure the left quick-release snap hook has proper spring tension and that the gate is closed. Makes sure the quick-release is seated. Follows the left attachment strap around to the improved equipment attachment sling, making sure it is not routed under any portion of the MC-4 harness or rucksack frame.
- Makes sure the right quick-release snap hook has proper spring tension and that the gate is closed. Makes sure the quick-release is seated. Follows the right attachment strap around to the improved equipment attachment sling, making sure it is not routed under any portion of the MC-4 harness or rucksack frame.
- Inspects the HPT lowering line assembly at its point of attachment on the right V-ring. Makes sure the gate on the quick-ejector release is closed and that the locking arm is locked. Checks the routing of the tubular nylon to the nylon duck container (stow pocket), making sure it is routed free of any portion of the MC-4 parachute system or the rucksack frame and is located between the parachutist's leg and the shoulder strap of the rucksack.
- Checks the running end of the HPT lowering line for proper attachment. Makes sure it is attached between the lateral locking straps where the diagonal straps cross. Checks that the running end of the lowering line passes through its own loop and is tightened down.
- Grasps both shoulder straps and pulls to the outside of the parachutist's legs to make sure they are attached correctly and that the parachutist has a leg through each shoulder strap.

G-11. Jumpmaster returns to the JMPI inspection sequence at the left main lift web large equipment ring.

Appendix G

JMPI WITH NIGHT VISION GOGGLES (WITH AND WITHOUT OXYGEN)

G-12. The jumpmaster will conduct a general visual inspection of the helmet and NVGs to ensure overall serviceability, suitability, and proper fit.

G-13. The jumpmaster will ensure the NVGs are properly mounted and secured, as well as verify that the NVGs function as designed; for example, power on and off, and flip up and down freely. If the NVGs have an external battery pack and power cables, the jumpmaster will inspect to ensure the battery pack and power cables are secured to the helmet and properly stowed.

G-14. The jumpmaster will trace the bungee cord to ensure it is routed through the retention loops with a not less than 4-mm-diameter bungee cord, but not larger than a 6-mm-diameter bungee cord. The jumpmaster will also ensure the hook is no more than 2 inches long. The jumpmaster will continue to trace the bungee cord to the termination point where it will be attached to the 550 cord loop on the NVG. The jumpmaster will then complete the bungee inspection by verifying that the bungee hook is closed with one turn of paper masking tape with a quick release. The chin strap and oxygen system will be inspected IAW current JMPI standards.

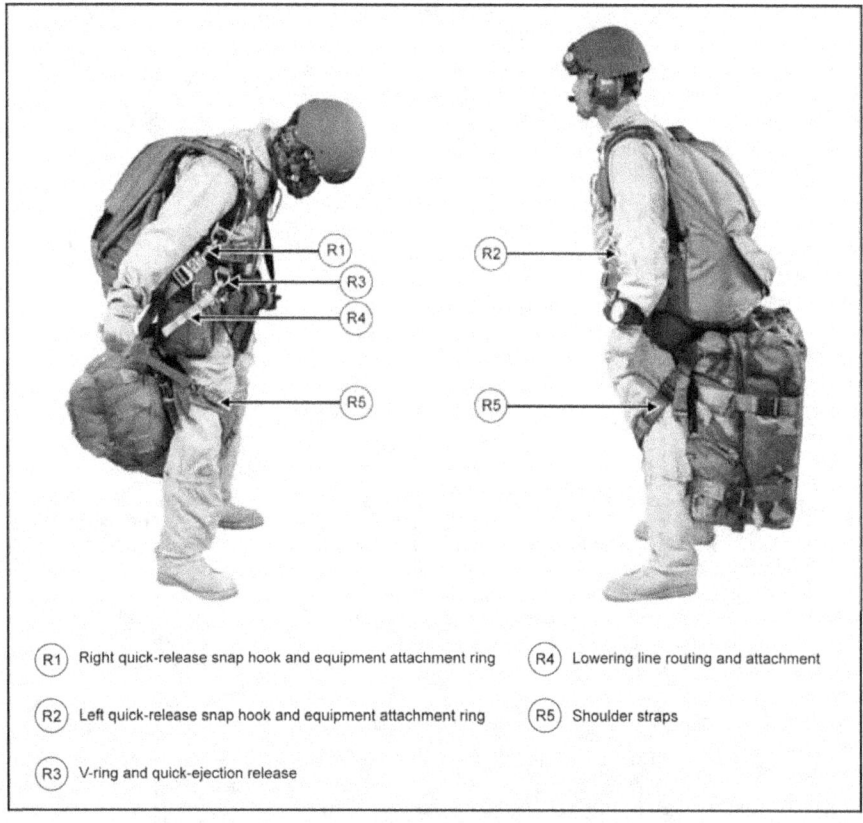

R1 Right quick-release snap hook and equipment attachment ring	R4 Lowering line routing and attachment
R2 Left quick-release snap hook and equipment attachment ring	R5 Shoulder straps
R3 V-ring and quick-ejection release	

Figure G-4. JMPI with the rear-mounted rucksack/parachutist drop bag

Appendix H
Sample Aircraft Inspection Checklist

Special Forces operational detachments primarily use USAF troop carrier aircraft when conducting MFF operations and proficiency training. The preparation of the aircraft for parachute operations is an aircrew responsibility. The jumpmaster, accompanied by the aircraft loadmaster, inspects the aircraft and coordinates any activities particular to the airborne operation (for example, loading and placement of oxygen consoles). At a minimum, the jumpmaster checks the exterior and interior areas of the aircraft directly related to the airborne operation. FM 3-21.220 contains the specific items that must be inspected and the peculiarities of certain aircraft. Figure H-1 contains a sample aircraft inspection checklist.

AIRCRAFT EXTERIOR
(Vicinity of the Jump Doors or Ramp)
- Projections.
- Sharp Edges.

AIRCRAFT INTERIOR
- Seats and Safety Belts.
- Jump Caution Lights.
- Cabin Lighting, if Required.
- Jump Doors:
    - Sharp or protruding edges.
    - Door latches.
    - Jump platforms.
    - Air deflectors.
    - Floors.
    - Clean.
    - Excess equipment secured.
    - Roller system removed or reversed.
- Oxygen Equipment:
    - Secured.
    - Operational.
    - Jumpmaster and spare console stations.
    - Safety Equipment:
        * Alarm bells.
        * Intercom system.
        * Fire extinguishers.
        * Emergency exits.
        * First-aid kits.
        * Overwater flight equipment.
- Troop Facilities:
    - Airsickness bags.
    - Latrine/head.
    - Walk-around bottle filler stations operational.

Figure H-1. Sample aircraft inspection checklist

This page intentionally left blank.

Appendix I
Jumpmaster Aircrew Briefing Checklist

The jumpmaster briefs the aircrew as a part of his duties at the departure airfield. He uses the following checklist (Figure I-1, pages I-1 and I-2) to brief the aircrew.

- Free-Fall Operation Concept.
- Aircrew Troop Safety Briefing (Time, Location).
- Marshaling Plan.
- Drop Zone:
    - Designation and location.
    - Desired impact point.
    - Proposed HARP location.
    - Elevation.
    - Major obstacles.
    - Marking/identification.
    - Strike reports.
- Flight Route/Checkpoint Warnings/Altitudes.
- Drop Heading.
- Racetrack (Turnoff Direction, Turnaround Time).
- Drop Altitude AGL and MSL.
- Number of Passes.
- Drop Speed.
- Formation or Interval (Multiple Aircraft).
- Number of Parachutists/Safety/Static Personnel.
- Command of Personnel Remaining Onboard the Aircraft.
- Time Warnings: Relayed From Crew to Jumpmaster.
- Jump Caution Lights: When Turned On and Off.
- Confirmation of Load/Station/TOT Times.
- Aircraft Inspection.
- Aircraft Configuration.
- Call Signs and Frequencies.
- Intercom.
- Cabin Lighting.
- Opening/Closing of Troop Doors/Ramp.

Figure I-1. Sample jumpmaster aircrew briefing checklist

Appendix I

- Aircraft Emergencies:
    - Load jettison.
    - Fuselage fire.
    - Abandon aircraft.
    - Emergency bailout.
    - Crash landings.
    - Ditching.
- Movement in the Aircraft.
- Smoking Restrictions.
- Airsickness.
- Latrine Facilities.
- Forecasted Weather Conditions.
- In-Flight Rigging.
- Oxygen Procedures:
    - Pressurized/depressurized flight.
    - Prebreathing requirement.
    - Oxygen emergencies.
- Automatic Ripcord Release:
    - Arming/disarming altitude.
    - Activation altitude.
- Free-Fall Bundles:
    - Location and movement.
    - Ejection procedures.
- Visual Jumpmaster Release:
    - Spotting procedure.
    - Increments of correction.
    - Hand-and-arm signals.
- Manifest.

Figure I-1. Sample jumpmaster aircrew briefing checklist (continued)

Appendix J
Joint Precision Airdrop System

This appendix discusses the new JPADS, which uses the GPS, steerable parachutes, and an onboard computer to steer loads to a designated point of impact on a DZ.

AUTONOMOUS BUNDLE OPERATIONS

J-1. New and emerging technologies are now allowing for the deployment of GPS-guided bundles that can carry an ever-increasing load to pinpoint grid coordinates. This new technology will greatly increase mission capabilities and allow for more than just personnel to be delivered to the objective area via the RAPPS.

J-2. In recent years, GPS-guided RAPPSs render more flexibility to load drops. Since 2004, the U.S. Marine Corps has been using Sherpa TM/MC GPS-guided parachutes in Iraq, dropping one-ton loads within 70 meters from their designated target point. In August 2006, the first "Screamers" were tested in Afghanistan, dropping container-delivery-system bundles containing food, water, ammunition, and other supplies, weighing 500 to 2,200 pounds (0.2 to 1 ton) to troops on the ground.

J-3. Precision airdrop for special operations has been chosen among the ten highest priority areas defined for the North Atlantic Treaty Organization's (NATO's) defense against terrorism effort. The purpose of the defense against terrorism program is to develop new, cutting-edge technology to protect troops and civilians against terrorist attacks. Precision airdrop capabilities will enhance the capability of NATO forces to deliver personnel or equipment stealthily and precisely under all weather conditions wherever they may be needed. This capability will also support the increasing deployment of NATO troops to long-distance out-of-area contingency operations, which have lately become in vogue.

J-4. High-altitude precision airdrop is expected to be a key enabling technology for future forces' deployment. Such capabilities will facilitate rapid strategic and tactical deployment of forces, supported "just in time," with supplies delivered precisely to any location throughout the world. The increased accuracy and ability to drop to more than one location simultaneously means that Soldiers on the ground can recover the cargo quickly and know exactly where it will land. Such capability is providing military planners with the capability of strategically and covertly positioning equipment and supplies for rapidly moving ground and special operations forces.

J-5. Aimed at supporting ground troops with essential supplies, the U.S. Army Natick Soldier Center has teamed with the U.S. Air Force Air Mobility Command to develop new airdrop capabilities, first pushing immediately essential supplies such as ammunition, water, fuel, and medical supplies to forward-deployed troops. The medium and heavy systems will be fielded at a later phase and will enable precision airdrop of loads ranging from 20,000 to 60,000 pounds (9 to 27 tons) of cargo, more than enough to deliver the Army's eight-wheel 19-ton Stryker combat vehicle. As the Army transforms to the Future Combat System, JPADS will provide the just-in-time logistics needed. The ultimate goal is to resupply troops anywhere in the world within 24 hours with supplies directly flown from the U.S. bases. JPADS will satisfy four "gaps" identified in the current airdrop capability:

- Increased ground accuracy.
- Standoff delivery.
- Increased air carrier survivability.
- Improved effectiveness of airdrop mission operations.

J-6. The JPADS family of systems consists of "self-guided" cargo parachute systems with navigation aids for MFF parachute systems all linked to a common mission planning and weather system. The JPADS is intended to be deployed as a containerized-delivery-system bundle via static-line deployment. The JPADS uses a ram-air parachute canopy for the entirety of the descent, and is controlled by an Autonomous Guidance Unit (AGU). The canopy is deployed directly off the ramp of the aircraft and the opening is slider-controlled. The system is designed for operation up to 24,500 feet MSL and offset up to 25 kilometers from the target, providing a high standoff delivery and aircraft survivability. Minimum release altitude during combination drops is 5,000 feet AGL. The JPADS uses Military Global Positioning System (MIL-GPS) sensor data for navigation to waypoints and accurate soft landings within 150 meters from the designated impact point. The weight classes are as follows:

- *Micro lightweight (MLW)*: 10 to 150 pounds.
- *Ultra lightweight (ULW)*: 250 to 700 pounds.
- *Extra light (XL, also known as 2K)*: 700 to 2,200 pounds.
- *Light (L)*: 5,000 to 10,000 pounds.
- *Medium (M)*: 15,000 to 30,000 pounds.

J-7. All systems will be required to hit a preplanned GPS ground target within 50 meters, cleared for high-altitude drop from 24,500 feet MSL, and capable of being deployed from at least 8 kilometers horizontal offset from the ground target. Using a portable mission-planning tool and wireless communications, loadmasters will be able to update the mission plans uploaded to the rigged JPADS before the flight with last minute changes of DZ location, threats, and so on.

RIGGING

J-8. Because of the complexities of JPADS rigging, joint airdrop inspection is required for all JPADSs and is only conducted by qualified personnel.

J-9. As early as possible, personnel should try to identify the type of aircraft and style of rollers that they are using for delivery. It is crucial that the rollers are wide enough to accept the skid plate. If the rollers are too thin, then the load may roll off of the rollers during deployment. C-130 rollers are too thin to accept the TORDS (round) bundle, but work well with square loads.

J-10. Square loads are tied down with a minimum of two cargo straps ensuring the handles are forward of the load so the bundle safety can manipulate them without getting between the ramp and the bundle. A chain bridle with four cargo straps works best for the TORDS bundle.

PERSONNEL AND JPADS COMBINATION AIRDROP OPERATIONS

J-11. MFF parachutists exiting directly following JPADS bundles are restricted to HAHO operations only. Combination drops will only be conducted with extra light, ultra lightweight, and micro lightweight systems.

J-12. Minimum pull altitude for jumpers conducting combination MFF/JPADS drops is 1,000 feet from drop altitude.

J-13. The limiting factor for combination airdrops is the jumper. The JPADS generally has a greater glide distance and should easily make the DZ during combination airdrops using the more restrictive personnel release point.

J-14. Jumpers must realize the JPADS is robotic in nature and will normally have an increased forward airspeed and a slightly increased rate of descent compared to a typical MFF parachutist. Figure J-1, page J-3, depicts personnel and JPADS combination airdrop operations. The following paragraphs discuss considerations that must be taken into account when conducting combination airdrops using JPADS.

Joint Precision Airdrop System

Figure J-1. Personnel and JPADS combination airdrop operations

PREJUMP CONSIDERATIONS

J-15. Jumpers should note the size and weight of the JPADS they will be following. As the JPADS weight increases, the airspeed and descent rate of the JPADS bundle also increase. This information should give the parachutist an idea how difficult the JPADS will be to fly with.

J-16. Personnel jumping with the JPADS should adjust their planned exit weight to ensure they do not exceed the descent and glide rate of the JPADS bundle. Under no circumstances should parachutists intentionally perform maneuvers that place them lower than the JPADS bundle.

J-17. The jumpmaster will brief the JPADS procedures for grouping on the bundle during canopy flight. Units may consider having the low man concentrate primarily on maintaining visual contact with the bundle while other jumpers in the stick monitor the low man's pattern and planned flight routes.

Appendix J

Canopy Flight Considerations

J-18. The JPADS will normally be the lowest canopy in the stack. All parachutists should use prebriefed grouping procedures to get into a stacked formation and exercise added caution to ensure they maintain separation with the JPADS and other jumpers while joining the stack.

J-19. The jumpers will accomplish standard postopening procedures, build the initial canopy formation behind the bundle, and begin flying the intended flight route.

J-20. Jumpers should become familiar with and consider using trim tabs to stay with the bundle and to alleviate arm fatigue from excessive front riser use. Jumpers must avoid radical canopy maneuvers while following a JPADS.

J-21. In the event that jumpers lose visual contact with the bundle, or cannot acquire the AGU after exit, they should keep scanning and attempt to locate the bundle system. After completing postopening procedures, all jumpers should turn to the briefed heading and continue scanning, keeping a sharp lookout during the entire descent in an attempt to visually reacquire the bundle system.

Collision or Entanglement With the Bundle

J-22. These procedures are intended to correspond with personnel entanglement emergency procedures so individual jumpers will not have to memorize a different set of emergency procedures if entanglement occurs with a JPADS or bundle:

- If jumper's canopy is entangled with the JPADS and the JPADS has a good canopy, jumper should cut away no lower than 1,000 feet AGL.
- If jumper has a good canopy but the JPADS canopy is entangled with the jumper, the jumper should clear the canopy entanglement from himself and his equipment.
- If jumper and JPADS are entangled and neither has a good canopy, jumper should clear himself from the entanglement and cut away regardless of position in the entanglement.
- If impact with the ground is imminent, jumper should deploy the reserve in an attempt to slow his descent.

Duty Positions

J-23. The *bundle safety* is responsible for ensuring the bundle is programmed, rigged, loaded, and deployed correctly.

J-24. The *bundle pusher* is responsible for verifying the bundle is programmed, rigged, loaded, and deployed correctly. It is also his primary responsibility to deploy and follow the load out of the aircraft and lead the stack to land in close vicinity of the bundle by using proper canopy control techniques.

Bundle Operations

J-25. During night operations, the bundle will be marked IAW a published unit SOP. It is recommended that all MFF jumpers conducting combination drops with JPADS utilize available NVG technology to maintain continuous visual contact with the system. For night operations, it is recommended that the bundle be marked as follows:

- One strobe light on each side and one on the top.
- All strobe lights should be secured with hook-pile tape.
- Overt or covert markings should reflect mission profile.

J-26. The bundle should be placed as close as possible to the ramp hinge, with ratchet portion of cargo strap facing the ramp, so the safety releasing the cargo strap does not interfere with the jumper controlling the bundle. The bundle should be rigged with the static line on the port side of the aircraft, but the overall rigging will depend on the aircraft.

J-27. Time warnings for bundle operations are as follows:
- *10 Minutes*: All jumpers will put rucksacks on and perform buddy checks of equipment. The safety turns on the JPADS GPS system.
- *6 Minutes*: Aircraft will be flat and level on heading for jump run.
- *4 Minutes*: All jumpers stand up and perform pin checks.
- *3 Minutes*: Lead jumper controls bundle and Safety disconnects cargo strap. All jumpers are ready with eyes on the jumpmaster. The jumpmaster will call for the ramp after the safety's thumbs-up signal. The ramp will not be opened until both jumper and safety have positive control of the bundle.
- *2 Minutes*: The bundle will be moved to the ramp hinge. Lead jumper and safety will maintain control.
- *1 Minute*: With the ramp open, the bundle will be moved by the lead jumper with the assistance of the Safety halfway to the ramp edge.
- *Stand By*: Lead jumpers will move the bundle to the ramp edge. The safety at this time will control the static line of the bundle and assist the lead jumper in any way necessary.
- *Go*: Lead jumper will push the bundle. After the bundle is pushed, the lead jumper will delay 5 to 10 seconds to ensure the bundle has a good canopy and the GPS guidance system has found its heading.

J-28. In the event of a mishap with the bundle guidance unit, it is recommended that the stack follow the bundle until the low man decides the bundle is not flying to its intended landing point.

JUMPER ACTIONS ON EXIT, UNDER CANOPY, AND LANDING

J-29. Exit procedures for jumpers are as follows:
- After the bundle is pushed, the lead jumper will delay 5 to 10 seconds ensuring the bundle has a good canopy and the GPS guidance system has found its heading.
- On exit, all jumpers will exit the aircraft immediately, turn toward the DZ, and deploy the main canopy.
- Pull procedures are as normal with NVGs; these pull procedures are as follows:
  - The jumper will "ARCH," look under the NVGs at the rip cord housing and rip cord, "TRACE" rip cord housing, "GRAB" rip cord, "PULL" rip cord, and "RAISE" right arm and shoulder.
  - The jumper's counter hand is moved past the jumper's head, exaggerating the counter.
  - To clear the partial vacuum, the jumper will raise his right arm and shoulder to disrupt the vacuum.
- After deploying the main canopy, jumpers will orient in the direction of the DZ and check in on the radio in exit number order.

J-30. The lead jumper's responsibility is to maintain visual contact with the bundle and set up a safe landing pattern for the remainder of the formation. The second jumper and the last jumper (jumpmaster) should be primary for navigation. Additionally, these jumpers should use the GPS tracking system (Parachutist Navigation System [ParaNav]) mounted to the jumpers' equipment.

J-31. The last jumper (jumpmaster) should be the stack commander; his responsibility is to maintain stack integrity and to relay any directions to all jumpers in the stack.

J-32. All jumpers not using the computer GPS should use compass boards with a marine compass and a commercial-brand GPS.

J-33. At 2,000 feet AGL, the lead jumper should call by radio the landing pattern to the remainder of the stack.

J-34. All jumpers will land as normal; once on the ground, jumpers will turn off strobe lights and rally on the low man.

This page intentionally left blank.

Glossary

SECTION I – ACRONYMS AND ABBREVIATIONS

AAUL	Approved and Authorized for Use List
ACH	advanced combat helmet
ACH-ARC	advanced combat helmet accessory rail connector
AF	Air Force
AFI	Air Force instruction
AFWA	Air Force Weather Agency
AGL	above ground level
AGU	Autonomous Guidance Unit
ALICE	all-purpose, lightweight, individual carrying equipment
AO	area of operations
ARR	automatic rip cord release
ASFS	American Safety Flight System
ATTP	Army tactics, techniques, and procedures
AWR	Airworthiness Release
BMNT	beginning morning nautical twilight
C	Celsius
CFLOS	cloud-free line of sight
CMWH	center-mounted weapons harness
CO_2	carbon dioxide
CRU	connector regulator unit
CYPRES	Cybernetic Parachute Release System
DA	Department of the Army
DACO	departure airfield control officer
DAF	departure airfield
DCS	decompression sickness
DSN	Defense Switched Network
DZ	drop zone
DZCO	drop zone control officer
DZSO	drop zone safety officer
DZSTL	drop zone support team leader
EAAD	electronic automatic activation device
F	Fahrenheit
fps	feet per second
ft	foot/feet
GPS	Global Positioning System
HAHO	high-altitude high-opening

Glossary

HALO	high-altitude low-opening
HARP	high-altitude release point
Hg	mercury
hPa	hectopascal
HPT	hook-pile tape
HRPO	highest release point obstacle
HSPR	harness, single-point release
IAW	in accordance with
ID	identification
in/Hg	inches of mercury
IOH	improved oxygen harness
IPB	intelligence preparation of the battlefield
IR	infrared
JMPI	jumpmaster personnel inspection
JPADS	Joint Precision Airdrop System
km	kilometer(s)
kph	kilometer(s) per hour
lb	pound(s)
LBE	load-bearing equipment
LCD	liquid crystal display
LED	light-emitting diode
LPU	life preserver unit
m	meter(s)
mbar	millibar(s)
MBITR	multiband inter/intrateam radio
METT-TC	mission, enemy, terrain and weather, troops and support available, time available, and civil considerations
MFF	military free fall
MICH	modular integrated communication headset
mm	millimeter(s)
MMPS	Multimission Parachute System
MO	malfunction officer
mph	miles per hour
MSL	mean sea level
MTTB	Military Tandem Tethered Bundle
NATO	North Atlantic Treaty Organization
NCO	noncommissioned officer
NGA	National Geospatial-Intelligence Agency
NLT	not later than
nm	nautical mile(s)
NSN	national stock number

Glossary

NVG	night vision goggle
OP	opening point
PAS	Personnel Airdrop Systems
PBOS	portable bailout oxygen system
PC	personal computer
PDA	Personal Digital Assistant
PDB	parachutist drop bag
PLF	parachute landing fall
POM	Parachutist Oxygen Mask
PRP	preliminary release point
psi	pounds per square inch
RAPPS	ram-air personnel parachute system
RFII	request for intelligence information
RSL	reserve static line
SARPELS	Single-Action Release Personal Equipment Lowering System
SCAR	Special Operations Forces Combat Assault Rifle
SEAD	suppression of enemy air defenses
SLR	Standard Lapse Rate
SOF	special operations forces
SOP	standing operating procedure
STANAG	standardization agreement
STT	special tactics team
SWO	staff weather officer
TFSS	Tactical Flotation Support System
TM	technical manual
TO	technical order
TORDS	Tandem Offset Resupply Delivery System
TOT	time on target
UDT	underwater demolition team
U.S.	United States
USAF	United States Air Force
USAJFKSWCS	United States Army John F. Kennedy Special Warfare Center and School
USASOC	United States Army Special Operations Command
USMC	United States Marine Corps
USN	United States Navy
USSOCOM	United States Special Operations Command
VDZ	virtual drop zone
VWT	vertical wind tunnel
ZAR	Zone Availability Report

Glossary

SECTION II – TERMS

abort
    To terminate a mission for any reason other than enemy action. It may occur at any point after the beginning of the mission and prior to its conclusion.

above ground level
    The actual distance of the aircraft above the ground, normally expressed in feet. Also called **AGL**.

altimeter setting
    The setting applied to an aircraft altimeter's barometric correction counter that results in an altimeter reading equal to the MSL ground elevation of an airport, or other location, when the aircraft is sitting on the ground at that location.

automatic rip cord release
    A mechanical device designed to automatically extract the rip cord pin(s) at a predesignated altitude. Also called **ARR**.

aviation (master) altimeter
    An altimeter of the type installed in all aircraft. This instrument has an adjustment mechanism to provide offset corrections (altimeter settings) for local deviations from standard barometric pressure conditions.

barometric correction counter
    The adjustment mechanism that allows an altimeter to be corrected for local deviations from standard pressure.

body stabilization
    A movement made in free fall to attain and maintain a stable body position.

body turn
    A movement made in free fall to effect a turn by moving the upper torso either to the right or left.

control lines
    The lines that connect the toggles and turn slots, and by which the parachutist may control the action of his canopy.

crabbing
    A movement made in free fall to maneuver the canopy at an angle to the direction of the wind.

cutaway
    A term used for jettisoning the main canopy in the event of a malfunction.

departure airfield
    An airfield on which troops and/or materiel are enplaned for flight. Also called **DAF**. (JP 1-02)

desired impact point
    A desired spot for parachute landings on the drop zone.

dogleg
    A term used to describe calculations when the directional difference in winds is 90 degrees or more at two consecutive altitudes.

drop zone
    A specific area upon which airborne troops, equipment, or supplies are airdropped. Also called **DZ**.

drop zone elevation
    The actual physical (ground) altitude of the DZ, in feet MSL, obtained from topographic maps or a global positioning system (GPS).

drop zone safety officer
   The officer responsible for the conduct of operations on the drop zone. Also called **DZSO**.

equivalent altitude
   The MSL altitude for a location, derived from the standard atmosphere pressure-altitude curve, which corresponds to the prevailing ambient pressure at that location. This parameter is computed by the barometric calculator and is used to bring locations with different atmospheric pressure variations to the same reference point. Depending on the prevailing pressure, equivalent altitude may be higher or lower than the actual ground elevation. Equivalent altitudes may be negative for locations at or near sea level. The equivalent altitude of any location may also be determined using an aviation altimeter by adjusting its barometric correction counter to the standard setting of 29.92 (no correction) and reading the altitude directly.

glide
   A position used to permit forward movement to prevent collision with other parachutists. Parachutists bring the hands toward the shoulders. They do not break the arch in their back. They extend their legs slightly.

grouping
   A technique used to enable parachutists to fall together in the air, remain together under canopy, and land as a compact tactical unit.

guide ring
   A ring attached to the rear risers through which the control lines pass.

holding
   A term used when the canopy is pointed directly into the wind (as opposed to crabbing or running).

impact point
   A point on the ground where the parachutist should land.

jump commands
   The commands given by the jumpmaster to the parachutists on his sortie to control the parachutists' actions between the 2-minute warning and exit.

jumpmaster
   The assigned airborne-qualified individual who controls parachutists from the time they enter until they exit.

jumpmaster personnel inspection
   An inspection by the military free-fall jumpmaster similar to that of the static-line jumpmaster to ensure all safety requirements have been met. Also called **JMPI**.

loadmaster
   An Air Force technician qualified to plan loads, to operate auxiliary materials handling equipment, and to supervise loading and unloading of aircraft. (JP 1-02)

local barometric reference data
   Information describing the atmospheric conditions presently surrounding the barometric calculator. When this data is accessed, the equivalent altitude of the calculator is displayed. The equivalent altitude of a calculator located at the DZ is used during operation to compute wrist altimeter settings.

lowering line
   A cord designed to allow a parachutist to lower a rucksack or a piece of equipment to the ground prior to his own impact.

Glossary

malfunction
A discrepancy in the deployment or inflation of the parachute that can create any faulty, irregular, or abnormal condition increasing the parachutist's rate of descent, or a condition in which the canopy is uncontrollable.

mean sea level
The average height of the sea for all tidal conditions; the zero altitude reference point used in determining ground elevations. Also called **MSL**.

millibar(s)
A unit of measurement of barometric pressure used when setting the ARR. Also called **mbar**.

military free fall
Methods of delivering personnel, equipment, and supplies from a transport aircraft at a high altitude via free-fall parachute insertion. The parachute (RAPPS) can be manually deployed during free fall or with the assistance of a static line depending on mission and jumper capabilities. Also called **MFF**.

nonoxygen jump
A parachute jump, normally below 10,000 feet, that does not require the use of oxygen equipment.

nonoxygen procedures
The signals given by the jumpmaster to control the action of the parachutists between takeoff and the 2-minute time warning when oxygen is not used.

opening point
The point on the ground over which the parachutist deploys his canopy. Also called **OP**.

oxygen check
A visual check made by the jumpmaster to see that each parachutist is receiving oxygen.

oxygen jump
A free-fall parachute jump requiring the use of oxygen, normally at any altitude above 10,000 feet.

oxygen mask
A face mask that may be connected to an oxygen supply, allowing parachutists to operate above nonoxygen altitudes.

oxygen procedures
The procedures used by parachutists and the jumpmaster when they jump using oxygen equipment.

partial malfunction
A situation in which the canopy does not fully deploy.

physiological training
The training conducted by the Air Force to enable parachutists to identify oxygen equipment and systems and explain the effects of high-altitude physiology, cabin pressurization, and hazardous noise and stress.

pilot briefing
A briefing the jumpmaster gives the pilot to clarify any points related to the airborne operation, such as drop signal, time, or alternate drop zone.

prebreathing time
The time spent prior to a high-altitude drop when the parachutists and jumpmaster breathe 100-percent oxygen.

preliminary release point
The point above the ground at which the initial vector stops and the free-fall drift factor begins. Also called **PRP**.

Glossary

release point
: The point over which parachutists exit the aircraft.

running
: A technique used for pointing the canopy in the direction of the wind.

safe-to-arm altitude
: An altitude 5,000 feet AGL or 2,500 feet above the ARR activation altitude, whichever is higher.

special tactics team
: A team consisting of combat control, special operations weather, and pararescue personnel. Also called **STT**.

spotting
: A technique used by the jumpmaster to visually align the aircraft and release the parachutists at the proper release point.

standard atmosphere
: An idealized description of the average properties of the earth's atmosphere, including the relationship between atmospheric pressure and altitude, based on a pressure of 29.92 Hg or 1,013 millibars, at MSL. All aviation altimeters are calibrated to the standard atmosphere pressure-altitude curve to permit uniform adjustment under variable atmospheric and geographic conditions.

terminal velocity
: The velocity at which a falling object attains its maximum, constant speed, normally about 125 miles per hour for a free-fall parachutist.

time warnings
: The warnings given by the jumpmaster, in minutes, to alert the parachutist to the time remaining before exiting the aircraft.

toggles
: The nylon loops attached to lines that control the forward speed of the canopy and left and right maneuvering, mounted on the front side of the front risers.

total malfunction
: A type of malfunction in which the parachute remains in the pack tray.

walk-around bottle
: A large, low-pressure oxygen cylinder that may be used by either the jumpmaster or safety personnel not connected to the oxygen console or the aircraft oxygen system.

wind-cocked
: The claimed tendency of some parachute designs to face downwind automatically and to maintain this facing throughout descent.

wind drift formula
: A formula used to locate the proper release point.

wind line
: An imaginary line extending upwind from the target area to the opening point.

wind/weather-cocking
: The environmentally (for example, wind, air pressure/density, altitude, and so on) induced "downwind" flight path of an aerodynamic projectile/parachute. By virtue of design, the ram-air parachute will, if unmanipulated, establish a downwind glide path and fly a predictable route to a calculated point of impact.

Glossary

wrist altimeter adjustment
In general, the practice of setting a wrist altimeter to zero on the ground at the DZ to permit determination of a jumper's height AGL. The barometric calculator provides the capability for adjusting wrist altimeters at a remote location to obtain a zero reading on the ground at the DZ.

wrist altimeter setting
The number of feet, AGL, to which a wrist altimeter at a remote location should be set to obtain a zero reading on the ground at the DZ.

References

SOURCES USED
These are the sources quoted or paraphrased in this publication.

AIR FORCE PUBLICATIONS
AF Form 702 (Individual Physiological Training Record).
AF Form 1274 (Physiological Training).
AF IMT Form 3823 (Drop Zone Survey).
AFI 11-409, *High-Altitude Airdrop Mission Support Program*, 1 December 1999.
AFI 11-410, *Personnel Parachute Operations*, 4 August 2008.
AFI 13-217, *Drop Zone and Landing Zone Operations*, 10 May 2007.
AFJ 13-210(I), *Joint Airdrop Inspection Records, Malfunction/Incident Investigations, and Activity Reporting*, 8 April 2008.
USAF Warfare Center Air Combat Command Project 07-169A Final Report, *Free-Fall Night Vision Devices Tactics Development and Evaluation*, November 2008.

ARMY PUBLICATIONS
AR 59-4, *Joint Airdrop Inspection Records, Malfunction/Incident Investigations, and Activity Reporting*, 8 April 2008.
AR 70-62, *Airworthiness Qualification of Aircraft Systems*, 21 May 2007.
DA Form 285-AB (U.S. Army Abbreviated Ground Accident Report [AGAR]).
DA Form 1306 (Statement of Jump and Loading Manifest).
DA Form 2028 (Recommended Changes to Publications and Blank Forms).
DA Pamphlet 385-40, *Army Accident Investigations and Reporting*, 6 March 2009.
FM 3-05.210, *Special Forces Air Operations*, 27 February 2009.
FM 3-21.38, *Pathfinder Operations*, 25 April 2006.
FM 3-21.220, *Static-Line Parachuting Techniques and Tactics*, 23 September 2003.
GTA 31-01-003, *Detachment Mission Planning Guide*, 1 March 2006.
TM 10-1670-287-23&P, *Unit and Direct Support Maintenance Manual Including Repair Parts and Special Tools List for MC-4 Ram-Air Free-Fall Personnel Parachute System*, 30 July 2003.
TM 10-1670-329-13&P, *Operator and Field Maintenance Manual Including Repair Parts and Special Tools List for Parachutist Oxygen Mask (POM)*, 21 December 2009.
USASOC Form 1099-E (Military Free Fall Jump Record).
USASOC Policy 20-10, *Wear of Night Vision Devices During Military Free-Fall Operations*, 1 September 2010.
USASOC Regulation 350-2, *Airborne Operations*, 4 June 2008.

DEPARTMENT OF DEFENSE PUBLICATIONS
DD Form 1748-2 (Joint Airdrop Malfunction Report [Personnel Cargo]).
DODD 5100.1, *Functions of the Department of Defense and its Major Components*, 21 December 2010.

References

JOINT AND MULTI-SERVICE PUBLICATIONS

JP 1-02, *Department of Defense Dictionary of Military and Associated Terms*, 8 November 2010.

TM 1-1680-377-13&P-5, *Operator's Unit and Direct Support Maintenance Manual, Including Repair Parts and Special Tools List for Helicopter Oxygen Systems (UH-60)*, 9 October 2009.

TM 1-4220-252, *Maintenance Instructions With Illustrated Parts Breakdown USAF Flotation Equipment LRU-1/P, F-2B, 20-Man VPLR, and 25-Man Life Rafts LPU-3/P, LPU-6/P, LPU-10/P, A-A-50652, and MB-1 Life Preservers*, 31 January 2005.

TM 10-1670-287-23&P, *Unit and Direct Support Maintenance Manual, Including Repair Parts and Special Tools List for MC-4 Ram Air Free-Fall Personnel Parachute System*, 30 July 2003.

TM 10-8470-204-10, *Operator's Manual for Advanced Combat Helmet (ACH)*, 17 May 2010.

TM 55-1660-247-12, *Operation, Fitting, Inspection and Maintenance Instructions With Illustrated Parts Breakdown for MBU-12/P Pressure-Demand Oxygen Mask*, 1 April 1981.

TO 14S-1-102-21, *Maintenance Instructions With Illustrated Parts Breakdown USAF Flotation Equipment LRU-1/P, F-2B, 20-Man VPLR, and 25-Man Life Rafts LPU-3/P, LPU-6/P, LPU-10/P, A-A-50652, and MB-1 Life Preservers*, 31 January 2005.

TO 15X5-3-6-1, *Operation, Fitting, Inspection and Maintenance Instruction With Illustrated Parts Breakdown MBU-12/P Pressure-Demand Oxygen Mask*, 1 April 1981.

MARINE CORPS PUBLICATIONS

MCO 13480.1D, *Joint Airdrop Inspection Records, Malfunction/Incident Investigations, and Activity Reporting*, 8 April 2008.

MCO 3120.11, *Marine Corps Parachuting and Diving Policy and Program Administration*, 4 May 2009.

USMC TM 10121A-12&P, *Single-Action Release Personal Equipment Lowering System (SARPELS)*, 30 September 1997.

USMC TM 11019-12A&P, *Operation and Organizational Maintenance With Repair Parts List for the Cybernetic Parachute Release System (CYPRES) Automatic Opening Device*, September 2006.

USMC TM 70244A-OI, *Tactics, Techniques, and Procedures Manual for U. S. Marine Corps Military Free-Fall Operations*, July 2006.

NAVY PUBLICATIONS

OPNAVINST 4630.24D, *Joint Airdrop Inspection Records, Malfunction/Incident Investigations, and Activity Reporting*, 8 April 2008.

U.S. Navy Diving Manual, 15 April 2008.

OTHER PUBLICATIONS

National-Geospatial Intelligence Agency, *List of Lights, Radio Aids, and Fog Signals*, 2011.

USSOCOM Directive 10-1, *Organization and Functions, Terms of Reference Roles, Missions, and Functions of Component Commands*, 25 August 2008.

STANDARDIZATION AGREEMENT (STANAG)

STANAG 3570, *Drop Zones and Extraction Zones—Criteria and Markings*, 26 March 1986.

Index

A

aircraft jump commands, 6-2, 6-9 through 6-12
  abort, 6-2, 6-12
  go, 6-2, 6-11
  move to the rear, 6-2, 6-10
  stand by, 6-2, 6-10
  stand up, 6-2, 6-10
aircraft procedure signals, 6-1 through 6-9
  check oxygen, 6-1, 6-2, 6-6
  don helmets, 6-1, 6-3
  emergency bailout, 6-4, 6-5
  mask, 6-1, 6-5
  time warnings, 6-2, 6-7
  unfasten seat belts, 6-1, 6-3
  wind speed, 6-2, 6-7, 6-8
altimeter
  MA2-30/A, 2-12
  MA-10, 2-13 through 2-18
  PA-200, 2-12
  setting computation, 2-12, 2-13
aviator's kit bag, 2-22

B

body stabilization skills, 7-1 through 7-10
  body turn, 7-4
  box man method, 7-2
  diving exit, 7-3
  gliding, 7-4
  main ripcord pull, 7-5, 7-6
  poised exit, 7-1
  stable free fall, 7-4
  tracking position, 7-7

C

calculations
  canopy drift, F-4
  doglegs, F-5
  drift and direction, F-3, F-4
  forward throw, F-4
  high-altitude release point (HARP), F-1, F-2, F-5 through F-8

canopy control
  crabbing, 8-12, 8-13
  flat turn, 8-18
  full brakes, 8-15
  full flight, 8-14
  glide angles, 8-13
  half brakes, 8-14, 8-15
  holding maneuver, 8-11
  running maneuver, 8-11
  spiral turn, 8-17
  stall, 8-15, 8-16
canopy entanglement procedures, 9-9
checklists
  aircraft inspection, H-1
  canopy control, 8-11
  jumpmaster briefing, I-1, I-2
  jumpmaster responsibilities, 13-1, 13-2
  portable bailout oxygen system preflight inspection and operational, 4-29
  prebreather preflight inspection and operational function, 4-30, 4-31
controllability check, 9-5, D-9
cutaway handle, 2-7, 5-59, 8-5, G-3, G-9, G-10
Cybernetic Parachute Release System (CYPRES)
  CYPRES Military Absolute Adjust Circular Calculator, 3-25 through 3-28
  Military CYPRES Calculator, 3-27 through 3-32
  PDA computer with Military CYPRES Absolute Adjust Model Calculator, software download, 3-27

E

emergency procedures
  free fall, 9-4
  in-flight, 9-2, 9-3
  landing, 9-10
  ram-air personnel parachute system, (RAPPS), 9-1 through 9-10

H

harness, single-point release (HSPR)
  attaching HSPR and all-purpose, lightweight, individual carrying equipment (ALICE) pack to parachutist, 5-24 through 5-27
high-altitude release point (HARP) calculation, F-1 through F-8
high-altitude high-opening (HAHO) parachutist requalification, B-2

I

inspections
  aircraft, H-1
  jumpmaster personnel, G-1 through G-12

J

jumpmaster
  briefing checklist, I-1, I-2
  cardinal rules, 13-2
  qualifications, 13-2, 13-3
  responsibilities, 13-1, 13-2
jumpmaster personnel inspection, G-1 through G-12

L

landing approaches
  base leg, 8-21
  downwind leg, 8-21
  final approach, 8-21
  glide angles, 8-22
landing procedures, emergency, 9-10
life preservers
  B-7, 5-54 through 5-55
  LPU-10/P, 5-57, 5-58
  UDT, 5-59
  TFSS-5326, 5-55 through 5-57

M

main parachute, 2-12, 3-1, 7-1, 8-2, 8-3, 9-6, D-11

malfunctions
- bag lock, 9-6, D-12
- broken control lines, 8-19, 9-7, D-13
- broken lines, 9-7
- closed end cells, 9-7, D-13
- floating rip cord, 9-6, D-12
- hard pull, 9-6
- horseshoe, 7-5, 9-6, 10-16, D-12
- hung slider, 9-7, D-13
- line twists, 9-8, D-14
- main parachute, 9-8 through 9-10
- pack closure, 9-6, D-12
- partial (RAPPS), 9-5, D-12 through D-14
- pilot chute
  - hesitation, 9-6, D-12
  - over the nose, 9-7, D-13
  - tears, 9-8, D-14
  - tension knots, 9-8, D-14
- total (RAPPS), 9-5, D-12

masks, oxygen
- types, 4-9 through 4-15

N

night vision goggles, 10-12 through 10-17

O

oxygen
- AIROX VIII assembly, 4-16 through 4-20
- equipment-related emergencies, 4-4
- handling and safety, 4-32
- MA-1 portable oxygen assembly, 4-25, 4-26
- masks, 4-6 through 4-15
- MBU-12/P, 4-7, 4-8, 4-10, 4-12, 4-13, 4-16
- Parachutist Oxygen Mask (POM), 4-6 through 4-16, 4-20, 4-21
- portable bailout oxygen system (106-cubic-inch), 4-16 through 4-20
- prebreather attachment, 4-26
- pressure, regulator, indicator, connections, and emergency equipment (PRICE) check, 4-27
- requirements, 4-5, 4-6
- safety personnel, 4-28, 11-4, 13-1
- six-man prebreather portable oxygen system, 4-21, 4-22

oxygen fitting block, 2-9, 4-19, G-4, G-6

P

postopening procedures, 9-5

W

water operations, deliberate
- drop zone (DZ) pickup procedures, 12-5, 12-6
- DZ requirements and markings, 12-4
- equipment recovery boats, 12-2
- equipment requirements, 12-2, 12-3
- night water parachute operations, 12-6
- parachutist procedures, 12-4, 12-5
- parachutist recovery boats, 12-1
- parachutist requirements, 12-2
- safety swimmers, 12-2
- water jumps with combat equipment, 12-6

winds
- high and low patterns, 8-23
- change in wind direction, 8-23, 8-24

ATTP 3-18.11 (FM 3-05.211)
14 October 2011

By Order of the Secretary of the Army:

RAYMOND T. ODIERNO
General, United States Army
Chief of Staff

Official:

JOYCE E. MORROW
Administrative Assistant to the
Secretary of the Army
1130602

DISTRIBUTION:

Active Army, Army National Guard, and United States Army Reserve: Not to be distributed; electronic media only.

PIN: 102456-000